All-in-One Electronics Guide

A comprehensive electronics overview for electronics engineers, technicians, students, educators, hobbyists, and anyone else who wants to learn about electronics

Your complete practical guide to understanding and utilizing modern electronics!

By: Cammen Chan

C & C Group of Companies LLC.

Published by C & C Group of Companies LLC.
Copyright 2015. All rights reserved.

No part of this publication may be reproduced, stored in a retrieval system, or transmitted in any form or by any means, electronic, mechanical, photocopying, recording, or otherwise, without the prior written permission of the publisher.

Website: http://www.ALLinOneElectronicsGuide.com

E-mail: ALLinOneElectronicsGuide@gmail.com

Facebook: http://www.facebook.com/ALLinOneElectronicsGuide

Twitter: http://www.twitter.com/ai1_electronics

All trademarks mentioned herein are property of their respective companies.

Book Cover Editor: Flora Gillis
Book Editor: Priscilla P. Flores
Book Cover Designer: Kristin Fleming http://www.kristinfleming.com/

ISBN-10: 1479117374
ISBN-13: 978-1479117376

Printed in the United States of America

About the Author

Cammen Chan has been working in the electronics industry since 1996. After receiving his bachelor of science degree in electronic engineering technology from the Wentworth Institute of Technology and master of science degree in electrical engineering from Boston University, he began his engineering career at IBM Microelectronics, then worked at Analog Devices Inc., National Semiconductor, and several technology startups. He has one US patent invention in the area of nanotechnology. Since 2009, Cammen has also been an adjunct faculty member at a number of US colleges and universities including ITT Technical Institute, DeVry University, Western International University, University of Advancing Technology, Chandler Gilbert Community College, Remington College, and Excelsior College. He teaches electronics engineering technology, information technology, mathematics, and emerging technologies. Cammen has taught all the subjects in this book in various formats such as on-site, online, and blended classes. Currently, Cammen is a technical training engineer at Microchip Technology in the Phoenix area.

Introduction

The semiconductor industry is a big business. The electronics industry is even bigger. The semiconductor industry alone was a US $300 billion plus industry in 2012. The long-term trend of electronics is bright and promising. With increasing use of electronic devices in consumer, commercial, and industrial products and systems, the electronics industry is always growing. If you are considering becoming an electronics engineer, this book gives you the technical skills needed to "pass" the technical parts of interviews and the confidence to increase your chances of getting employed. If you are already an electronics technician or engineer, this book improves your ability to perform at the highest level at work in the electronics field. If you want to be a microelectronics engineer or are already one, you will find the microelectronics-related contents in this book applicable to your work. If you are an educator teaching electronics, this book is the perfect reference for you and your students with step-by-step technical examples and quizzes. If you are an electronics hobbyist, this book offers sampled electronic circuits (electronic components connected with each other by wires or traces) you can apply to your design. For everyone else interested in learning about electronics, this book provides a strong foundation of what you need to know when working with electronics.

The chapters are divided into various electronic principles levels, from basic to advanced, along with practical circuits and quizzes. Answers provide step-by-step explanations of how and why the answers were derived. Examples and circuits in later chapters build upon previous chapters, thus creating a consistent flow of learning and a gradual accumulation of knowledge. The level of mathematics is moderate without tedious and complicated math models and formulas. For students majoring in electrical engineering, this book is more than your typical academic electronics textbook that overwhelms you with excessive theories, formulas, and equations. Instead, the material covered in this book is easy to read, with plenty of diagrams, pictures, waveforms, and graphs, and is easy to understand. Accurately representing our non-ideal world, this book's technical contents greatly differ from most academic textbooks' false "ideal" perspective. The content is injected with real world quantities and characteristics. For experienced electronics professionals, educators, and hobbyists, this book affords a good reality check and comprehensive review to assist your career or your students, to better prepare for your next job interview, and to inspire your next electronics projects.

How This Book Is Organized

Chapter 1: Direct Current (DC)
First, learn direct current (DC) theories. Then, apply them in practical circuits. Basic electrical parameters, concepts, and theories are covered. This chapter closes with practical DC circuits.

Chapter 2: Diodes
Zero in on diode, the building block of transistors. This chapter explains not only what a diode is made of but also the real world characteristics of diode and some practical diode circuits.

Chapter 3: Alternating Current (AC)
After comprehending DC and diodes, learn about AC, another critical electronics concept. From high-power electric plants to computers and wireless communications, AC operations take place in countless electronic systems. Get a good hold on AC definitions, common AC parameters, capacitors, inductors, and simple AC circuits.

Chapter 4: Analog Electronics
Analog electronics use a substantial amount of analog quantities. Transistors and operational amplifiers (op-amp) are the building blocks of mainstream electronic circuits and systems. Bipolar and Complementary-Metal-Oxide-Semiconductor (CMOS) are the most common types of transistors. Bipolar transistors consist of two diodes. On the other hand, CMOS does not contain any active diodes. Although germanium, gallium, and arsenide can be used to build transistors, both bipolar and CMOS transistors primarily use silicon as the raw material. Performance differences between raw materials types must be considered to choose the correct transistor type. CMOS and bipolar transistors have similar voltage and current characteristics with major differences in fundamental operation. A solid understanding of these differences is essential for analyzing and designing transistors and op-amp circuits.

Chapter 5: Digital Electronics
Basic digital electronics require an in-depth understanding of digital quantities, high (1) and low (0) logic level, logic gates, and circuits. It is considerably the best semiconductor technology choice for high-speed design and operations. In comparison to analog quantities, the simple two levels (1 and 0) offer distinct advantages over analog technology such as lower noise. For cost reasons, digital electronics present a good case for using CMOS transistor technology in digital systems. CMOS transistors are made in deep sub-microscopic scale with advanced chip manufacturing capability, while manufacturing throughputs continues to increase exorbitantly. For high speed, high-density digital designs such as Application Specific Integrated Circuit (ASIC), Field Programmable Gate Array (FPGA), or microprocessors, digital designers often use software to write programs/code for generating CMOS design. Using VHDL or Verilog, instead

of manually placing transistors individually in schematics as in analog design, digital circuits are generated to represent the functional and behavioral models and operations of the target CMOS design. In recent years, BiCMOS process has gained popularity. As its name implies, this process combines both bipolar and CMOS devices, offering the best of both.

Chapter 6: Communications

Electronic communications are the foundation of wired and wireless communications technology. It is an enormous industry with its market covering both consumers and businesses. Radios, cell phones, home and business computers connected to the internet by using either wired or wireless connections are just some examples. The vast majority of this technology is only possible due to the advanced development of electronic communication systems. Additionally, amplitude modulation, frequency modulation, and phase locked loops will be discussed in this chapter. Understanding basic communication theories, techniques, and parameters will greatly assist your work in the communications engineering field.

Chapter 7: Microcontrollers

Microcontroller silicon chips have found their way into a variety of electronic products. One automobile alone has an average of eighty microcontrollers controlling the engine, steering wheel controls, GPS, audio systems, power seats, and others. Microcontrollers are embedded in many consumer and industrial electronics including personal computers, TV sets, home appliances, children's toys, motor control, security systems, and many more. The final products that use microcontrollers are embedded systems. These devices are field programmable: they allow system designers to program the chip to the needs of a specific application, while letting end users perform a limited amount of modification. For example, an end user turning on a microwave oven is actually "programming" the timer. However, the end user does not have access to the source code on the microcontroller, hence the name "embedded systems." Moreover, the same microcontroller can be used in multiple designs. For instance, dishwashers and refrigerators use the same microcontroller with each design having its own specific code downloaded to the microcontroller, resulting in two completely different applications. The microcontroller's field programming capabilities allows many applications to be designed at a very low cost. Comprehending microcontroller architecture and basic programming techniques will prepare you to excel in this field.

Chapter 8: Programmable Logic Controllers

Programmable Logic Controllers (PLCs) are widely used in industrial and commercial applications. Thus, it is worthwhile to study them in addition to consumer-based systems. Types and uses of PLCs are covered first, followed by an inside look at PLCs. Ladder logic programming, a graphical programming technique, is the heart of PLCs. In addition, after exploring practical PLC programs and applications, the chapter closes with PLCs troubleshooting techniques and future development.

Chapter 9: Mental Math

If you have to use a calculator to solve **1 / 1 k = 1 m**, you are probably not making a good impression on interviewers or even coworkers. Using mental math to decipher simple arithmetic answers demonstrates solid mathematic, analytic, and problem solving capabilities. You can learn simple techniques to improve your mental math ability for calculating electronics arithmetic.

Chapter 1: Direct Current (DC) — 1 —

- Current — 1 —
- Resistor — 1 —
- Voltage — 5 —
- Definition — 5 —
- Ohm's Law — 6 —
- Power — 7 —
- Voltage Source and Schematic — 7 —
- Current Source and Schematics — 8 —
- Electrons — 8 —
- Current versus Electrons — 9 —
- Kirchhoff's Voltage Law (KVL) — 9 —
- Kirchhoff's Current Law (KCL) — 11 —
- Parallel Circuit — 11 —
- Parallel Resistor Rule — 12 —
- Series Resistor Rule — 13 —
- Current Divider Rule — 15 —
- Voltage Divider — 16 —
- Superposition Theorems — 19 —
- DC Circuits — 22 —
- IC Packages — 24 —
- Summary — 33 —
- Quiz — 33 —

Chapter 2: Diodes — 37 —

- P-N Junctions — 37 —
- Forward-Biased and Reverse-Biased — 40 —
- Diode I-V Curve — 42 —

X Table of Contents

Diode Circuits	- 43 -
Summary	- 47 -
Quiz	- 48 -

Chapter 3: Alternating Current (AC) — - 49 -

Sine Wave	- 49 -
Frequency and Time	- 50 -
Peak Voltage vs. Peak-to-Peak Voltage	- 52 -
Duty Cycle	- 52 -
Vrms	- 54 -
Impedance, Resistance, and Reactance	- 54 -
Capacitors	- 55 -
XC versus Frequency	- 56 -
Simple Capacitor Circuit	- 57 -
I (Δt) = C (ΔV)	- 59 -
Capacitor Charging and Discharging Circuit	- 60 -
Parallel Capacitor Rule	- 63 -
Series Capacitor Rule	- 63 -
Power Ratio in dB	- 64 -
R C Series Circuit	- 64 -
– 20 dB per Decade	- 65 -
Low-Pass Filter	- 68 -
Phase Shift	- 69 -
Radian	- 70 -
ICE	- 71 -
Inductors	- 73 -
XL versus Frequency	- 74 -
V (Δt) = L (ΔI)	- 75 -

ELI	- 77 -
Q Factor	- 77 -
Parallel Inductor Rule	- 78 -
Series Inductor Rule	- 79 -
High-Pass Filter	- 80 -
Real L and C	- 83 -
Practical AC Circuits	- 85 -
Ringing and Bounce	- 86 -
Inductive Load	- 87 -
Diode Clamp	- 88 -
Series R L C Circuit	- 89 -
LRC Parallel (Tank) Circuit	- 91 -
Transformers	- 93 -
Half-Wave Rectifier	- 95 -
Switching versus Linear Regulators	- 97 -
Buck Regulator	- 97 -
Summary	- 100 -
Quiz	- 101 -

Chapter 4: Analog Electronics — *- 105 -*

What Is Analog?	- 105 -
Analog IC Market	- 106 -
What Are Transistors Made Of?	- 107 -
NPN and PNP	- 108 -
NPN and PNP Symbols	- 109 -
Transistor Cross-Section	- 110 -
Bipolar Transistor Terminal Impedance	- 111 -
IC, IB, IE, and Beta (β)	- 111 -

Table of Contents

- VBE ... - 113 -
- IE = IC + IB ... - 113 -
- IC versus VCE Curve ... - 114 -
- Common Emitter Amplifier ... - 115 -
- Common Collector Amplifier (Emitter Follower) ... - 118 -
- Common Base Amplifier ... - 120 -
- Single-Ended Amplifier Topologies Summary ... - 121 -
- Tranconductance (Gm), Small-Signal Models ... - 121 -
- Common Emitter Amplifier Input Impedance ... - 123 -
- Common Emitter Amplifier Output Impedance ... - 124 -
- Common Collector Amplifier Small-Signal Model ... - 127 -
- Common Base Amplifier Small-Signal Model ... - 128 -
- Single-Ended Amplifier Summary ... - 129 -
- NMOS and PMOS ... - 130 -
- 3D NFET ... - 131 -
- Drain Current and Threshold Voltage ... - 132 -
- NFET and PFET Symbols ... - 132 -
- IC Layout ... - 134 -
- VHDL and Verilog ... - 135 -
- MOSFET Cross Section and Operations ... - 136 -
- MOSFET On-Off Requirements ... - 137 -
- ID versus VDS Curve ... - 139 -
- CMOS Source Amplifier ... - 139 -
- MOSFET Parasitic ... - 142 -
- Common Drain Amplifier (Source Follower) ... - 143 -
- Common Gate Amplifier ... - 145 -
- Bipolar versus CMOS ... - 147 -
- Differential Amplifiers ... - 148 -

- Common Mode — 149
- CMRR and Differential Gain — 150
- Current Mirror — 152
- Op-Amp — 153
- Op-Amp Rules — 155
- Inverting Amplifier — 158
- Non-Inverting Amplifier — 160
- Op-Amp Parameters — 162
- LM741 — 164
- Current Mirror Inaccuracies — 165
- Wilson Current Mirror — 166
- Bipolar Cascode — 167
- Darlington Pair — 168
- CMOS Cacosde — 170
- Buffer (Voltage Follower) — 171
- Summing Amplifier — 172
- Active Low-Pass Filter — 174
- Circuit Simulator — 176
- Hysteresis — 179
- Positive Feedback (Oscillation) — 182
- Instrumentation Amplifier — 184
- Linear Regulator — 185
- Low Drop-out (LDO) Regulator — 186
- Summary — 189
- Quiz — 190

Chapter 5: Digital Electronics — *- 195 -*

- 1s and 0s: The Inverter — - 196 -
- NMOS Inverter — - 197 -
- NFET and PFET Inverter — - 197 -
- Inverter Action — - 198 -
- Shoot-Through Current — - 199 -
- Ring Oscillator — - 200 -
- OR Logic Gate — - 202 -
- OR Gate Schematic — - 202 -
- Three-Input OR Gate — - 203 -
- LSB, MSB — - 204 -
- NOR Gate — - 204 -
- AND and NAND Gates — - 205 -
- XOR Gate — - 206 -
- Combinational Logic — - 206 -
- Boolean Algebra — - 207 -
- Latch — - 208 -
- Flip-Flop — - 210 -
- D and J-K Flip-Flops — - 211 -
- Frequency Divider — - 211 -
- Shift Register — - 213 -
- Parallel Data Transmission — - 214 -
- Multiplexer — - 215 -
- Mixed-signal — - 216 -
- Level Shifter — - 217 -
- Multi-Layer Board — - 217 -
- Digital Voltage Levels — - 219 -

- Analog-to-Digital Converter — 219 -
- Nyquist Frequency — 221 -
- ADC Gain and Offset Errors — 222 -
- Digital-to-Analog Converter — 224 -
- Binary-Weighted DAC — 225 -
- 555-Timer — 226 -
- Summary — 230 -
- Quiz — 230 -

Chapter 6: Communications — *- 231 -*

- Time versus Frequency Domains — 232 -
- Harmonics, Distortion, and Inter-modulation — 234 -
- Modulation — 236 -
- Bit Rate, USB, and Baud — 236 -
- C = F λ — 237 -
- Amplitude Modulation — 238 -
- Modulation Index and Bessel Chart — 239 -
- AM Transmitter — 240 -
- Frequency Modulation — 241 -
- Phase Lock Loop (PLL) — 242 -
- Summary — 245 -
- Quiz — 245 -

Chapter 7: Microcontrollers — *- 247 -*

- MCU Parameters — 248 -
- Harvard Architecture — 251 -
- Data and Program Memory — 251 -
- MCU Instructions — 255 -
- Instruction Clock — 257 -

XVI Table of Contents

- Internal Oscillator .... - 258 -
- Interrupt .... - 260 -
- Special Features .... - 261 -
- Development Tools .... - 262 -
- Debugger .... - 263 -
- Design Example: Comparator .... - 265 -
- Design Example: Timer .... - 269 -
- Summary .... - 271 -
- Quiz .... - 271 -

Chapter 8: Programmable Logic Controllers _____ - 273 -

- History .... - 273 -
- PLC Benefits .... - 275 -
- PLC Components .... - 276 -
- PLC Programming and Ladder Logic .... - 278 -
- PLC Programming Example .... - 283 -
- PLC Programming Syntax .... - 286 -
- Timers .... - 292 -
- On-Timer .... - 293 -
- On-Timer Application .... - 294 -
- Off-Timer .... - 295 -
- Off-Timer Application .... - 296 -
- Counter .... - 297 -
- Counter Application .... - 298 -
- Program Control Instructions .... - 300 -
- Jump to Label Instructions .... - 300 -
- Jump to Subroutine Instructions .... - 301 -
- Nested Subroutines .... - 303 -

Temporary End .... - 304 -
Data Manipulation Instructions .... - 304 -
PLC Data Structure .... - 305 -
MOV Instruction .... - 306 -
MOV Instruction Application .... - 307 -
Data Compare Instructions .... - 308 -
Math Instructions .... - 311 -
Sequencer Instructions .... - 315 -
Trends .... - 317 -
Summary .... - 317 -
Quiz .... - 318 -

Chapter 9: Mental Math .... *- 319 -*

Multiples and Submultiples of Units .... - 319 -
Decimal Numbers .... - 320 -
Whole Numbers .... - 320 -
Multiples Number Conversion .... - 320 -
Submultiples Number Conversion .... - 321 -
One-Over Reciprocal with Multiples and Submultiples .... - 323 -
Multiplication and Division with Multiples and Submultiples .... - 325 -
Percentage to Decimals .... - 326 -
Log to Real Number .... - 326 -
Summary .... - 328 -
Quiz .... - 328 -

Abbreviations and Acronyms .... *- 329 -*

Index .... *- 335 -*

Chapter 1: Direct Current (DC)

Students majoring in electronics always start with a Direct Current (DC) class. DC is a basic electronic theory that you must learn and understand well. This is the first step to a successful career in electronics. Let's first define some DC parameters.

Current

Electrical current is quantified as change (Δ or delta) of electron charge (Q) with time. Think of it as flow rate in plumbing. Electrical current is a measure of the number of electron charges (ΔQ) flowing through a point (node) with time (see figure 1.0). Current's unit is amperes (A) with "I" being its symbol.

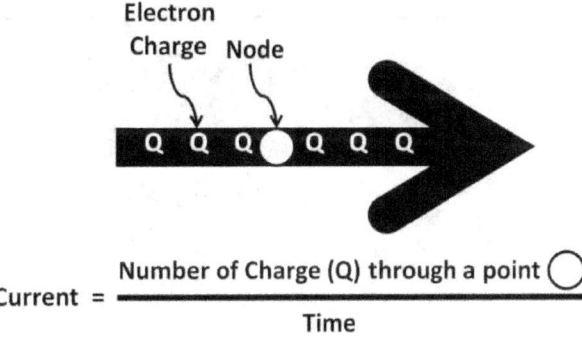

$$\text{Current} = \frac{\text{Number of Charge (Q) through a point}}{\text{Time}}$$

Figure 1.0: ΔQ / time

Current = ΔQ / Time

Resistor

All materials possess resistance, which is a measure of the amount of resistor value. A resistor is a passive electronic device made exclusively for electronic systems. Resistors resist current flow for a given electrical voltage (voltage will be defined shortly). A passive device by definition does not generate energy but rather stores and/or dissipates energy. The most abundant materials used in resistors are copper (Cu) and aluminum (Al). Carbon, thin-film, metal film, and wire-wound are popular resistor types. Resistor size (resistance) is measured in unit Ohms (Ω) with "R" as the symbol. Resistors come in many physical forms. Wire-leads, surface-mount, integrated circuits (ICs) package are popular ones. Figure 1.1 on the next page shows a graphical view of a copper (Cu) wire bundle with a certain length and area exhibiting a finite resistance amount. Internet wires and cables found in residential and commercial dwellings are largely made of copper with a plastic shield on the outside. A resistor can be discrete (one device per item) or manufactured via an IC process housed in an IC package. We will explore more on semiconductor packages later in the chapter. Resistance for a given material strongly depends on the resistor dimension, where resistivity is unique to the materials type:

$$\text{Resistance} = \text{Resistivity} \times \frac{\text{Length}}{\text{Area}}$$

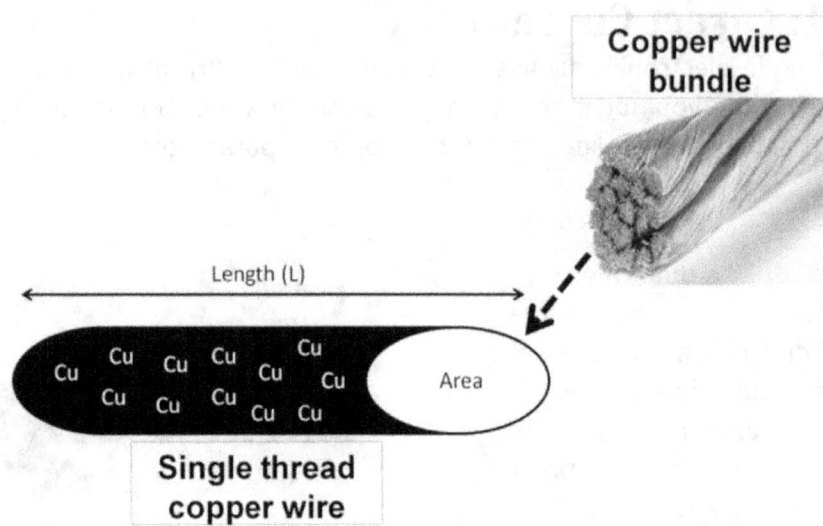

Figure 1.1: Copper wire

Common carbon resistors are measured in the order of several centimeters (see figure 1.1a).

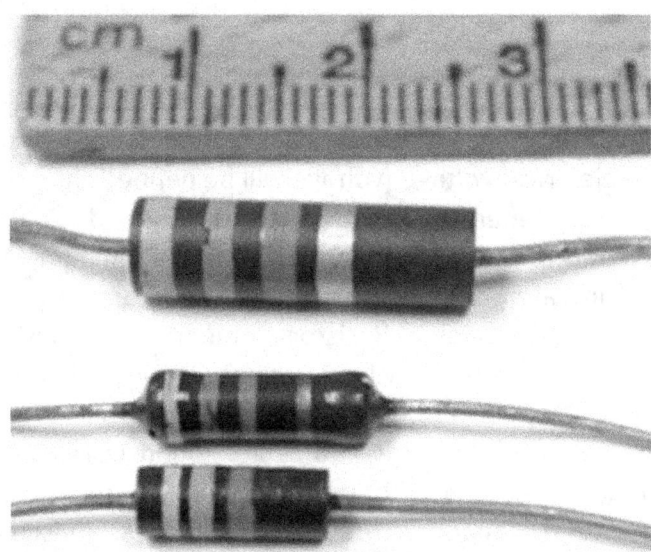

Figure 1.1a: Carbon resistors

Due to the small carbon resistor sizes, color bands are used to indicate resistance values instead of printing them on the resistors. There are four bands. The first band on the left represents the first significant resistance digit. The second band is the second significant digit. The third band is the multiplier, and the last is tolerance. Tolerance determines the maximum percentage change in resistance from its nominal value. Table 1-1 shows the details among band color, digit values, multiplier, and tolerances.

Band Color	First-band Digit	Second-band Digit	Third-band Multiplier	Fourth-band Tolerance
	0	0	1	
Brown	1	1	10	1 %
Red	2	2	100	2 %
Orange	3	3	1000	3 %
Yellow	4	4	10000	4 %
Green	5	5	100000	
Blue	6	6	1000000	
Violet	7	7	10000000	
Gray	8	8	100000000	
White	9	9	1000000000	
Gold				5 %
Silver				10 %
None				20 %

Table 1-1: Resistor band color, digit values, multiplier, and tolerances

Let's apply this to an example. What is the resistance of the carbon resistor that has Brown, Orange, Red, and Gold bands? First, brown yields "1"; orange means digit "3"; red multiplier means "100"; gold represents 5% tolerance. The resistance is therefore calculated as:

$$13 \times 100 = 1,300 \text{ } \Omega \text{ or } 1.3 \text{ k}\Omega \text{ with 5\% tolerance}.$$

Figure 1.1b: Surface-mount resistor

Surface-mount resistors, on the other hand, are popular due to their miniature sizes. They are ideal for portable applications when small size is necessary. Figure 1.1b shows several surface-mount resistors. A surface-mount resistor can be measured as small as 0.2 mm (millimeter) X 0.4 mm (millimeter). Because surface-mount resistors are small, in order to determine their values, numbering codes are used instead of color bands. The numbers printed on the resistor are usually 3-digit numbers. The first two numbers represent the first two digits of the resistor values while the third digit represents the number of zeros. For example, a resistor marked with 203 means **20 X 1,000 Ω** or **20 kΩ**. A 105 resistor gives **10 X 10^5 Ω** or **1 MΩ**. Resistors manufactured by microelectronics technology use different methods to determine resistances. Depending upon the chip manufacturing process, there can be multiple resistor types, ranging from metal and thin-film to poly resistors. The resistances are determined by the vertical and horizontal dimensions in conjunction with the sheet rho (pronounced as row) resistance. Sheet rho's units are in **Ω per square (Ω / square)**. For example, a Bipolar-CMOS (BiCMOS) process thin-film resistor's sheet rho is specified as **1,000 Ω / square**. **Length / Width** defines the square numbers. If the resistor's length and width are drawn as 10 micrometers (um) by 10 micrometer (um) respectively, the number of square equates to **10 um / 10 um = 1**. The resistance is then calculated as:

$$\frac{\text{Length}}{\text{Width}} \times \text{Sheet Rho} = \frac{10 \text{ um}}{10 \text{ um}} \times 1{,}000 = 1 \; \cancel{\text{square}} \times 1{,}000 \frac{\Omega}{\cancel{\text{square}}} = 1{,}000 \; \Omega$$

Regarding the chip manufacturing process, in addition to sheet rho resistances, each process offers a slew of devices with a unique set of parameters. Below are some common ones you will likely encounter. Transistors' minimum geometries: CMOS uses gate length where bipolar transistors use emitter width. Transistors' maximum operating frequencies: capacitors' capacitance per unit area; temperature coefficient (it determines how much variations device parameter changes with temperature); maximum voltage supply and break down voltages; transistors' drawn versus manufactured dimensions, metal level numbers available, and many more. Further explanations of these parameters will be discussed later in this book. Full understanding of these parameters is necessary before deciding on a process to use for a particular chip design. Further details on microelectronic design will also be discussed in later chapters.

Voltage

Voltage is the potential difference (subtraction) between two points (nodes). The object of these points can be any material. The most common materials are electronic devices such as resistors, diodes, and transistors, which are the main focus of this book. Each electrical parameter has its own symbol and unit. They are summarized in table 2-1.

Name	Symbol	Unit
Current	I	Ampere (A)
Voltage	V	Volt (V)
Resistance	R	Ohms (Ω)

Table 2-1: V, I, R symbols; units

Definition

Direct current (DC) states that electrical current flows through a resistor without changes in amplitudes or frequencies. A waveform can be used to make clear such phenomenon. A waveform is a time (transient) domain graph that shows quantities such as voltage, current, or power on the vertical (Y-axis); time on the horizontal (X-axis) (see figure 1.2). In this waveform, the DC voltage level stays the same over time while the frequency of DC is zero. We will further define amplitude and frequency in chapter 3, AC.

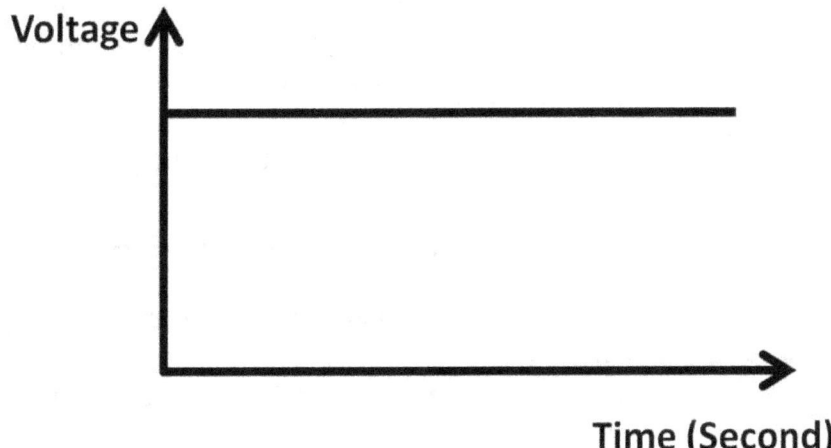

Figure 1.2: Voltage vs. time in DC

Ohm's Law

Ohm's law states that when there is a voltage developed (drop) across a resistor, i.e., voltage difference between two resistor ends (nodes), electrical current is bound to flow. The mathematical relationship between voltage (V), current (A or Amp), and resistance (Ω):

Voltage = Current X Resistance

For a given resistor size, increasing voltage causes current to increase linearly. Thereby, Ohm's law is simply a linear function (see figure 1.3). We can apply the above linear relationship among voltage, current, resistor, and slope concept to calculate resistance. A V-I graph is shown in figure 1.4. Any two points can be used to calculate slope (resistance). Because this is a linear function (straight line), slope (resistance) is fixed.

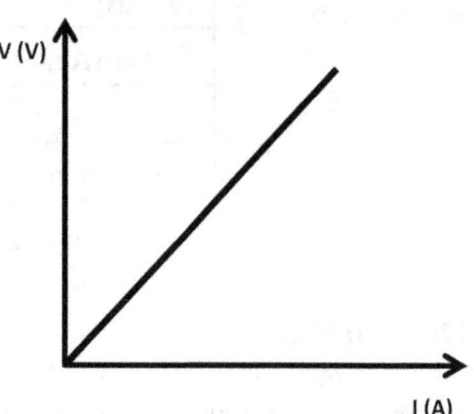

Figure 1.3: Ohm's Law, a linear graph

Resistors are usually in large sizes—thousands of Ωs, sometimes even more. This is because, for a given voltage, large resistance results in lower current (linear relationship). This is essential due to safety and power-saving reasons. Using Ohm's law, 1 V divided by 1 A equals 1 Ω resistance **(1 V / 1 A = 1 Ω)**. One ampere is a lot of current, in fact, current above 100 mA (milliamp) going through the human body is deemed lethal. To lower the current for a given voltage at a safe level, resistance needs to increase. For example, to lower the current to 1 mA, 1 V source yields:

$$R = (1 \text{ V} / 1 \text{ mA}) = 1{,}000 \text{ Ω or } 1 \text{ kΩ}$$

Note: $k = 1 \times 10^3 = 1{,}000$

Many portable electronic designs draw less than 1 mA of current to conserve battery life resulting in large values of R. This explains why thousands or even hundreds of thousands of Ω are frequently seen.

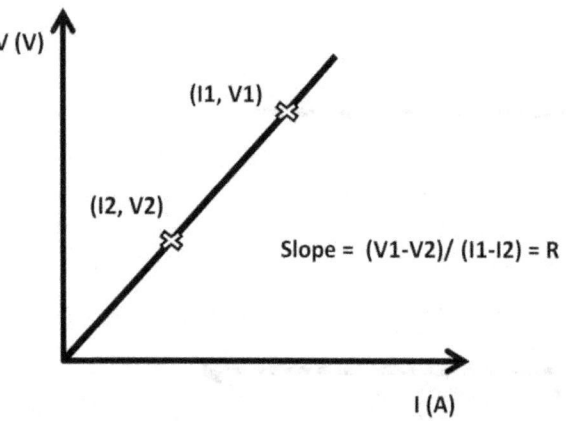

Figure 1.4: Slope equals resistance

Power

Power (P) definition:

$$P = I^2 \times R \text{ or } V^2 / R$$

The unit of power is Watts (W) and its symbol is "P". A modern smartphone power amplifier consumes about 300 mW (milliwatt) in idle mode. With 4 V lithium-ion battery (a popular cell phone battery type), antenna load resistance can be calculated:

$$300 \text{ mW} = 4^2 / R$$

$$R = 53.33 \text{ }\Omega$$

Voltage Source and Schematic

A voltage source is an electronic device that supplies voltage to an electronic load. The electronic load acts as an output that delivers or receives electrical energy to and from an input. Load examples are motors, electric fans, lights, etc. An ideal DC voltage source has zero internal resistance, capable of sourcing (sending) and sinking (receiving) infinite current amount to and from the load. A non-ideal voltage source contains finite (non-zero) internal resistance and cannot supply or receive infinite current amount. The most common DC voltage source is alkaline household battery commonly used in portable electronics. Figure 1.5 shows several popular alkaline battery types (Energizer brand). Most alkaline batteries are cylindrically shaped except the 9 V type, which is rectangular. They differ in sizes, voltage ratings, and mAh. mAh stands for milliamp-hour, which is equivalent to electron charge.

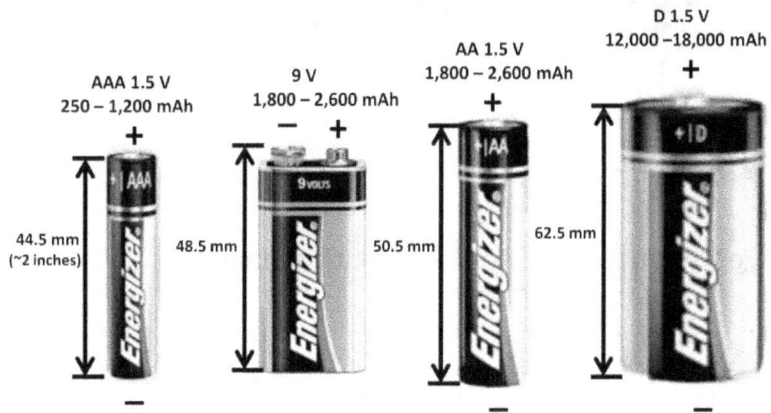

Figure 1.5: Alkaline battery types

It describes the electrical current capacity of a battery. Both AA, AAA, and D batteries and are rated at 1.5 V with different mAh ratings. A 9 V battery is rated at 9 V DC (1,800 – 2,600 mAh). If, for example, a portable device draws 100 mA discharge current to operate, the battery will last a minimum of 18 hours (1800 mAh / 100 mA = 18 hours). Other popular batteries are button-sized batteries (button cells) suitable for lightweight applications. They come in a wide range of types, sizes and voltage ranges. Button cells typically are rated at 1.5 V with less mAh (150 – 200 mAh).

Current Source and Schematics

A current source is an electronic device that supplies electrical current to a load. An ideal current source has infinite output resistance capable of supplying an infinite amount of current. Most electronic designs can be graphically expressed in the form of schematics (electronic circuits). Schematics include graphical V, I, and R symbols, plus various electronic components and wires. Figure 1.5a shows schematic symbols of voltage and current sources with ground connected at the other end. Ground is an electrical connection that is referenced to zero voltage potential (0 V). Schematics can be hand drawn on paper, although the majority of schematics are entered into computer software. This makes it very easy to design and modify electrical schematics. Popular electronic schematic software tools will be discussed in chapter 4, Analog Electronics. Ideally, ground is at absolute 0 V with zero resistance. Keep in mind real-world ground has non-zero resistance. The ground signal amplitude depends on multiple factors (mostly from electrical noise), which will be discussed later on. The current source symbol in figure 1.5a contains an arrow signifying the current flow direction. Both triangular and horizontal line ground symbols are interchangeable although some use the triangular symbol strictly for power ground; the horizontal symbol for signal ground. Triangular ground symbols are used throughout this book.

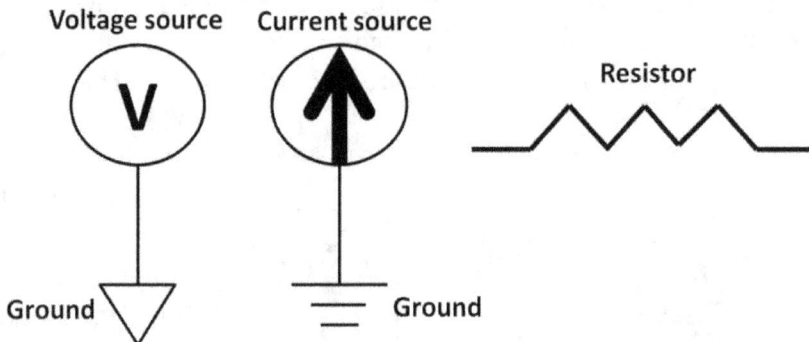

Figure 1.5a: Voltage, current, and resistor schematic symbols

Electrons

An atom is made up of tiny particles: protons (positive charge), neutrons (neutral), and electrons (negative charge). Protons and neutrons are in the center of an atom while electrons surround the nucleus. Electrons are ions (particles) containing negative charges. Difference in electron and proton numbers gives rise to various atom structures (chemical elements). In this book, we mainly focus on chemical elements that are used in microelectronics, such as silicon and germanium. The negatively-charged electrons are attracted to positive charges (terminals and polarities). The symbol "Q" quantifies electron charges. The unit of Q is coulomb (C). One electron charge holds:

$$\text{One Electron Charge} = 1.6 \times 10^{-19} \text{ C}$$

Current versus Electrons

In figure 1.6, a positive voltage source (positive signs) is connected to a resistor with a wire. The other end of resistor connects to ground (negative polarity) creating a loop. Due to a positive charge at the voltage source, according to Ohm's law, a current is bound to flow through the resistor in clockwise direction (inner arrow) while electrons (E-) are flowing towards the positive charges arriving at the voltage source. Keep in mind the electron and current flow in reverse directions.

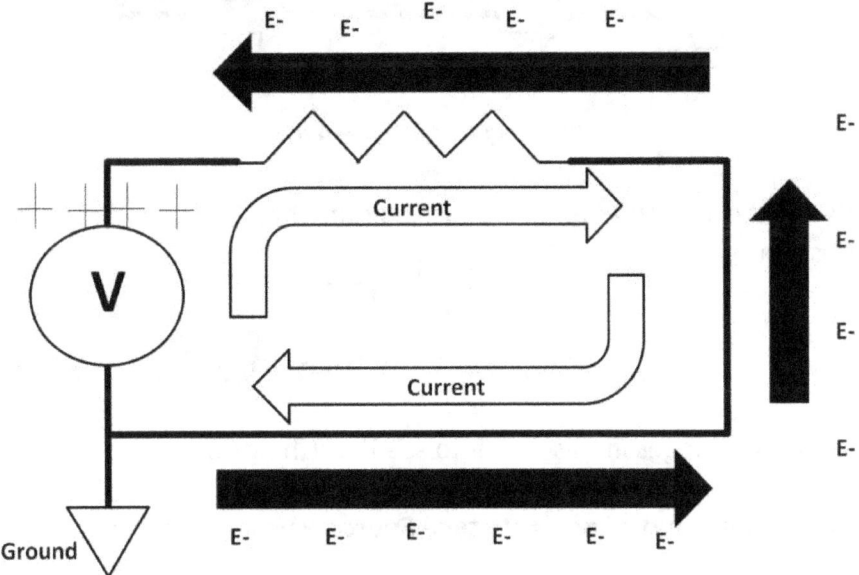

Figure 1.6: Electron vs. current flow

Kirchhoff's Voltage Law (KVL)

KVL states that the **sum of all voltages around a loop = 0**. A simple circuit in figure 1.7 applies and explains this theory. There is only one theory to apply: Ohm's law and we will use it twice. This circuit contains a 5 V voltage source connects to a 10 Ω resistor. We use Ground to close the loop. By using Ohm's law, current can be evaluated:

$$V = I \times R$$

$$I = V / R$$

$$I = (5\ V) / (10\ \Omega) = 0.5\ A$$

This circuit is a series circuit. There is only one branch the current could go. We will visit more series circuits in a moment.

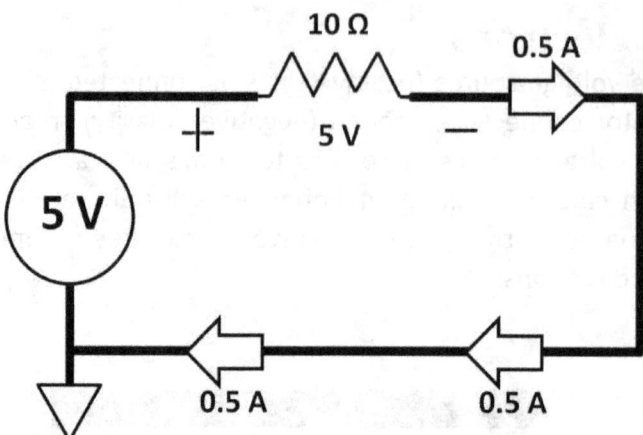

Figure 1.7: Series circuit

By using Ohm's law the second time, we could find out what the voltage drop is across the 10 Ω resistor, V_resistor:

V_resistor = I X R

V_resistor = (0.5 A) X (10 Ω) = 5 V

Now, we could use all voltages in this circuit to see if KVL holds up.

Sum of All Voltages Around a Loop = Voltage Source + Voltage Drop Across the Resistor.

5 V + (− 5 V) = 0 V

It checks out! Notice that the voltage drop across the resistor contains a negative sign (polarity). The reason is that the voltage on the left-hand side of the resistor was higher (+) than the voltage on the right-hand side of the resistor (−). The positive resistor sign "opposes" the positive polarity of voltage source, hence the negative sign in the KVL calculations (see figure 1.7). The importance of this circuit is twofold. First, it demonstrates how simple it is to apply and explain the circuit using Ohm's law and KVL. Secondly, despite the circuit's simplicity, any electronic circuit regardless of its complexity can always be explained by Ohm's law and KVL. Sometimes, you will hear statements such as; there is a "short" in an electronic circuit that caused damages. Applying Ohm's law easily explains it. In figure 1.7, if the 10 Ω resistor were "shorted" (zero resistance) and we applied Ohm's law, I = V / R, where V = 5 V, R = 0. I = 5 V / 0 Ω = **infinite**. Current becomes infinitely large causing damage to the system. Realistically, any electronic system, no matter how shorted it becomes, possesses a finite amount of resistance.

Kirchhoff's Current Law (KCL)

KCL states that current going into (passing through) a point (node) is equal to current coming out of the same node. We could use the same circuit in figure 1.7 to examine this theory. This is a series circuit. The current goes into left-hand side (node) of the resistor, and thus is equal to the right side of the resistor. We will use a parallel circuit in the next segment to further explain KCL.

Parallel Circuit

Series circuit states that current only flows in one direction. In parallel circuits however, current flows in more than one direction (see figure 1.8).

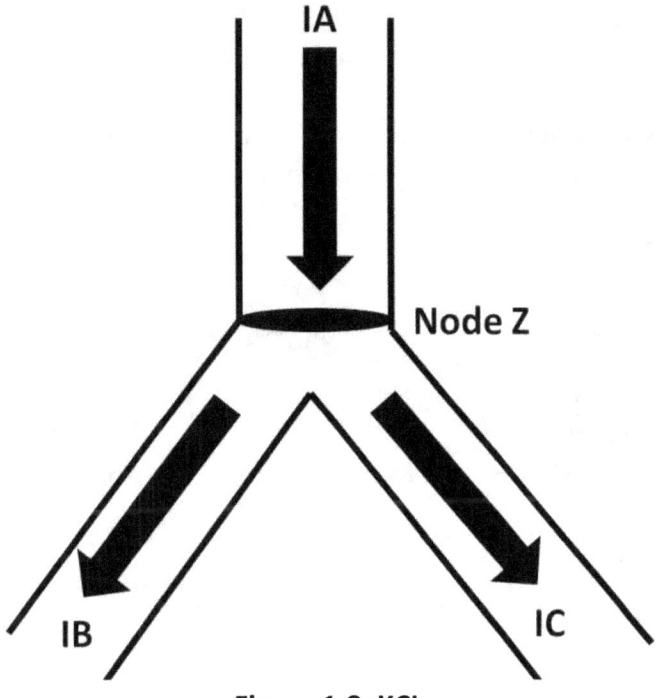

Figure 1.8: KCL

Current A (IA) goes into node Z and is equal to sum of both currents IB and IC, coming out of the same node (node Z). Mathematically, it's simply:

$$IA = IB + IC$$

Parallel Resistor Rule
Equivalent resistance (R_ equivalent) of two resistors (see figure 1.9):

$$R\_equivalent = \frac{Product\ of\ Resistors}{Sum\ of\ Resistors}$$

Figure 1.9: Parallel resistor rule

If the parallel (||) resistors number is two or more, the equivalent resistance is equal to the reciprocal of the sum of individual reciprocal resistances (see figure 1.10).

Figure 1.10: Multiple parallel resistors

If A = 1 Ω, B = 2 Ω, C = 5 Ω,

$$R\_equivalent = \left(\frac{1}{A} + \frac{1}{B} + \frac{1}{C}\right)^{-1}$$

$$= \left(\frac{1}{1} + \frac{1}{2} + \frac{1}{5}\right)^{-1}$$

$$= \frac{10}{17} \Omega = 0.58\ \Omega$$

You may notice that the equivalent resistance of multiple resistors is always slightly less than the smallest resistor among the resistor groups. From the above example, the equivalent resistance of 1 Ω, 2 Ω, and 5 Ω is 0.58 Ω. It's less than the smallest resistor value 1 Ω. This gives you a quick way of knowing if the equivalent resistance you come up with makes sense or not. Note that if the resistor numbers in parallel are exactly the same sizes, the equivalence resistance is calculated as resistance of one resistor divided by the total resistor number, **e.g., 10 || 10 || 10 = 3.33 Ω (10 / 3 = 3.33 Ω)**. This rule, however, doesn't apply to parallel resistors that have different sizes.

Series Resistor Rule

Equivalent resistance = Sum of all resistances (see figure 1.11).

Figure 1.11: Series resistor rule

Let's use a simple parallel circuit to explain series and parallel resistor configurations. (See figure 1.12).

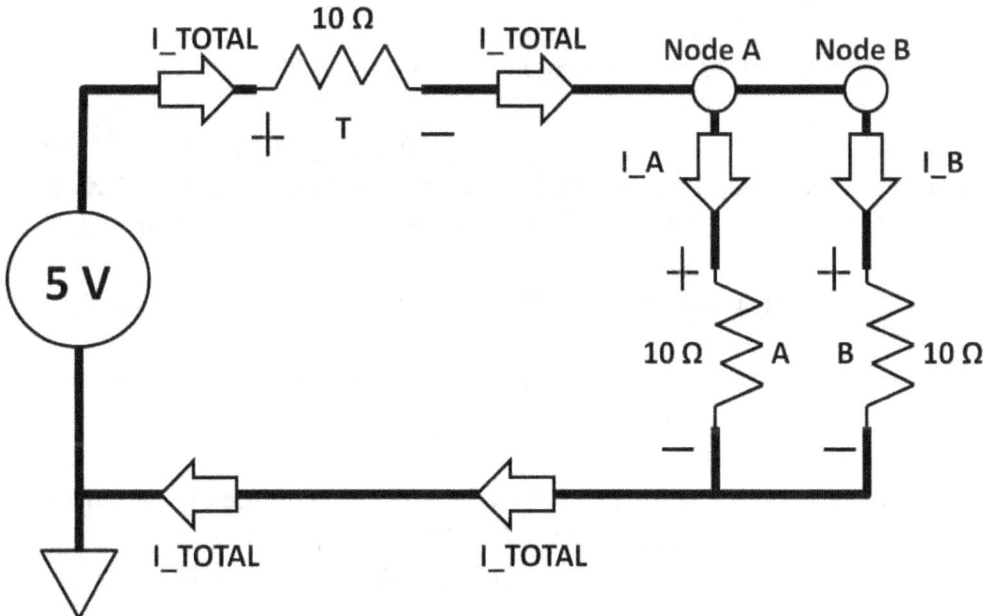

Figure 1.12: Simple parallel circuit

In this illustration, resistor A, B forms a parallel circuit. Total current (I_TOTAL) going towards node A, B is divided into two separate branches, according to KCL. To calculate I_A, I_B, we first calculate the total resistance of the entire circuit. I_TOTAL can then be found. The idea is to consolidate all three resistors T (10 Ω), A (10 Ω), and B (10 Ω) into one resistor (Equivalence) and one voltage source. We can then use Ohm's law to calculate I_TOTAL. According to figure 1.9, resistor A, B can be combined into one resistor, R_eq:

$$A \parallel B = R\_eq = \frac{10 \, \Omega \times 10 \, \Omega}{10 \, \Omega + 10 \, \Omega} = 5 \, \Omega$$

We then further consolidate the two resistors (T and R_eq) into one resistor. We will call it R_total. Using series resistor rule, **R_total = T + R_eq = 10 Ω + 5 Ω = 15 Ω**. The consolidated one voltage source, one resistor circuit is shown in figure 1.13.

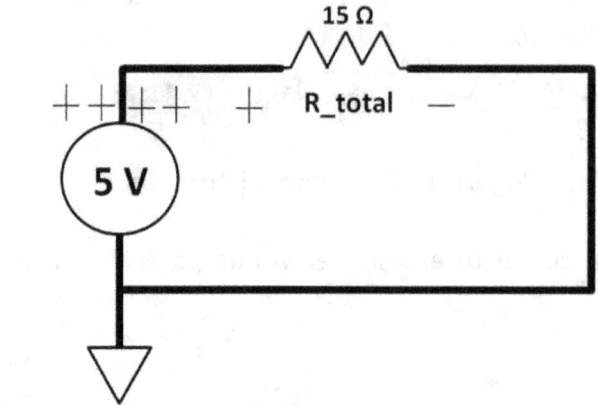

Figure 1.13: Simplified one voltage, one resistor circuit

We now can simply use Ohm's law to calculate **I_TOTAL. I_TOTAL = 5 V / 15 Ω = 0.33 A**. At this point, we can apply KVL to make sure the analysis is valid. **5 V + (− 0.33 A X 15 Ω) = 0 V**
To figure out I_A, I_B, we first calculate the voltage drop across T using Ohm's law:

Voltage Drop across T = (I_TOTAL) X (T) = 0.33 A X 10 Ω = 3.33 V

Since voltage at left side of T is 5 V and the voltage drop across T is 3.33 V, the voltage at the right-hand side of T (node A and B) is:

**Voltage Drop across T = (Voltage at Left Side of T) − (Voltage at Right Side of T)
5 V − (Voltage at the Right Side of T) = 3.33 V
(Voltage at the Right Side of T) = 5 V − 3.33 V = 1.67 V**

It's crucial to recognize that voltage across a device means the difference (subtraction) between two nodes. Now we can use Ohm's law again to calculate I_A and I_B. Because voltage at right side of T is common to node A and B (**Voltage at Node A = Voltage at Node B**):

**I_A = (Voltage at Node A) / A = 1.67 V / 10 Ω = 0.167 A
I_B = (Voltage at Node B) / B = 1.67 V / 10 Ω = 0.167 A**

To prove the analysis is correct, simply use KCL which states that **I_TOTAL = I_A + I_B**

I_A + I_B = 0.167 A + 0.167 A = 0.33 A = I_TOTAL, it checks out!

Current Divider Rule

The current divider rule states that the current on one branch is the total current multiplied by the ratio of total current. When seeking current A, the numerator contains resistor B and vice versa.

$$I\_A = I\_total \times \left(\frac{B}{A+B}\right)$$

$$I\_B = I\_total \times \left(\frac{A}{A+B}\right)$$

Using figure 1.12 in the previous example, I_A and I_B can be easily calculated:

$$I\_A = (0.33\ A) \times \frac{10\ \Omega}{10\ \Omega + 10\ \Omega} = 0.167\ A$$

$$I\_B = (0.33\ A) \times \frac{10\ \Omega}{10\ \Omega + 10\ \Omega} = 0.167\ A$$

Notice if both resistor sizes are the same on each branch, the current amount will be equally divided in a parallel circuit. If the resistors A and B are different sizes, the current is less on the branch that has the larger resistor and vice versa. This concept is illustrated in figure 1.14.

In this example, **I_total = 2 A, A = 20 Ω, B= 10 Ω**:

$$I\_A = I\_total \times \left(\frac{B}{A+B}\right) = (2\ A) \times \left(\frac{10\ \Omega}{10\ \Omega + 20\ \Omega}\right) = 0.66\ A$$

$$I\_B = I\_total \times \left(\frac{A}{A+B}\right) = (2\ A) \times \left(\frac{20\ \Omega}{10\ \Omega + 20\ \Omega}\right) = 1.33\ A$$

This shows that I_A is less than I_B **(A's resistance > B's resistance)**. To further prove this is correct, apply KCL:

I_total = I_A + I_B = 0.66 A + 1.33 A = 2 A
It checks out!

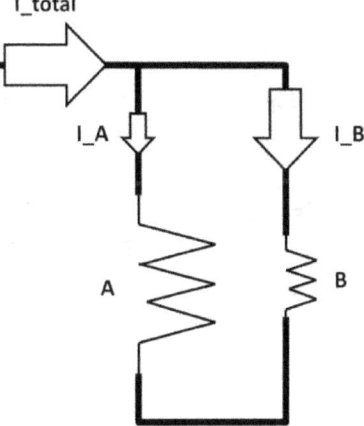

Figure 1.14: Resistor size vs. current amount

Voltage Divider

The voltage divider is used all too often. We will start with the definition then use simple circuits to explain it. Just like it sounds, a voltage divider "divides" voltage. The word "divides" does not mean there is a mathematical division; it means the voltage is "reduced" by the resistors. Below is a simple series circuit (see figure 1.15) to explain voltage divider.

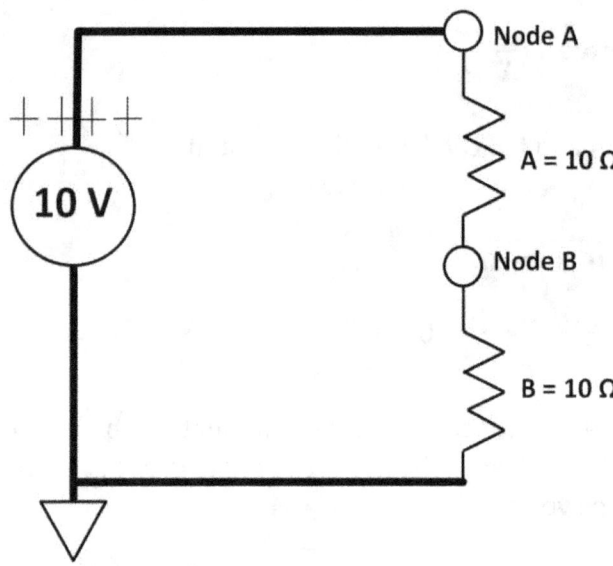

Figure 1.15: Simple series resistor circuit

The explanation of this circuit is simple, not surprisingly, using Ohm's law. There is only one current branch in this series circuit. The current can be calculated using Ohm's law and the series resistor rule:

$$\frac{10\text{ V}}{10\text{ Ω} + 10\text{ Ω}} = 0.5\text{ A}$$

The voltage at Node A is 10 V (connected to a 10 V voltage source). The voltage across (I R drop) resistor A is the potential difference between node A and B, i.e., **Voltage at Node A – Voltage at Node B** or it can be calculated using Ohm's law: **0.5 A X 10 Ω = 5 V**

Once again, it's important to realize that voltage drop across a resistor is the potential difference between two nodes. Knowing that voltage at node A is 10 V, and voltage drop across resistor A is 5 V, voltage at node B can be found using voltage definition:

(Voltage Drop across A) = (Voltage at Node A) – (Voltage at Node B)

5 V = 10 V – Voltage at Node B

Voltage at Node B = (10 V – 5 V) = 5 V

For voltage drop across resistor B, it would be **Voltage at Node B – ground (0 V) = 5 – 0 = 5 V**. All voltage drops (I R drops) are shown in figure 1.16.

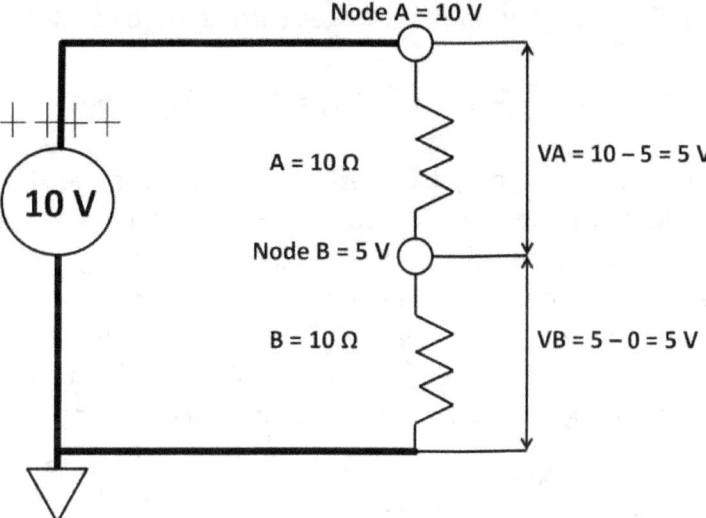

Figure 1.16: Voltages across A and B

There are voltage drops across each resistor. Voltage was reduced (divided) from the 10 V voltage source. In other words, voltages across each resistor cannot exceed the 10 V voltage source. Some use this formula when it comes to voltage divider:

$$VA = (10\ V) \times \frac{RA}{RA + RB}$$

$$VB = (10\ V) \times \frac{RB}{RA + RB}$$

RA is in the numerator when calculating VA. RB is in the numerator when calculating VB. VA and VB are simply the ratio of individual resistance (RA, RB) over the sum of all resistances (RA + RB) in the circuit. If you look closely, the VA, VB formula comes from Ohm's law and series circuit rule. We know that the current going through A and B are the same (series circuit rule). **VA / 10 Ω = VB / 10 Ω**. This current can be calculated from the 10 V source in series with **RA + RB** (Ohm's law):

$$\frac{10\ V}{10\ \Omega + 10\ \Omega} = 0.5\ A$$

$$\frac{VA}{10\ \Omega} = \frac{VB}{10\ \Omega} = \frac{10\ V}{10\ \Omega + 10\ \Omega} = 0.5\ A$$

Thus, **VA = VB = (10 Ω) X 10 V / (10 Ω + 10 Ω)**. This is essentially the voltage divider formula.

Although the voltage divider formula does come in handy, one formula will not and cannot fit all because the resistor configurations may be totally different from one circuit to the next. It's much more intuitive to apply basic principles to analyze voltage divider circuits, in fact, any circuits. To see if we come up with the voltages correctly, we use KVL to prove it.

$$10 \text{ V} + (-5 \text{ V}) + (-5 \text{ V}) = 0 \text{ V}, \text{ it checks out!}$$

In the above example, there are only two resistors. Their sizes are the same. In real life, voltage dividers could have more than two resistors exhibiting a variety of sizes and connection configurations. Despite different voltage divider configurations, the method of determining voltages on any node, voltage drop across any resistor, and current through each branch is the same: by using Ohm's law, KVL and KCL. One interesting fact is that if the resistance is larger than the other(s), such resistor would have the most voltage drop across it. This is demonstrated in figure 1.17, where resistor B value is larger than A, thus voltage drop across B is larger than A. This observation is exactly opposite to the current divider rule where larger R sought smaller I and vice versa. In figure 1.17, let's assume

$$RA = 5 \text{ }\Omega, RB = 10 \text{ }\Omega$$

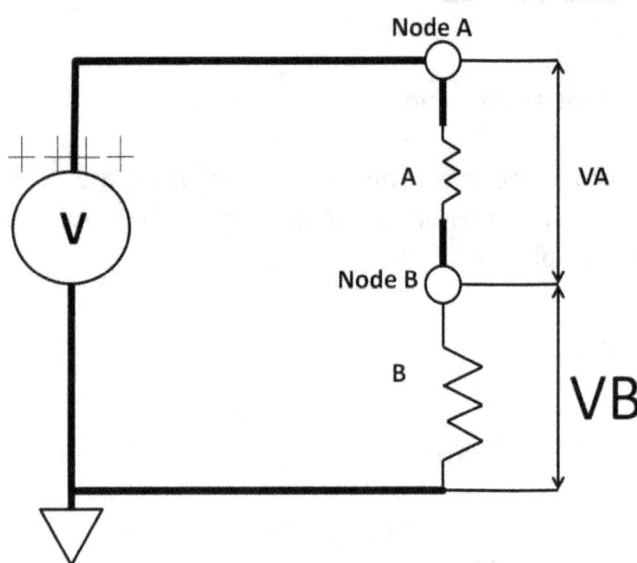

Figure 1.17: Voltage vs. resistor size

To seek the voltage drops across RA and RB, we use the voltage divider formula:

$$VA = (10 \text{ V}) \times \frac{RA}{RA + RB}$$

$$VA = (10 \text{ V}) \times \frac{5 \text{ }\Omega}{10 \text{ }\Omega + 5 \text{ }\Omega} = 3.33 \text{ V}$$

$$VB = (10 \text{ V}) \times \frac{RB}{RA + RB}$$

$$VB = (10 \text{ V}) \times \frac{10 \text{ }\Omega}{10 \text{ }\Omega + 5 \text{ }\Omega} = 6.67 \text{ V}$$

Check with KVL:

$$10 \text{ V} + (-6.67 \text{ V}) + (-3.33 \text{ V}) = 0 \text{ V}, \text{ it checks out!}$$

This example shows that, in a series circuit, if resistance (RA) is higher, there is more voltage drop (VA) across RA than RB. Regardless of resistor values or circuit configurations, KVL and Ohm's law always hold true.

Superposition Theorems

So far, we've focused only on one voltage source circuit. Practical circuits have more than one voltage and/or current source. Numerous theories exist which attempt to explain how the circuits are analyzed in academic textbooks (Thevenin, Norton, and Mesh, just to name a few). I decided to use superposition because of its simplicity. By definition, superposition states that if a circuit contains multiple voltage or current sources, any voltage at a node within the circuit is the algebraic voltage sum found by calculating individual voltage one at a time. Furthermore, any voltage source will be seen as a short to ground when calculating other voltages in the remaining circuit. Any current source will be seen as open circuit. Let's use a simple example to understand superposition (see figure 1.18).

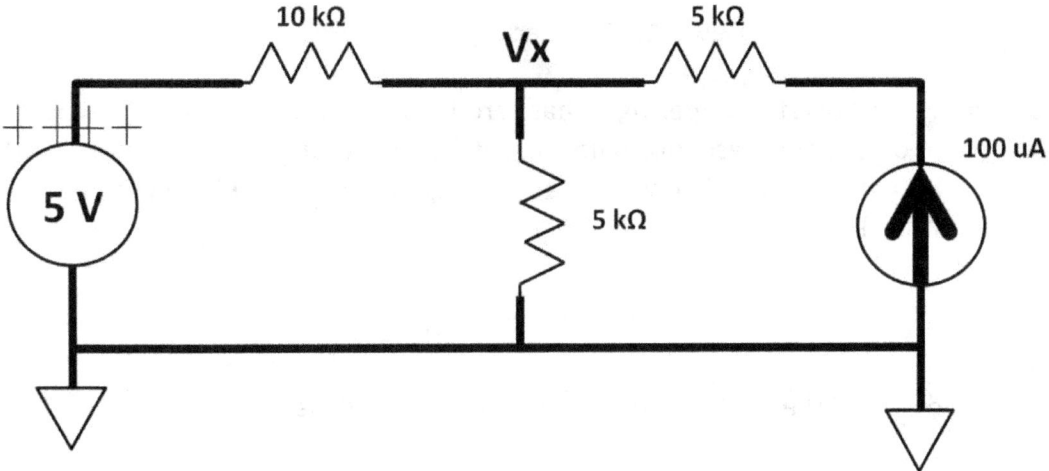

Figure 1.18: Superposition circuit example

The goal is to find out what the voltage is at Vx if the current source pushes out 100 uA (100 microamperes, 100×10^{-6} A) current and DC voltage source is 5 V.

Steps:
1) Isolate the circuit into two separate ones.

2) Start with the voltage source on the left; force the current source open. Then calculate Vx_1. The individual circuit is shown in figure 1.19.

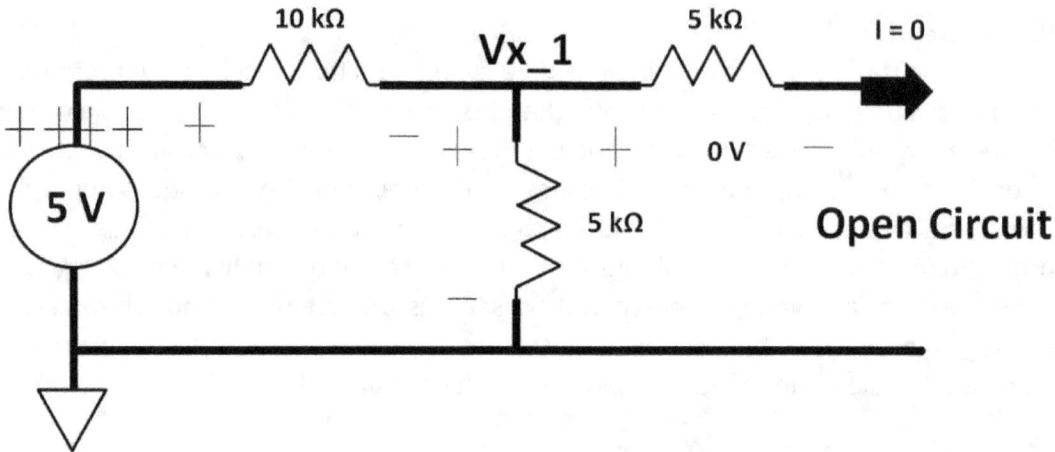

Figure 1.19: Superposition circuit 1

Noticed the 5 kΩ resistor (upper right) has zero I R drop (voltage across it) because of the open circuit on the right resulting in no current flowing through it. Ohm's law says, **V = I X R = 0 X 5 kΩ = 0 V**. Vx_1 is then viewed as a voltage between 10 kΩ and the vertical 5 kΩ (voltage divider):

$$Vx\_1 = (5\ V) \times \frac{5\ k\Omega}{5\ k\Omega + 10\ k\Omega} = 1.67\ V$$

3) Second circuit: The 5 V DC source is shorted to ground (see figure 1.20).

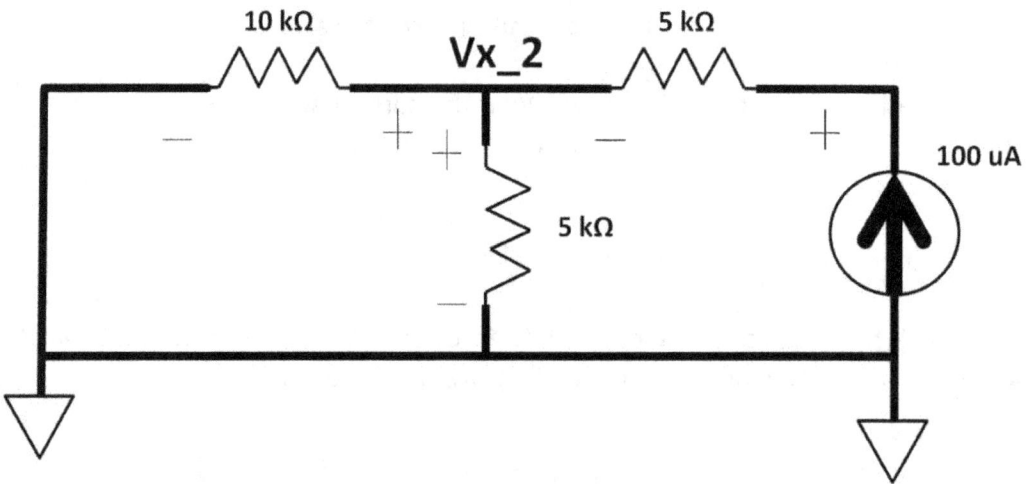

Figure 1.20: Superposition circuit 2

Use parallel resistor rule, 10 kΩ and the vertical 5 kΩ can be combined:

$$\frac{5\ k\Omega \times 10\ k\Omega}{5\ k\Omega + 10\ k\Omega} = 3.33\ k\Omega$$

By inspection, figure 1.20 is transformed to figure 1.21. This is a series circuit where the voltage drop across 3.33 kΩ is between Vx_2 and ground (0 V). Ohm's law states that **Vx_2 = 100 uA X 3.33 kΩ = 0.333 V** when a 100 uA fixed current source flows through 3.33 kΩ.

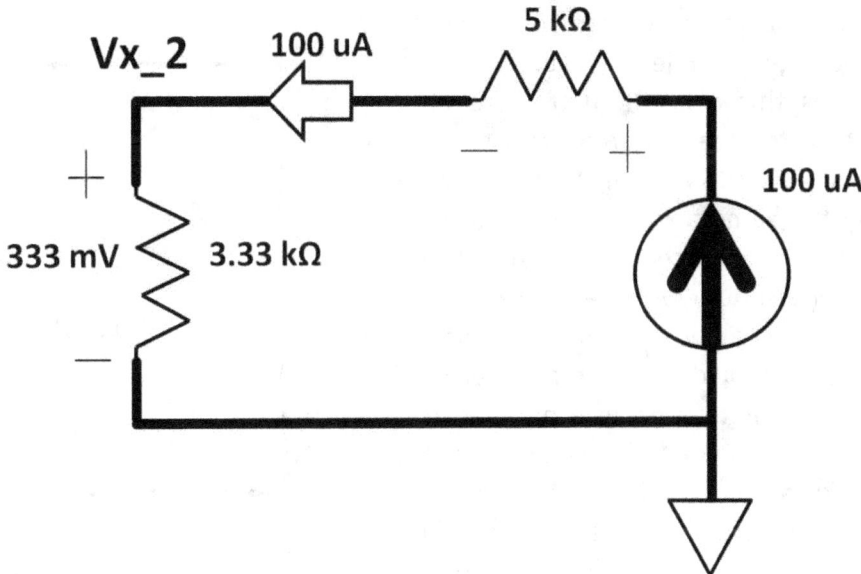

Figure 1.21: Circuit 2 transformation

The resulting Vx can now be found by summing Vx_1 and Vx_2:

Vx_1 + Vx_2 = 1.67 V + 0.33 V = 2 V

DC Circuits

1) What is the difference between an ideal and non-ideal voltage source?

   This question leads to the understanding of voltage source and voltage divider non-ideal characteristics.

   Rules:
   - Ideal voltage source: Zero internal resistance
   - Non-ideal voltage source: Non-zero (finite) internal resistance
   - Ideal current source: Infinite internal resistance
   - Non-ideal current source: Non-zero (finite) internal resistance

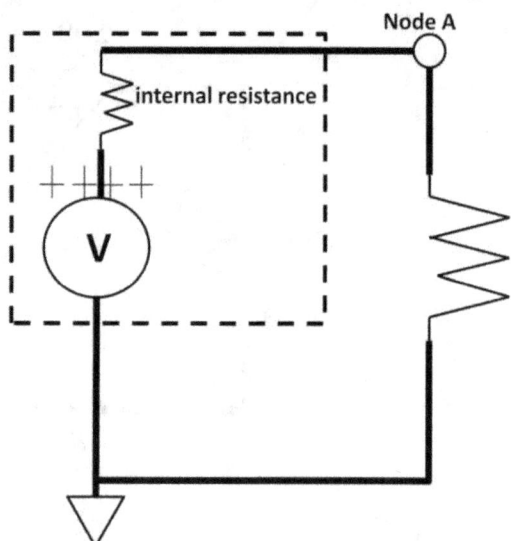

Figure 1.22: Non-ideal voltage source

A non-ideal voltage source can be viewed as a voltage divider. Figure 1.22 demonstrates this concept. If it were an ideal voltage source, internal resistance would be zero Ω. The voltage at node A will be exactly the same as voltage originating from the voltage source. If node A is the output voltage, input would be the same as the output. In a not-so-perfect world, voltage source would have finite internal resistance. This finite resistance originating from the voltage source makes the circuit look just like a voltage divider. Voltage at node A is no longer the same as the original voltage source. In a non-ideal world, when you connect a voltage source to a resistor, the output will not be exactly the same as the input. High quality power supplies offer extremely low internal resistance (still non-zero), and your output is "almost" the same as the input. It's for this reason voltage divider is seldom used as a constant voltage source. Using Figure 1.22 as an example, if the original voltage source on the left is 10 V, the intended voltage output is 5 V at node A. By design, we set both resistors to have the same values (voltage divided by half) so that 5 V at node A can be obtained. In reality, the voltage at node A won't be constant at 5 V. Firstly, any changes from the original input source will change the voltage at node A (again by the voltage divider action). Secondly, any change in the resistances (e.g., caused by temperature variations) will also change the voltage at node A. To achieve a more stable voltage output, low drop-out and switching regulators are used, which will be discussed later in this book.

2) Draw V, I curve of a "real" resistor.

This question tests how much you know about non-ideal resistors behavior. Back in figure 1.3, Ohm's law is depicted as a linear function. In reality, it's a linear relationship only up to a certain point. This point is determined by how hot the resistor gets. Figure 1.23 shows this heating effect. As electrons (E-, current in reverse direction) pass through a resistor made of copper (Cu), the resistor heats up causing random copper ion movements by the electron bombardment. This random motion decreases the likelihood of available electrons passing through Cu atoms. This causes the resistance to go up. These random copper ions' movements are the electrical noise source. Noise is unwanted signals interfering with circuits. It comes from many different sources, adversely affecting circuit performance and corrupting ground signal. Noise is particularly apparent in AC systems. The V versus I resistor function (see figure 1.24), unlike an ideal resistor I-V curve, is a non-linear function with increasing slope, i.e., increasing resistance. This phenomenon is called temperature coefficient (TC) where resistors' TC is positive. This means resistances go up with temperature. The exponential part of the curve depends largely on the resistor's power rating. If it's within or below the rating, it may not show up in the datasheets. Product datasheets are documentation provided by the electronic device manufacturers detailing device features, functions, descriptions, and ratings, along with device symbols, conditioned parameters, and specifications (spec). They often include graphs, waveforms, sampled circuits, application notes, and package information. Thorough device datasheet understanding allows you to decide quickly if the part is right for your design. A datasheet is a document provided by the electronic system component manufacturers that details specific device name, number, features, functions, and parameters related to the device electrical performances. Many datasheets come with electrical test and characterization graphs along with device's dimensions. Some even provide sampled application circuits.

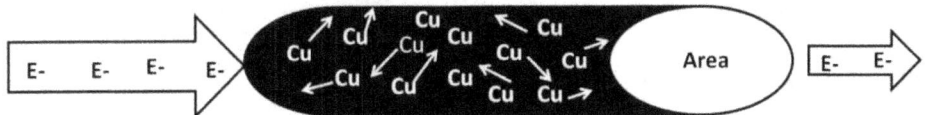

Figure 1.23: Resistor heating effect

Temperature is a major factor of consideration in most electronic systems. Many designs operate in a wide temperature range. The system you work with most likely have electronic parameters fluctuate with temperature. Pay special attention when design and analyze electronics products over temperature.

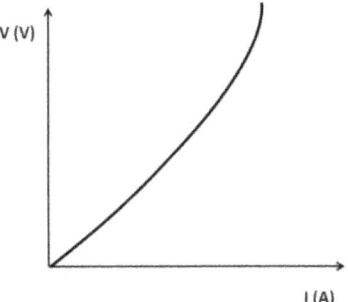

Figure 1.24: Resistor temperature coefficient (TC)

As for purchasing parts, most discrete (stand-alone) components are sold through third party wholesalers (distributors). Well-known ones are Digi-key, Mouser electronics, Arrow electronics, AVNET, and Future electronics. Some chip companies provide direct purchase systems to customers in conjunction with distributors. Analog companies such as Analog Devices, Texas Instruments, Freescale, STMicroelectronics, Maxim Integrated Circuits, Linear Technology, On Semiconductor, Intersil, International Rectifier and Microchip Technology have these systems in place.

IC Packages

Leading semiconductor companies all design and produce integrated circuits (ICs), which are microscopic electronics components manufactured on a piece of semiconductor chip (chip). Some chips are measured in several hundreds of square millimeters in area. The chip will then be housed inside an IC package. The semiconductor package is a crucial part of modern electronics devices. Many devices are manufactured at the sub-micron (less than a micrometer) level that requires an IC package to house the device inside. Figure 1.25 shows an IC that is placed in the middle of an IC package. Most ICs are small enough to place in the palm of a hand. The thin wires connecting the chip to the package pins (needle–shaped structure) are bond wires used to interface the chip with the outside world such as wires and traces on the printed circuit board (PCB). More on PCB in a moment.

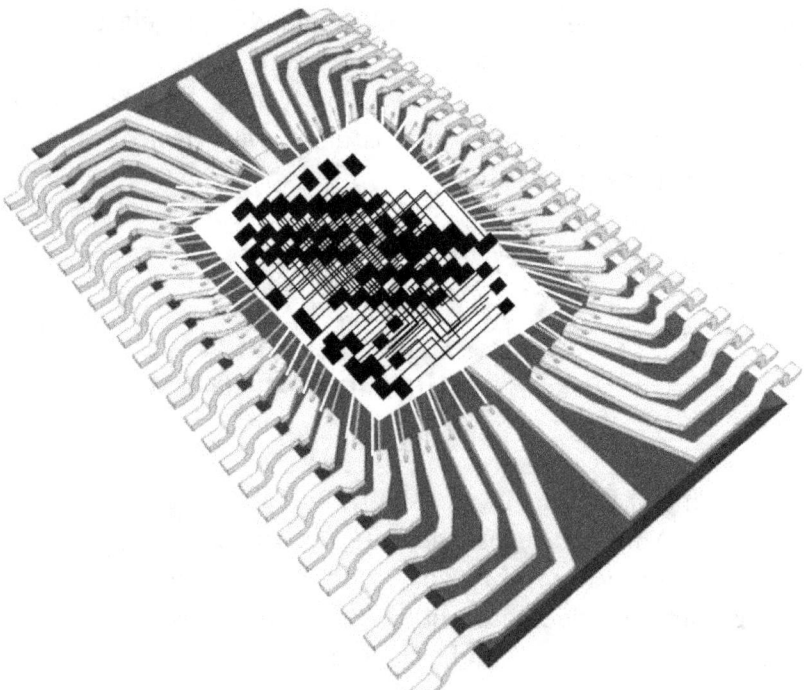

Figure 1.25: IC inside semiconductor package

The purpose of the IC package mainly is to protect the device from external damage, shock, and contaminants such as dust and moisture that could adversely affect device operation. The other purpose of the package is to provide a physical connection of the device itself to the outside world. The IC package is a vital part of the entire electronics industry. They come in many forms and sizes. From wire-bond, dual inline package, ball-grid-array to flip-chip, the advancement in IC packaging technology is always progressing. Semiconductor packaging is large enough that it is categorized as a separate industry. Major package manufacturers are Amkor, Advanced Semiconductor Engineering and Siliconware Precision Industries. The IC package could come with interface pins or ball-shaped bumps that connect to the other ICs or devices at the board level. Figure 1.25a shows a Microchip Technology Analog-to-Digital Converter (ADC) IC with a package length of about 10 mm long.

Figure 1.25a: Microchip Technology MCP3903, ADC IC

3) Show current flow direction and amount of current, if there is any in figure 1.26.
   This simple question tests the concept of potential difference and Ohm's law.
   The voltage drop across the 50 kΩ resistor is the difference between 10 V and 5 V,
   i.e., **10 V – 5 V = 5 V**. Apply Ohm's law:

$$\text{Current} = \frac{\text{Voltage Across Resistor}}{\text{Resistance}} = \frac{5\text{ V}}{50\text{ k}\Omega} = 0.1\text{ mA}$$

Figure 1.26 illustrates the current flow direction from higher potential to lower (left to right). Notice the two horizontal lines symbol representing voltage source symbol.

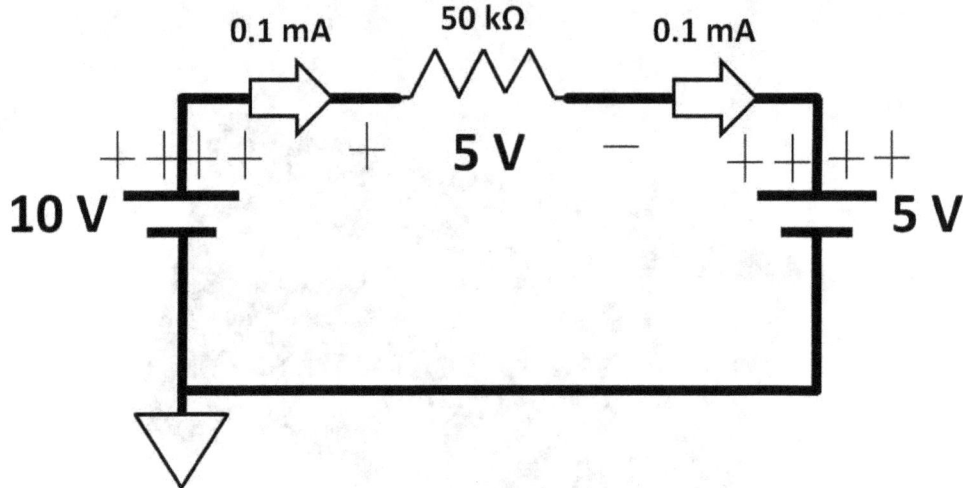

Figure 1.26: Current flow

4) What is the voltage at the ideal source, given the divider circuit in figure 1.27?
   This question tests your knowledge on a simple voltage divider where both resistors connected in series are same sizes. If voltage at node B is 2 V, the source would be twice as much, 4 V. The divider "divides" the source voltage by half with equal resistor sizes.

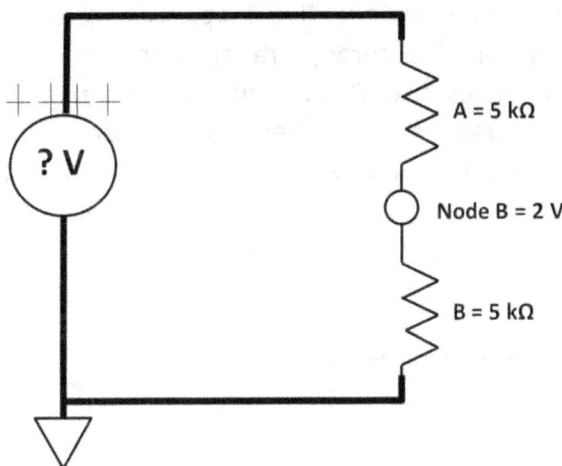

Figure 1.27: Voltage divider

5) There are many kinds of electronic measuring instruments. The most basic ones are multimeter, oscilloscope (scope), function generator, and DC power supplies. There are analog and digital multimeters (DMM). Both have the abilities to measure voltage and current. An analog multimeter has a needle to display the measurement results. Digital multimeters (DMM) come with 7-segment display. DMM prices vary depending on brands, features, and specifications such as accuracy and resolutions. Well-known DMM brands are Fluke, Agilent, and Tektronix. Figure 1.28 shows a Fluke DMM Model CNX3000 (Courtesy of Fluke Corporation). DMMs have become mainstream in recent years. Many are portable, designed to be lightweight and available with a wide range of features. In addition to voltage, current, and resistance measurements, some measure capacitance, inductance, frequency, temperature, and diodes. The center dial (see figure 1.28) allows users to switch from measuring voltage and resistance to current. Frequency, capacitance, and inductance will be discussed shortly in chapter 3, AC.

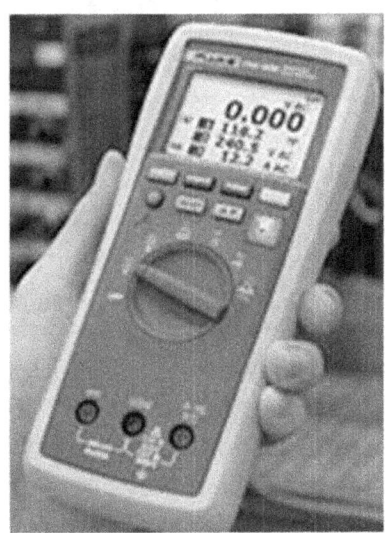

Figure 1.28: Fluke DMM Model CNX3000

A simplified graphical DMM view is shown in figure 1.29. "I", "V," and "COM" are terminals. Test cables and leads are plugged into these DMM terminals. The other ends of the cables connect to the device being measured. "COM" corresponds to common that should be connected to the lowest potential (ground or the most negative supply voltage) during the measurement. Size, accuracy, range numbers, resolutions (the smallest values the DMM could measure), maximum voltage, current ranges are criteria in choosing DMMs. Aside from knowing how DMM works, understanding how it measures voltages helps you troubleshoot your circuits much quickly. We use the voltage divider to expand this idea further (see figure 1.30).

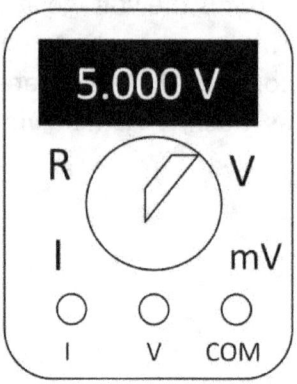

Figure 1.29: Simplified DMM view

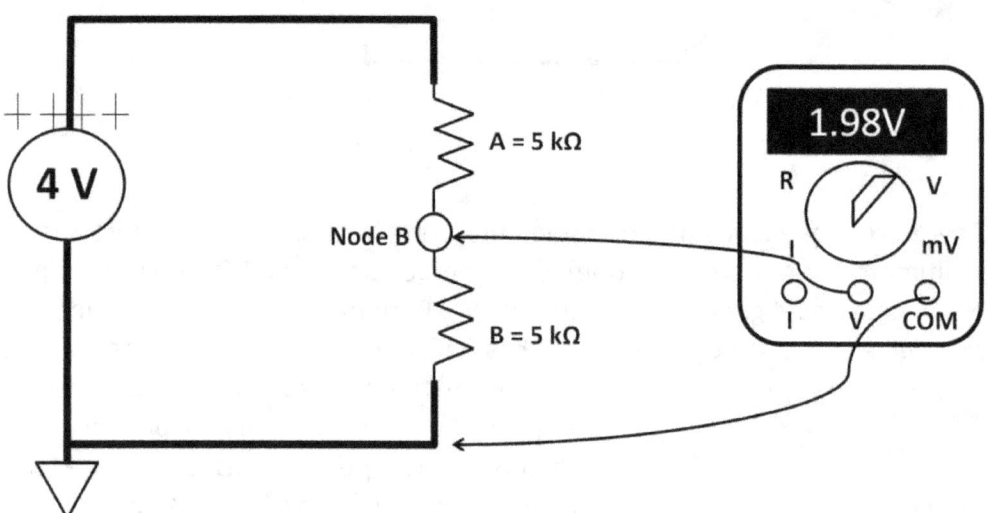

Figure 1.30: DMM measures voltage

We assume the 4 V power supply is an ideal voltage source. DMM is connected as a voltmeter measuring voltage. It measures only 1.98 V. According to the voltage divider rule, it should have measured 2 V. Why? There are two answers to this question. First of all, test leads (represented by the arrows) and plugs (connectors that go into the DMM terminals) consist of finite resistance adding additional resistances to the circuit affecting the measurement. Secondly, the DMM itself contains input resistance, and although very large by design, it's not infinite. DMM's input resistance hence determines the meter's resolution. We have little control over this parameter for a particular DMM. We do have control over leads. To achieve more accurate readings, short leads with the least resistance are more preferred.

What about measuring current? Figure 1.31 shows a simple current measurement using DMM. The DMM is set to measure current as an ammeter. It's connected in series with the circuit. Ideally, the ammeter resistance is infinite. This is the reason why the ammeter cannot be connected in parallel. If it were, no current would flow through the ammeter. Real current sources possess large internal resistance (non-zero) that would impact the overall resistance of the entire circuit. This internal resistance causes a small voltage drop across the ammeter. This "error" voltage is particularly important when measuring low, precise current, (e.g., microamperes (uA) and below). Leading test equipment suppliers such as Agilent offer many power supplies models. Figure 1.32 shows an Agilent DC power supply with multimeter, U3606A. It comes with a voltage supply, current, and resistance measurement with programming capabilities.

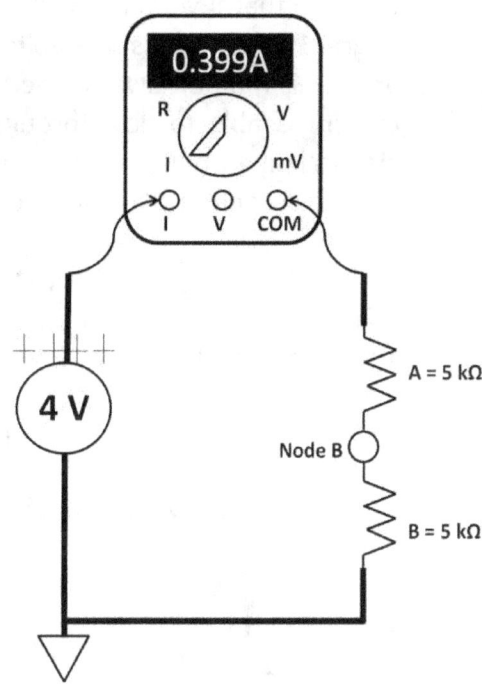

Figure 1.31: DMM measures current

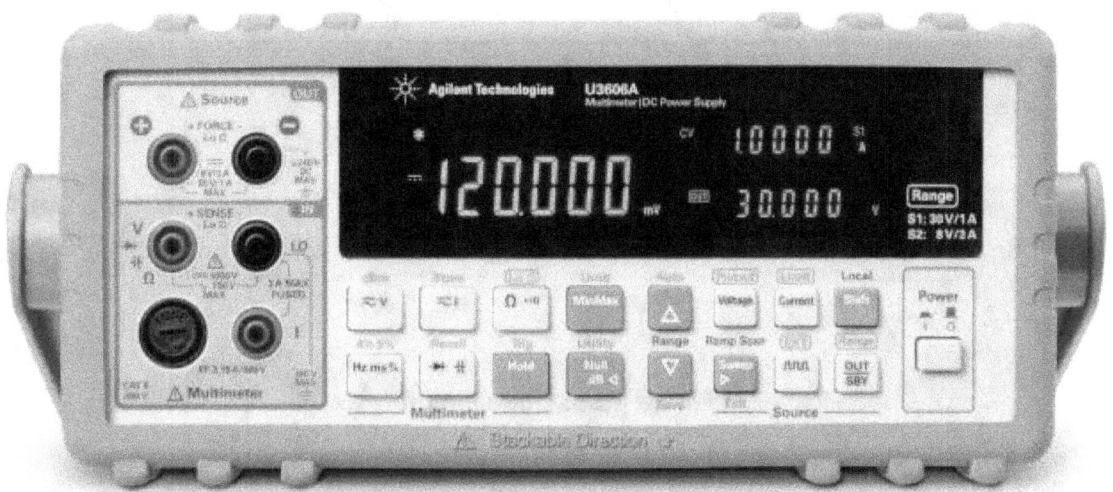

Figure 1.32: Agilent Power supply, multimeter, U3606A

6) Is it true that if you have voltage, you always have current? What about the circuit in figure 1.33? What is the voltage at node A assuming 9 V is an ideal source? The answer is no, not always. The circuit below is a series circuit with a broken loop. No current is able to flow through the loop due to infinite resistance from the open circuit. Using Ohm's law, there is 0 V drop across the resistor, **V = I X R = 0 X R = 0 V** The resistor potential difference can be derived:

(9 V – Voltage at Node A) = 0 V

Then, Voltage at **node A = 9 V**

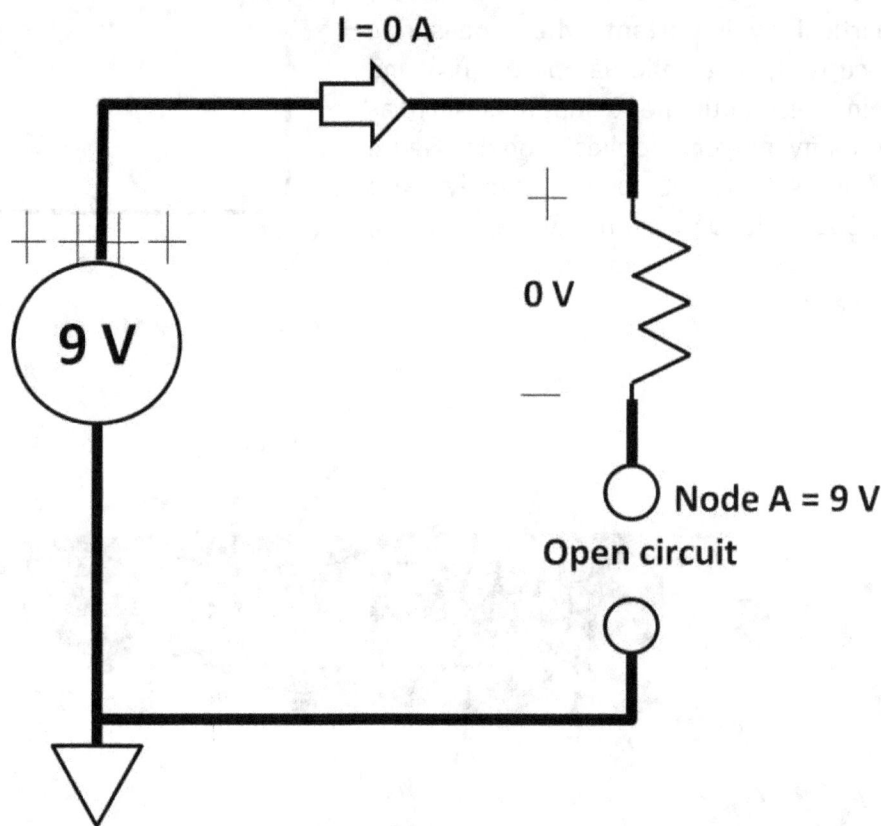

Figure 1.33: Open circuit

7) What is the equivalent resistance between node A and B in figure 1.34? We need to first consolidate this resistor network into one single resistor. Using series and parallel resistor rules, we start from the two parallel resistors D and E (5 Ω using the parallel resistor rule).

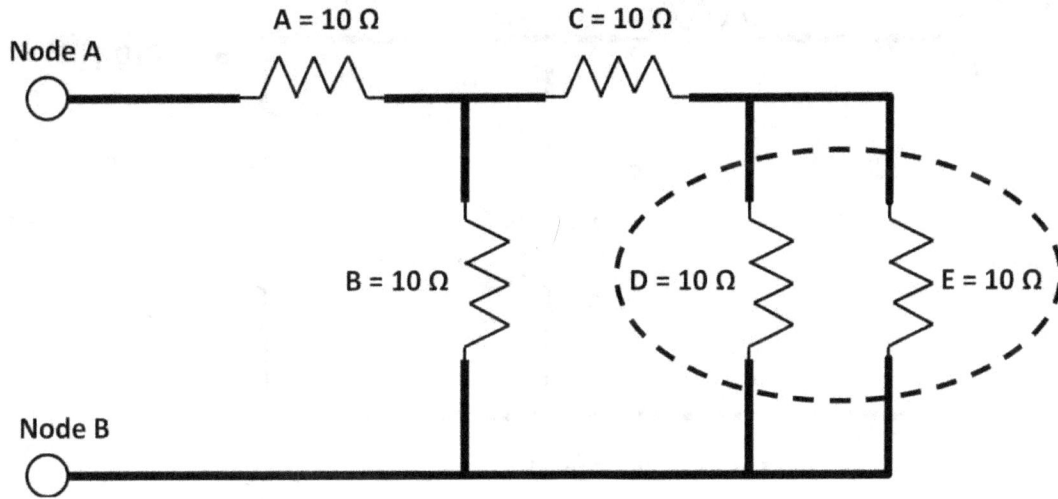

Figure 1.34: Equivalent D and E resistance

Then we further simplify it by combining C, parallel of D and E (5 Ω) below (see figure 1.35).

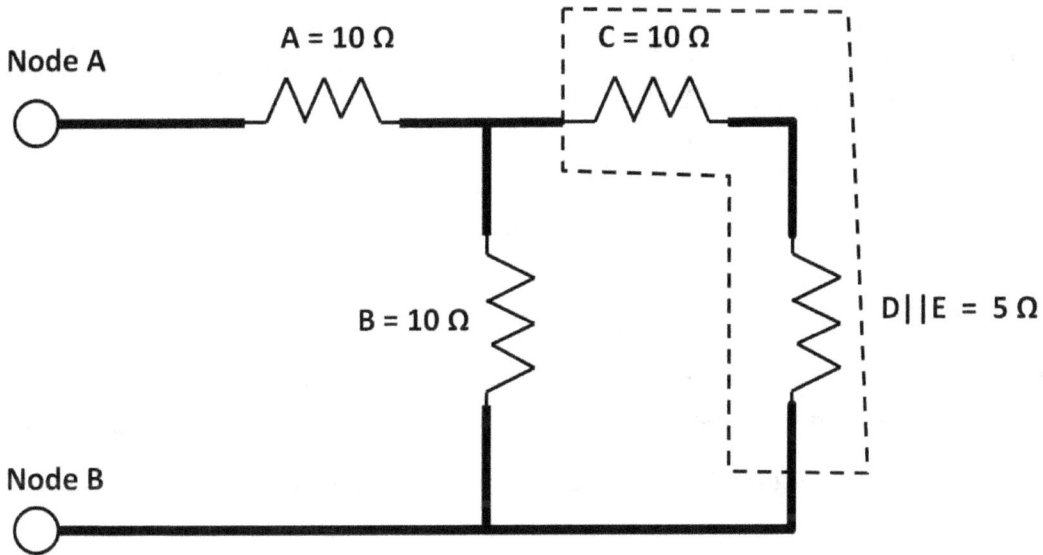

Figure 1.35: C + parallel of D and E

The result is a 15 Ω resistor. We then use the parallel rule to combine B and 15 Ω. This yields 6 Ω (see figure 1.36).

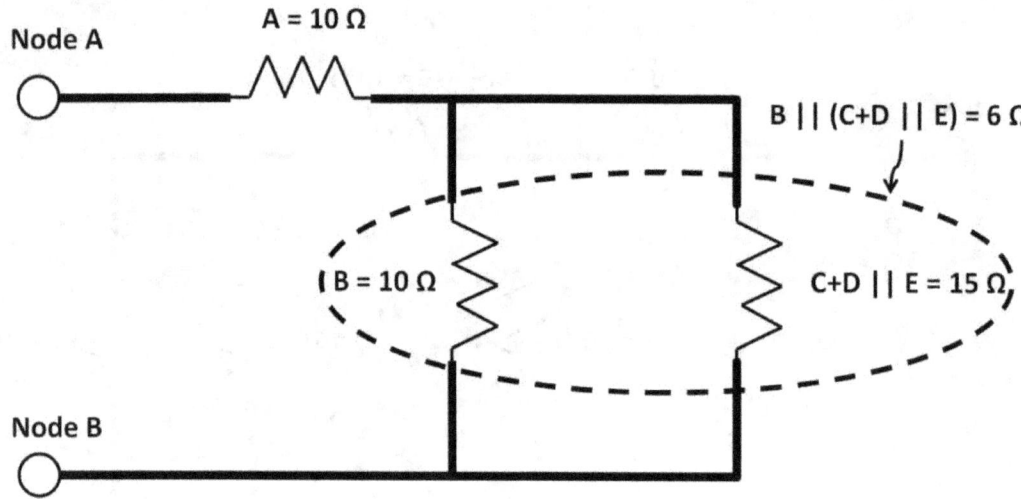

Figure 1.36: Combine B with 15 Ω in parallel

Finally, the result of the parallel combination is in series with 10 Ω resistor A. This gives rise to 16 Ω equivalent resistance between node A and B (see figure 1.37).

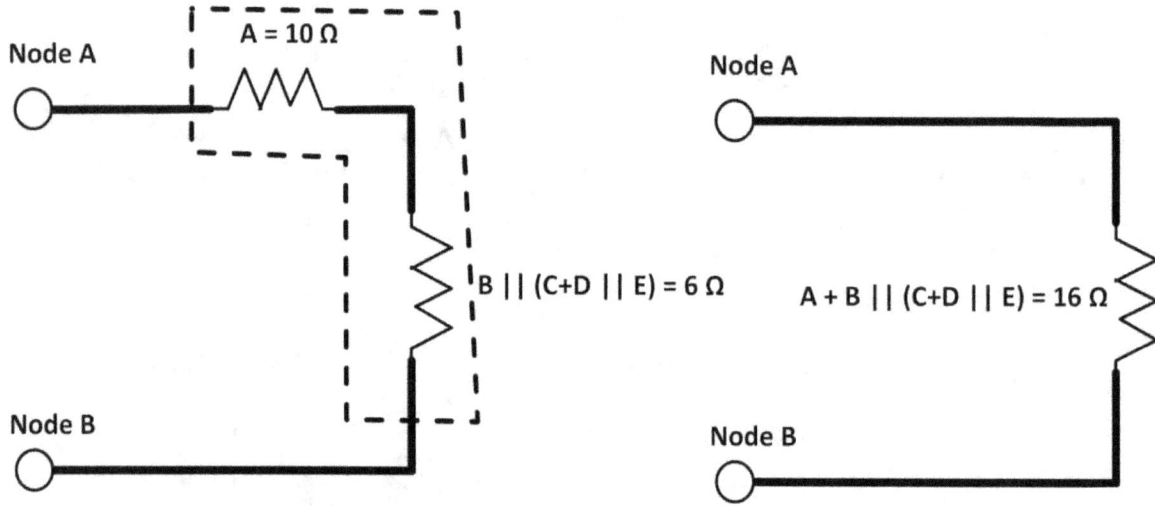

Figure 1.37: Final equivalent resistance value

Summary

DC electronics are the most basic, easy to learn electronic theory. The chapter started with basic electronic properties (voltage, current, and resistor). Basic electronic principles were then discussed: Ohm's law, KVL, and KCL. We then went over series and parallel resistors rules, and voltage-current divider rules explained by practical circuit examples. Superposition theorems, IC package, electronic measuring apparatus, non-ideal characteristics of voltage, current sources, and resistors were reviewed. Once you become proficient in basic electronics principles, you can then apply theories to explain and analyze any circuits with ease. This builds up a strong foundation for further study, use, and applications of more complex electronics.

Quiz

1) Show current flow direction and amount of current, if any (see figure 1.38).

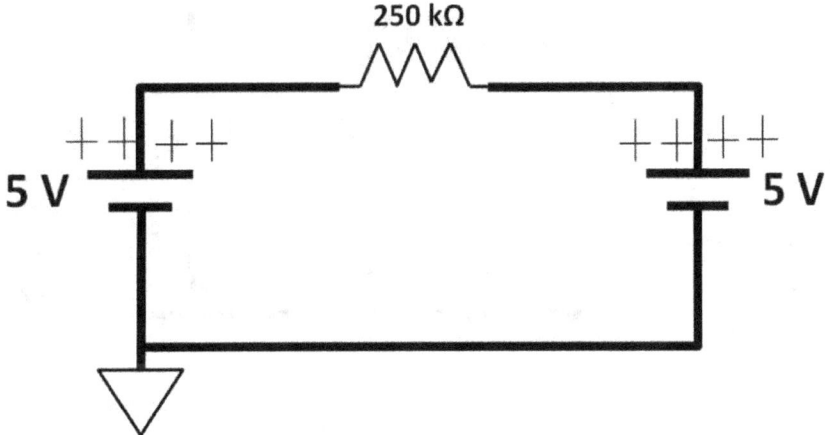

Figure 1.38: Current flow

2) The power supply was set to produce 5 V. When measuring using DMM, it only reads 4.95 V. Why?

3) Five parallel resistors sized from 1 Ω, 10 Ω, 100 Ω, 1 kΩ, and 10 kΩ. What is the approximate equivalent resistance by inspection?

4) Using figure 1.38, if the 5 V on the left is replaced with 10 V, what is the current flow direction? What is the amount of current, if any?

5) Design a voltage source that generates 3 V from a 12 V DC source. To save power, current is limited to 10 mA. Hint: Use a voltage divider.

6) What is the power in Watts from the circuit you designed in problem 5?

7) Refer to the circuit below in figure 1.39a. R1 is a variable resistor symbol (potentiometer or POT) in which users can adjust resistances by manually turning a knob. Figure 1.39b is a 10 kΩ POT with a body size of 9.5 mm X 9.5 mm X 4.9 mm. You can view the potentiometer as a resistive divider where the top, middle, and bottom pins are measured points. If you connect the top and bottom pins to your circuit, a full scale 10 kΩ is obtained. By connecting the top and middle pins to the circuit, resistance can be varied by turning the knob. The range of resistance would be from 0 Ω to 10 kΩ. Calculate current flow in each branch, assuming that the potentiometer is at mid scale.

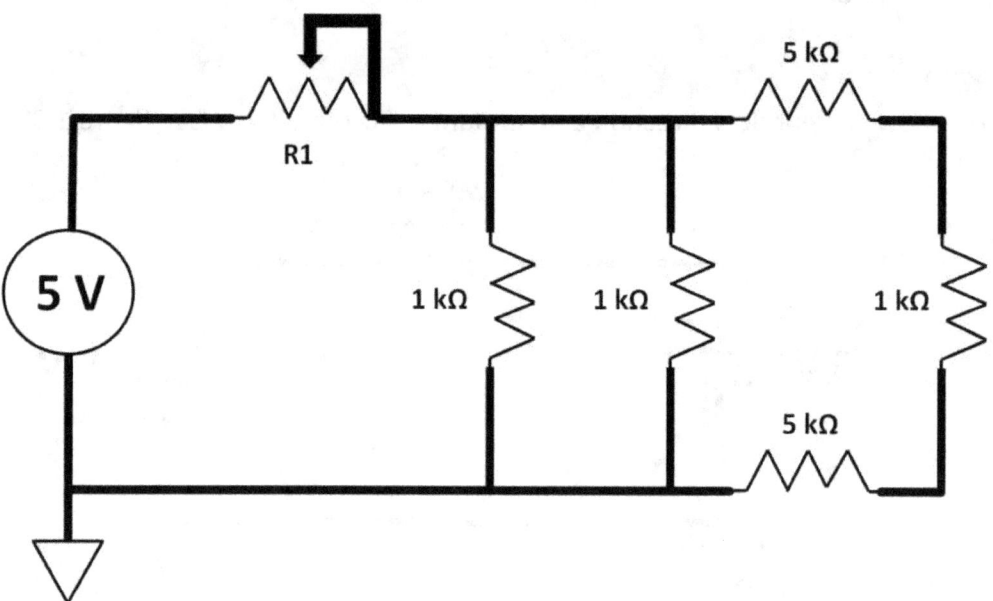

Figure 1.39a: Current flow in different branches

Figure 1.39b: Potentiometer

8) Use superposition to find Vx. Show steps (see figure 1.40).

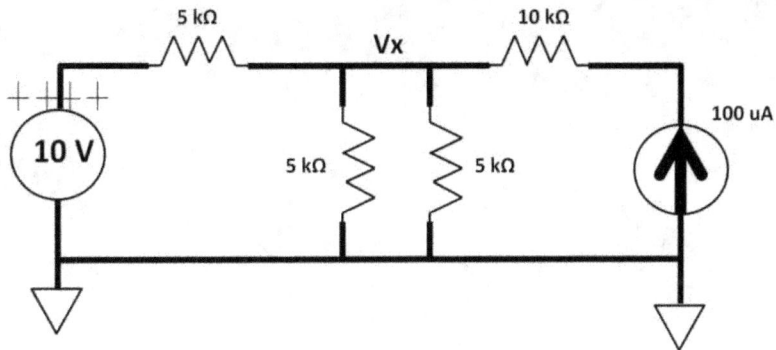

Figure 1.40: Superposition, voltage, current sources

9) Use superposition to find Vx. Show steps (see figure 1.41).

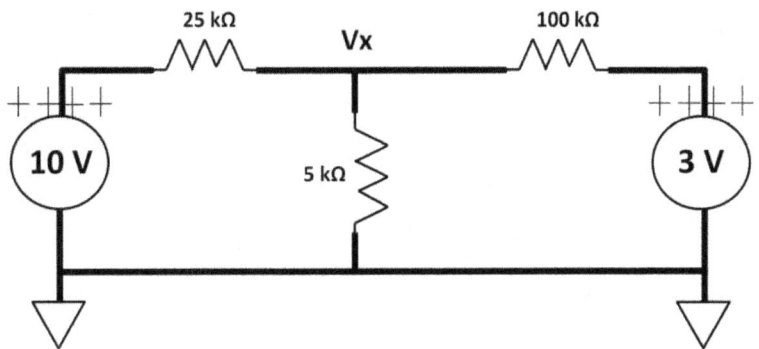

Figure 1.41: Superposition, two voltage sources

10) Two voltage sources are connected in series in figure 1.42. What are the voltages at node A and B?

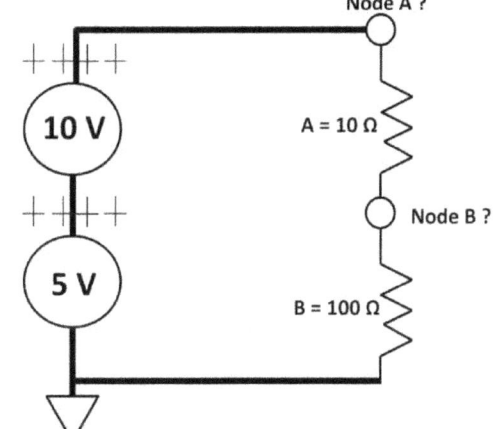

Figure 1.42: Two voltage sources in series

Chapter 2: Diodes

Diodes are passive electronic devices that do not generate electrical energy or power. Passive devices only dissipate or store energy. Resistors and diodes are examples of passive devices. Diodes are made of P (positive) and N (negative) type junctions. They are the building blocks of transistors. Transistors, by far, are the most widely used electronic components in electronic systems. Diodes are used in many electronic circuits that we encounter daily. Understanding diode structure, device physics, behavior, and diode circuits prepares you well to further understand transistors and complex electronic circuits.

P-N Junctions

Diodes are formed by merging two different types of materials. Silicon and germanium are the most popular material choices used in semiconductors. From a performance standpoint, germanium offers faster switching capability with lower reliability. With silicon's abundant supply and higher reliability, silicon is the most popular material in semiconductor technology. From chapter 1, DC, chemical materials (elements) are made of atoms. Each atom consists of electrons, protons, and neutrons. Silicon has total 14 electrons (dots) with 4 electrons in the outer shell (see figure 2.0). An atom is stable if the outer shell contains two or eight electrons. By bombarding silicon (Si) with chemicals, we can alter its properties to create P-N junctions. For example, to create a P-type junction in silicon, we bombard silicon with boron (Br), which has three electrons in its outer shell. By adding boron's three electrons to silicon, which currently has four electrons, seven electrons are now in the silicon atom's outer shell. Recall that the silicon atom wants to have eight electrons to fill up its outer shell. These seven electrons leave a net positive charge (hole) in the modified silicon atom outer shell (see figure 2.0a).

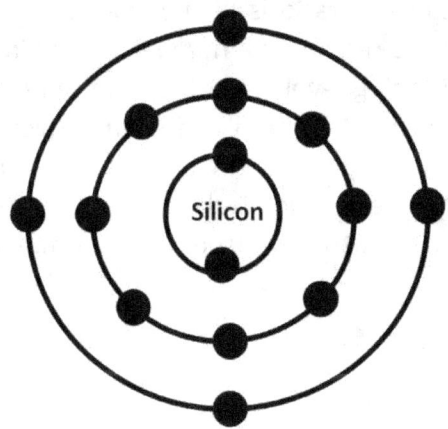

Figure 2.0: Silicon atom, 14 electrons, 4 electrons on outer shell

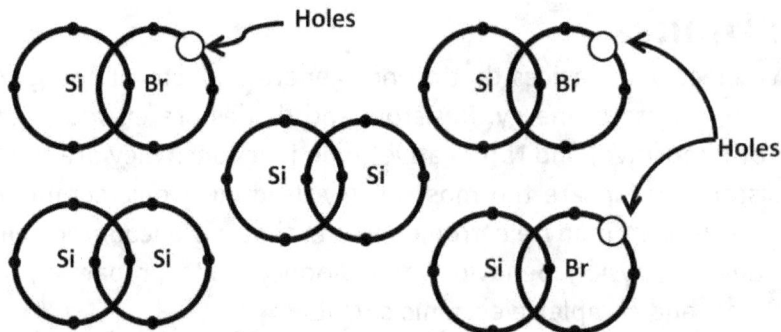

Figure 2.0a: Silicon implanted with boron, net positive charge

In other words, it's eager to seek one electron to fill the outer shell with total of eight electrons, which is the maximum number of electrons a shell could accept. This process leaves a net positive charge. P-type junction material means that the area is injected with more positive ions, namely holes. Precisely, the positive ion concentration and ratio is higher than with N-type. It does not mean there are no electrons at all in a P-type region. The number of ions in a given junction area is defined by its carrier concentration (doping levels). For P-type, the holes doping level is high. To create an N-type junction, phosphorus (P) is bombarded with silicon. Because phosphorus has 5 electrons on the outer shell, it nets a total of 9 electrons (e-). This extra electron results in net negatively charged silicon. An N-type junction is defined as the area that is dominated by negative ions, namely higher electron concentrations (see figure 2.0b).

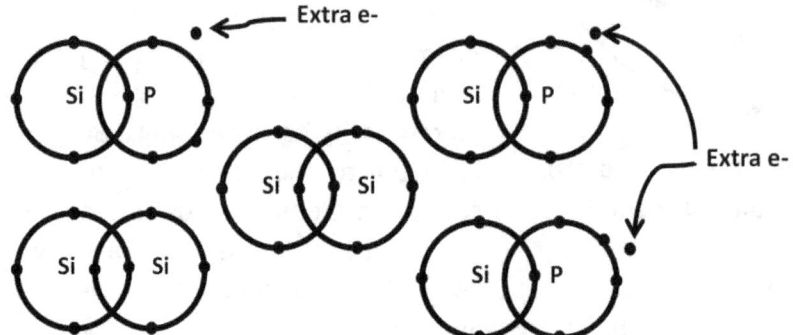

Figure 2.0b: Silicon implanted with phosphorus, net negative charge

The bombardment process mentioned above is called ion implantation, which is one of many IC manufacturing steps. The result of ion implantation gives the processed silicon unique properties so that it's not totally conductive but only semi-conductive, hence the name semiconductor. Electronic devices made by such process are called solid-state devices because the electrons and other charged carriers are confined in the solid materials. With appropriate voltage condition (bias), a semiconductor can be controlled by either turning it fully or partially on or off. This concept builds the foundation of diodes, which come in many forms in terms of junction carrier concentrations.

Figure 2.1 shows a graphical representation of a P-N junction (see top of figure 2.1). There is a region in between P-N junctions, called the depletion region (see bottom of figure 2.1). This region determines the amount of voltage across the diode needed in order to turn a diode on or off. The ability to turn a diode on and off gives limitless and powerful design possibilities. Electrons in the N junction diffuse into the P-type while the P-type migrates to the N region due to carrier concentration imbalance (see middle of figure 2.1). This difference in carrier concentration results in electrons diffusing into the P region, leaving the N-type with an extra hole. While the electron recombines with a hole in the P region, it leaves behind a negative ion. As this diffusion process continues (see bottom of figure 2.1), a wall of electrons accumulates near the P-type and wall of holes on the N-type edges. Finally, the diffusion process stops, reaching equilibrium. This process forms the depletion region. The reason for the end of the diffusion process is that as more holes recombine with electrons, the electron concentration starts to increase in the P region opposing additional electrons migration from N- to P-type. The same resisting force occurs at the P region. This is why there is a wall of electrons and holes on the edges of each type. These ions cannot diffuse anymore and are "stuck" at the depletion region.

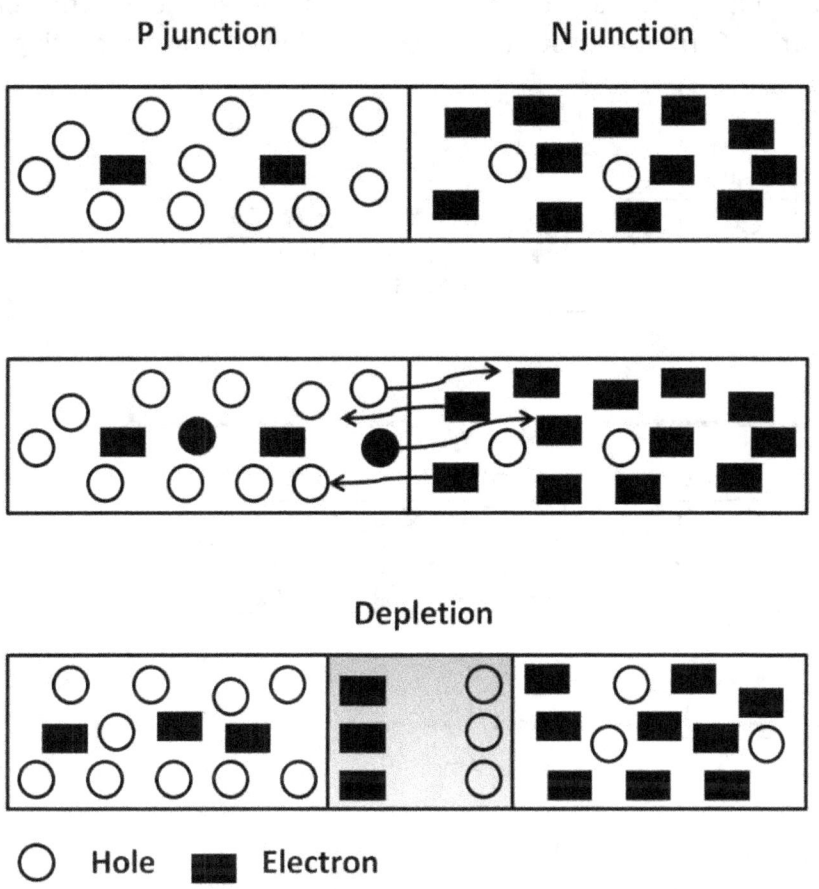

Figure 2.1: Graphical representation of P-N junction

Forward-Biased and Reverse-Biased

The implication of depletion region is significant. It sets the minimum voltage required to turn on the diode. Turning on the diode means forward-biasing a diode, in which case we say the diode is forward-biased. Many textbooks define the minimum built-in diode voltage potential to be 0.7 V. It is important to note that this number is only a typical number. Forward-biased voltage can have other values depending upon the diode types and many other factors. A datasheet would indicate the exact forward-bias voltages on any particular diode. Let's now go over the mechanisms behind forward-biasing a diode. In figure 2.2, there is a voltage source connected to a diode with the source's positive terminal connected to the P junction. The negative source terminal (polarity) connects to the N junction. Assume the forward voltage is 1 V. If we dial in the voltage source to 0.5 V across the diode, the positive charge from the source opposes the holes in the P junction causing the holes in the P junction to diffuse towards the depletion region. The same action takes place in the N junction where electrons are moving towards the depletion region. Since 0.5 V is less than 1 V, which is the minimum voltage required to turn on the diode, the diode is now reverse-biased. Modeling the diode as a switch, it's currently open (off). If we increase the voltage source to 1 V, the switch overcomes the built-in potential, causing holes and electrons to flow in the reverse direction breaking through the depletion barrier. Current then starts to flow. The diode is now conducting. As a switch, it's now closed (on).

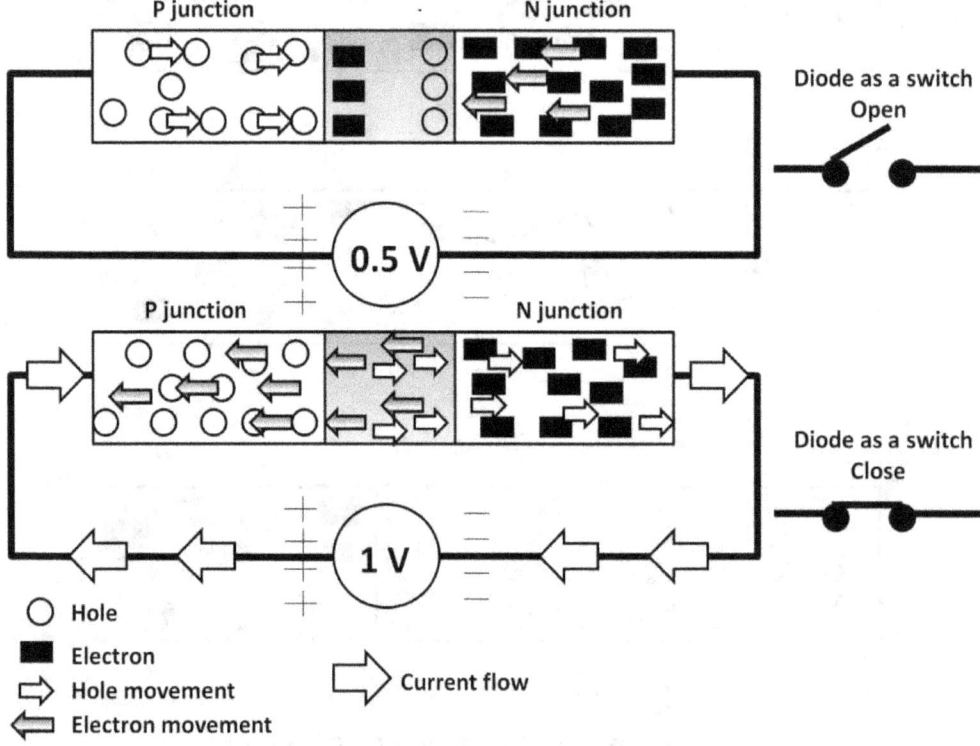

Figure 2.2: Voltage across diode

The diode schematic symbol includes a vertical line and a triangle (see figure 2.3). It may be obvious that the diode symbol looks like an arrow. The vertical line at the tip of the arrow end is a cathode. A cathode is simply the N junction of the diode in figure 2.1. The opposite side of the diode symbol is the anode (P junction). A diode is forward-biased when the voltage across anode and cathode is positive and at least equal to or above the forward-biased voltage, i.e., voltage at the anode is higher than voltage at the cathode. These conditions give rise to current flow from the anode to the cathode, just like an arrow moving from left to right. When a diode is forward-biased, current flows from anode to cathode. When the diode is reverse-biased, i.e., when voltage at cathode is larger than anode or the voltage at the anode and cathode is less than the minimum forward drop voltage, under these conditions, the diode is said to be reverse-biased (off or an open- circuit) without current flow.

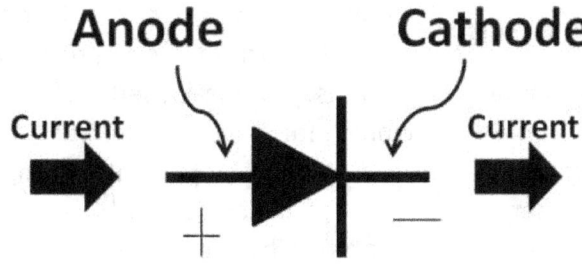

Forward bias

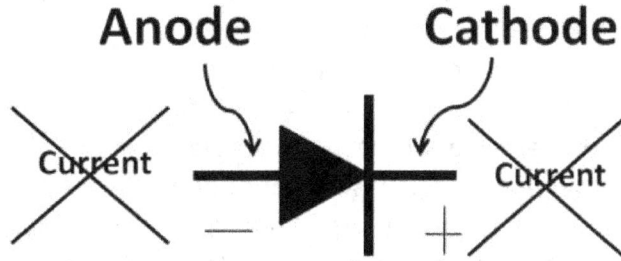

Reverse bias

Figure 2.3: Diode forward- and reverse-biased

Diode I-V Curve

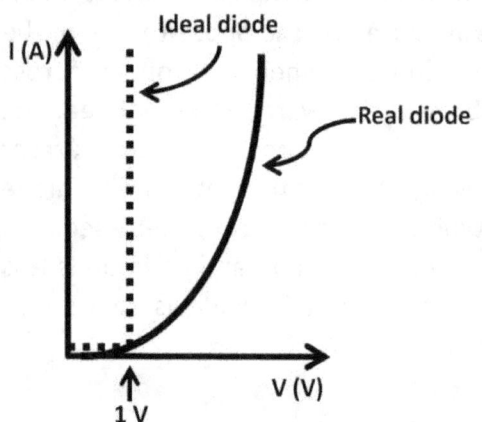

Figure 2.4: Diode voltage vs. current

As we continue to increase (sweep) the DC voltage source in figure 2.2, forward biasing the diode, the current continues to increase exponentially. Using current versus voltage of a 1 V diode (see figure 2.4), we can further examine diode behavior. There are two sets of curves in this figure. The dotted line is the ideal diode I-V characteristic. It shows that the current takes off infinitely once the diode is forward-biased. The non-ideal diode, however, shows the current is rising exponentially with voltage but not infinitely. This is because of the finite resistance in the real world diode that limits the current. Before the diode voltage reaches 1 V, the current is close to zero with the diode being off (reverse-biased). When an ideal diode is reverse-biased, it is an open circuit (infinite resistance). A real diode, however, would not have infinite resistance but extremely large resistance when it's off. It means that there would be current flowing through the diode when it's reverse-biased. This is characterized as leakage current, which is usually small and negligible but increases exponentially with temperature. The diode current transfer function is modeled as:

$$I = Io \times (e^{qV/KT} - 1)$$

Io: Leakage current; q: Electron charge (1.6×10^{-19} C); V: Voltage across diode; K: Boltzmann's constant (1.38×10^{-16}); T: Absolute temperature (Kelvin). The transfer function of temperature from Kelvin to Celsius is **K = °C + 273**; for room temperature, 27 °C, **K = 27 + 273 = 300 K**. This diode current model indicates that for a given temperature, increasing diode voltage increases diode current. Every diode has its own set of parameters (ratings). Maximum forward voltage is the maximum voltage a diode could withstand in forward-bias mode before it breaks down (shorts or blows open). Reverse voltage: This number determines the maximum reverse-biased voltage a diode could withstand before reverse breakdown. Diode output current defines the current level during forward bias and is approximately constant in high forward voltage. Maximum reverse current (leakage) defines the current amount through a diode during reverse bias. Maximum power dissipation describes the amount of power (Watts) allowed for a given diode voltage and current, **power = (I) X (V)**.

Diode Circuits

1) Many circuits utilize diodes. A diode can be used as a voltage regulator (see figure 2.5). A voltage regulator by definition is an electronic device that generates a constant voltage source. The ideal voltage regulator can source and sink infinite amounts of current. The tilted up-pointing arrow in the voltage source means it's a changing (sweeping) variable voltage source. The diode's forward-bias voltage is 200 mV in this sampled circuit. Voltage at node D is the thin trace (V-I graph on the right). As the voltage source sweeps from 0 V to 200 mV, there is no current flowing in this circuit due to the fact that the diode remains off (reverse-biased). Therefore, node D voltage follows variable voltage. Using Ohm's law, there is no voltage drop across the diode, i.e., **Input voltage − node D voltage = 0 V, and current = 0 A**. When variable voltage increases to 100 mV, node D follows at 100 mV. Once the variable voltage source reaches 200 mV, the diode starts to turn on. Node D voltage is now roughly fixed at 200 mV. Current continues to increase exponentially as the variable voltage source continues to go up.

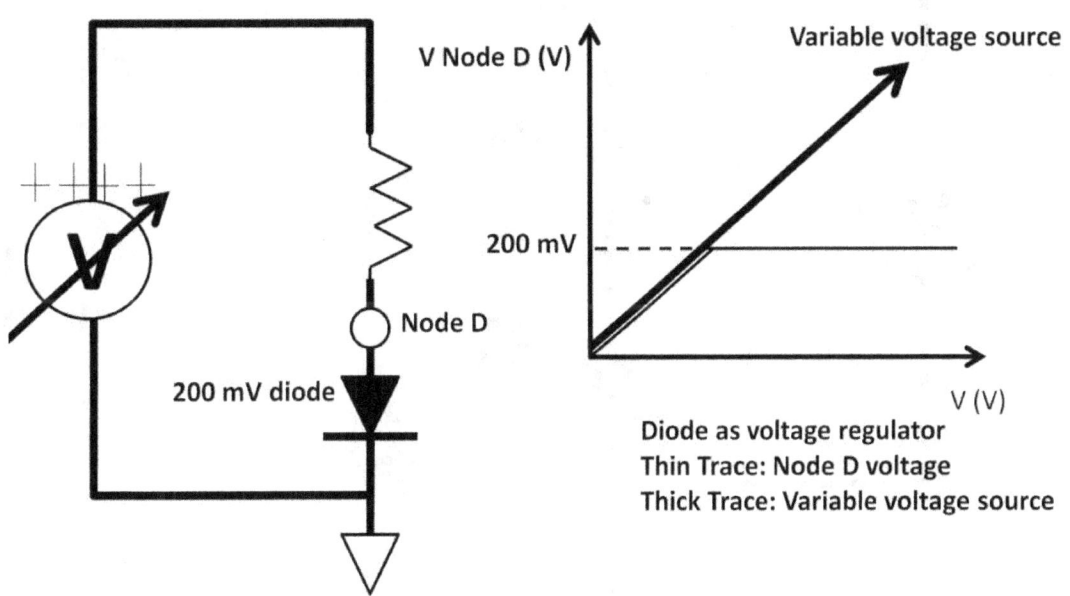

Figure 2.5: Diode as voltage regulator

2) A very popular diode type is the light emitting diode (LED). When the LED is forward-biased, it emits light. There are many color LED combinations (white, red, blue, yellow, orange, green, and violet are popular colors). Typical LED forward-biased voltage is between 2 V to 3 V. The intensity of the light is a strong function of current. Typical LED consumes 20 to 30 mA of forward current. Because of its small sizes, low power consumption, and long life (typically 10,000 hours), LEDs are suitable for lighting applications. As the prices of LEDs have continued to go down in recent years, they've found themselves further in automotive lighting applications. Figure 2.6 shows several LEDs that are in the order of 2 mm by 3 mm in dimensions (right) and a simple LED circuit in a series configuration (left).

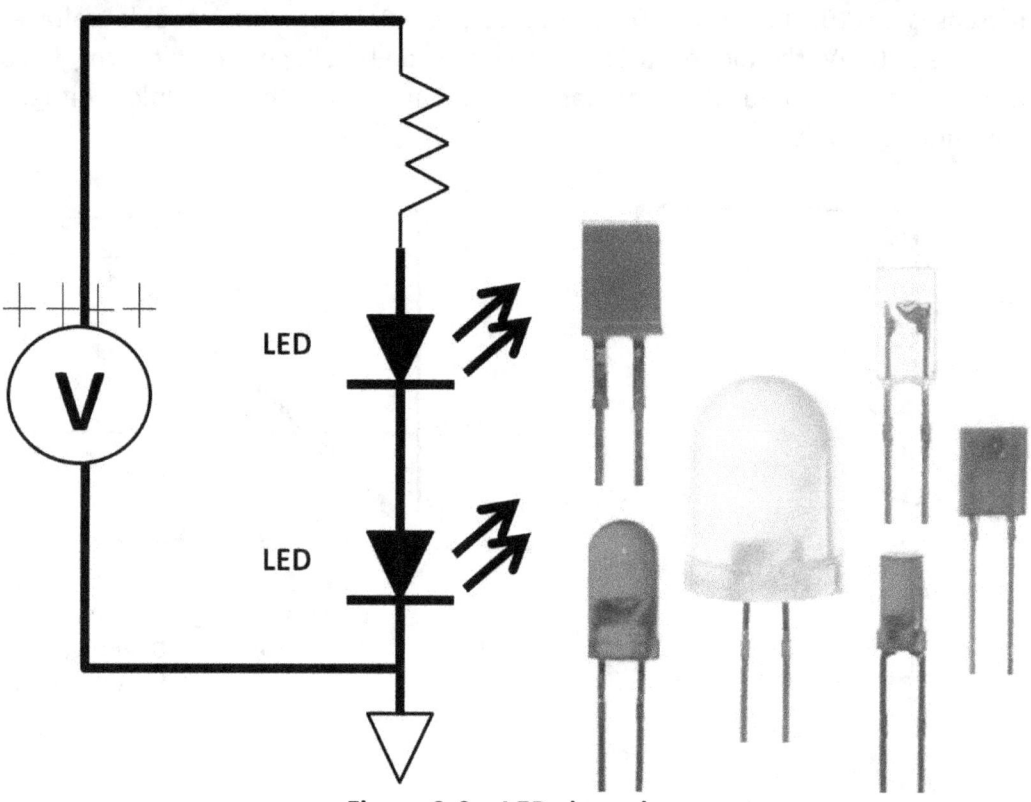

Figure 2.6a: LEDs in series

3) Another common LED application is constructed in parallel configuration (see figure 2.7). From chapter 1, DC, it is easily recognized that the trade-off between a series and a parallel LED application is that a series LED circuit requires higher voltage than the parallel one. A parallel circuit draws more current due to multiple LED branches as a result of the KCL rule.

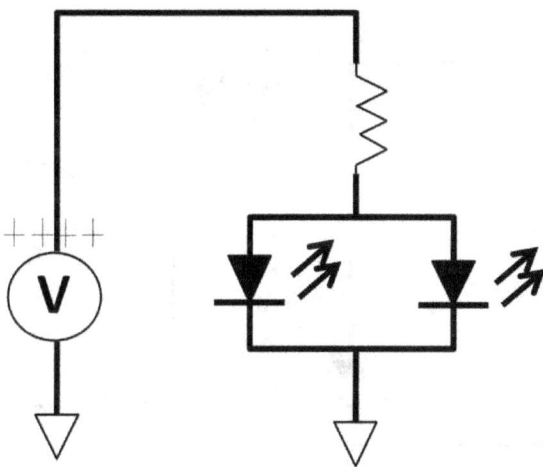

Figure 2.7: LEDs in parallel

4) As mentioned previously in this chapter, diodes are modeled as switches. Let's take a look at a practical circuit (see figure 2.8). An output is supplied by either one of the two voltage sources (V1 and V2). There are two assumptions. **1)** When V1 is present, V2 is not. **2)** When V2 is connected, it would be higher than V1. If **V1 = 5 V**, the forward diode drop is rated at 1 V. This forward-biases the diode causing the voltage at the output to be 4 V. If V2 is 10 V connected to the output, the diode is now reverse-biased (voltage at cathode > anode). The diode is off (switch is open), and V2 is then the only voltage supply to the output.

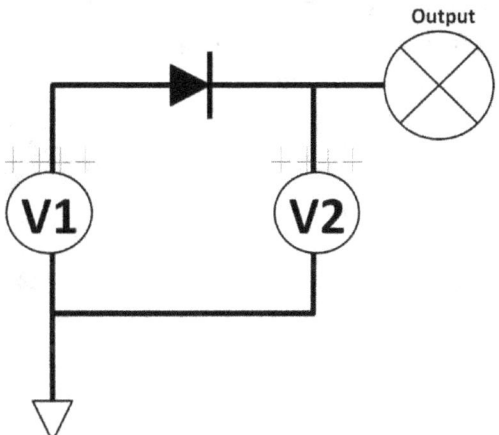

Figure 2.8: Diode application

5) A "real" diode does not behave the same as an ideal diode. Diode voltage is a strong function of temperature. The graph in figure 2.9 shows that diode voltage exhibits negative temperature coefficient with approximately **– 2 mV / °C**. Due to many diode types, you should refer to the specific diode datasheet for the correct temperature coefficient numbers.

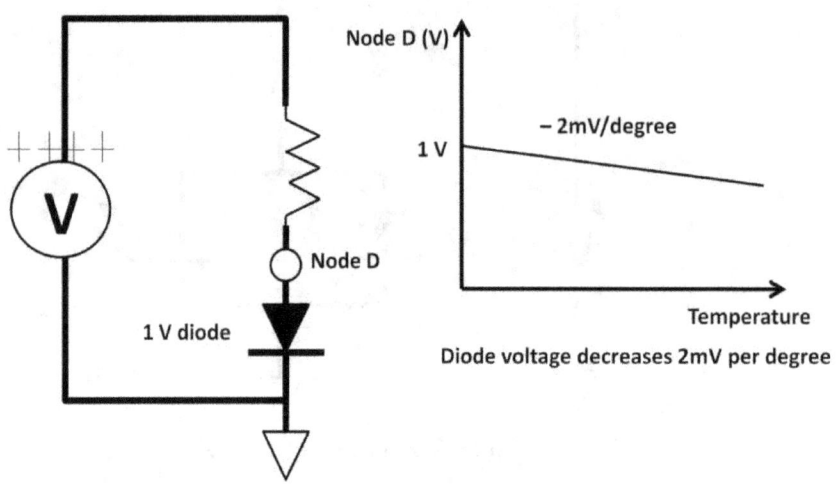

Figure 2.9: Diode negative temperature coefficient

6) A diode contains finite resistance when it's forward-biased. Additionally, there are resistances in a real diode due to physical leads. Figure 2.10 shows a simplified physical diode representation with leads. These leads present small finite resistance that may be significant in designs that are sensitive to noise. Surface-mount diodes are available in small footprints. Figure 2.11 shows surface-mount diodes from SEMTEX. Recall from chapter 1, DC, figure 1.21, that any resistor with current flowing through it generates heat and noise due to random ion bombardments. Noise in particular should be minimized at all costs, especially in highly accurate circuits.

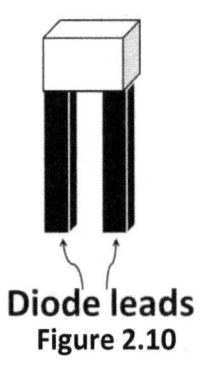

Diode leads
Figure 2.10

Figure 2.11 Surface-mount diode (Courtesy of SEMTEX)

7) A zener diode is a very popular diode type commonly used in linear regulator applications. Linear regulators are the building blocks of virtually all electronic power supplies. As opposed to switching regulators, linear regulators are on 100% of the time. Switching regulators means that devices that turn on and off periodically will potentially increase power efficiency. We will take a closer look at switching regulators in chapter 3, AC. Figure 2.12 demonstrates a simple zener diode implementation used as a voltage regulator. A zener diode operates in the reverse-biased region, i.e., the left-hand side of the I-V diode curve. When it reaches the rated reverse-biased threshold, 5 V, it behaves as a voltage source staying at 5 V. Once the zener diode starts conducting, it remains turned on as a linear regulator as long as variable voltage source stays at least or above 5 V.

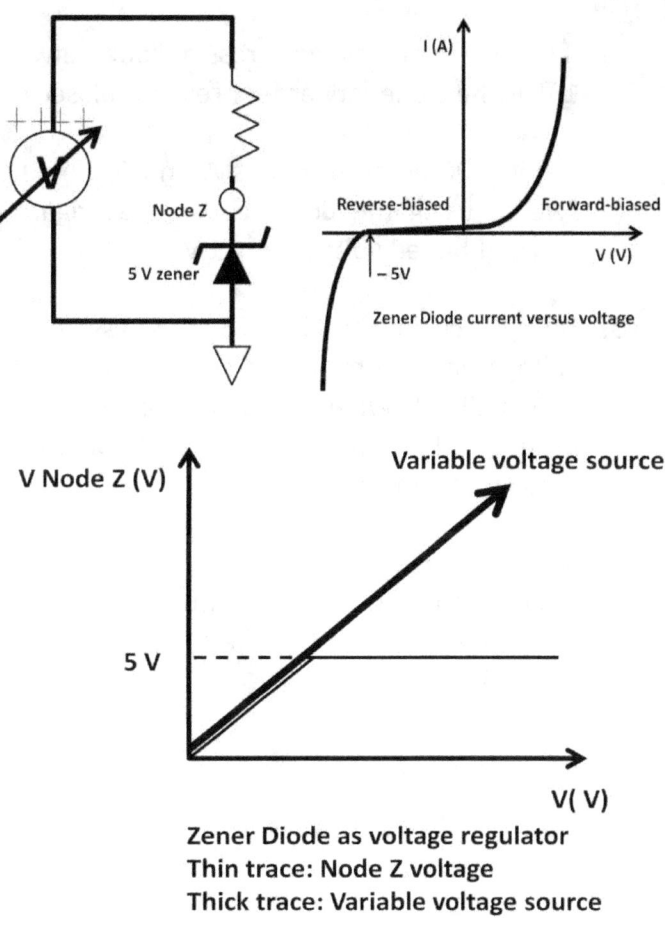

Figure 2.12: Zener diode circuit, V, I curve

Summary

Diodes are formed by P-N junctions. They are basic transistor building blocks. Powerful, practical electronic circuits can be built and designed by using diodes. The chapter started with silicon atomic structure, then basic diode formation process, followed by diode DC characteristics (I-V diode curve). We then examined forward and reverse-biased definitions as well as several practical diode applications. Ideal- and non-ideal diode characteristics were discussed. The chapter closes with LED, zener diodes, and the linear and switching regulator principle of operations and applications.

Quiz

1) If you measure voltage across a diode between the anode and cathode, the DMM reads 1 V. Is the diode forward- or reverse-biased?

2) Draw a DC graph of nodes A and B during 0 V to 5 V (DC sweep) using the diode circuit (see figure 2.13). Assume forward-biased voltages are 1 V.

3) Design a circuit that drives 5 LEDs. Assume 5 V is the supply voltage and the minimum current needed to turn on each LED is 10 mA. When the LED conducts, it drops 1.5 V. Hint: Include LED voltage drop. Decide if you should choose parallel or series configurations.

4) Using figure 2.13, at 5 V DC, draw a DC sweep graph of nodes A and B over temperature ranging from − 40 °C to + 125 °C (see figure 2.14). The temperature coefficient of both diodes is − 2 mV / °C.

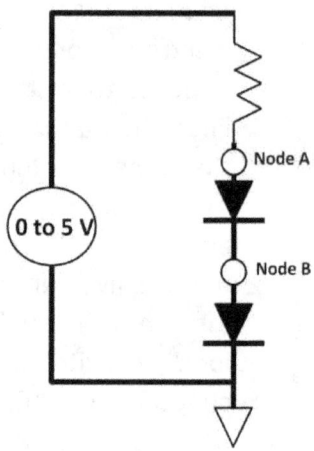

Figure 2.13: Diode circuit

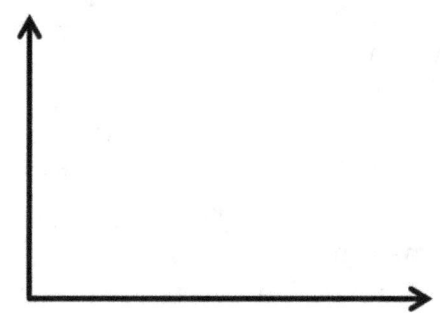

Temperature (C)
Figure 2.14: Diode voltage temperature sweep

5) 1N4001 is a popular general-purpose diode that is capable of handling up to 1 A of forward current (see figure 2.15). Its length is less than 10 cm; forward voltage drop is rated at 1.1 V, 27 °C room temperature; reverse voltage is specified at maximum of 50 V. Using DMM, 100 mA forward bias current is measured; the voltages at the anode and cathode of the diode are the same. What condition is the diode most likely to be? Would it mean it is shorted, open or working properly?

Figure 2.15: 1N 4001 general-purpose diode

Chapter 3: Alternating Current (AC)

Alternating current (AC) is not an isolated electronic theory but rather an extension of DC and diode. By definition, AC is an electrical signal (current, voltage, or power) that changes its amplitude over time. AC operations can be seen everywhere from electric power utilities, computers, Central Processing Unit (CPU) operations, radio broadcasting, wireless communications, etc. We first need to understand basic AC parameters, capacitors, and inductors before getting into more complicated AC electronics designs. Some AC parameters are listed in table 3-1.

Definition	Unit	Remarks
Frequency	Hertz (Hz)	Number of cycles occurs in one second
Period	Second (S)	$\frac{1}{Frequency} = 1 \text{ Period } (1 \text{ Cycle})$
Duty cycle	Unit-less	$\frac{On-Time}{On-Time + Off-Time}$

Table 3-1: AC parameters

Sine Wave

We will use sinusoidal wave and AC parameters to explain most AC operations. The most common AC waveform is the periodic sinusoidal wave. Sinusoidal (sine) wave comes from trigonometry in mathematics. Figure 3.1 shows a periodic sine voltage waveform in time (transient) domain. It means that the frequency is fixed while the waveform amplitude is changing. Other than the sine wave, the square wave and saw-tooth wave are also common AC signal sources. The schematic symbols of all three types are shown below.

Sine wave **Square wave** **Saw-tooth wave**

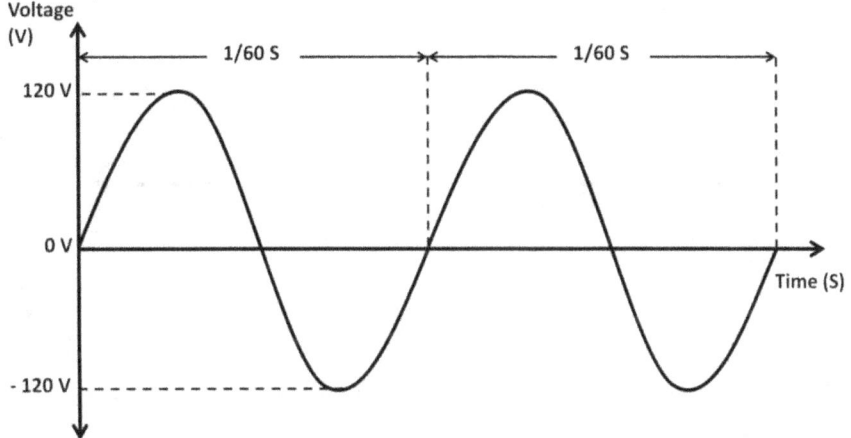

Figure 3.1: Periodic waveform

Frequency and Time

One operation cycle (period, unit in seconds) is defined as the total time it takes while the voltage stays above the X-axis (upper half of a sine wave cycle) plus the time the voltage stays below the X-axis (bottom half of one sine wave cycle). In figure 3.1, each cycle takes 1/60 second to complete. Furthermore, one period can be interpreted as: from one waveform peak (maximum point) to the next peak. It can also be measured from the rising (leading) or falling (trailing) edge of the waveform to the next rising or falling edge. From figure 3.1, one period is found by one rising edge to the next. By definition in table 3.1, frequency is defined as one over period (1 / Period):

$$\text{Frequency} = 1 / \text{Period}$$
$$\text{Or}$$
$$\text{Period} = 1 / \text{Frequency}$$

The frequency in figure 3.1 is:

$$\frac{1}{\frac{1}{60\,S}} = 60\ \text{Hz}$$

In other words, 60 Hz means that there are 60 cycles occurring in one second (see figure 3.2). The significance of this example is that 60 Hz is the US household power outlet frequency. A 3 gigahertz (GHz) signal (the typical CPU clock speed of today's desktop computers) runs 3 billion cycles in one second.

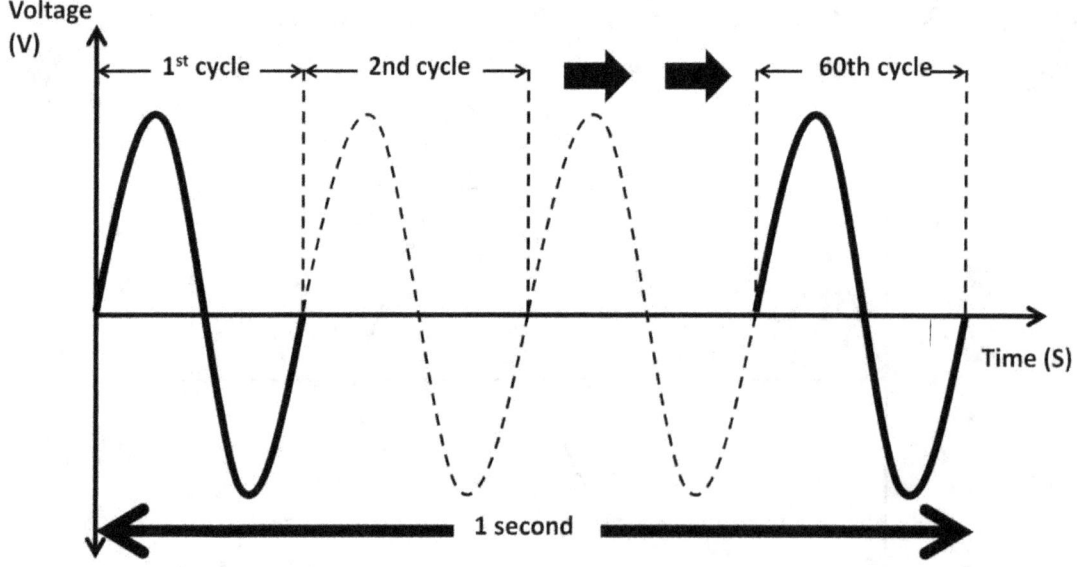

Figure 3.2: 60 Hz in time domain

We use AC in our daily lives. Figure 3.2a shows an electrical outlet (receptacle) commonly found in US residential households and commercial buildings. Each of the three terminals has a copper conductor connected it. The "hot" terminal provides 120 V AC source. "Neutral" is the return current path for the AC source. The *"ground"* terminal has a zero voltage potential and zero resistance. It provides a path for the current to the earth, which is a huge mass of conductive materials such as dirt, rock, ground water, etc. Since the earth is a superb conductor, it makes the ground terminal a great voltage reference for electrical systems as well as a safety measure by directing all unwanted buildup of electrical charge to the earth, thus preventing damage to the equipment and the user. Electrical equipment, such as computers, is often built with a chassis ground. This zero voltage connection provides a common point of voltage reference with respect to internal circuitries and for safety reasons.

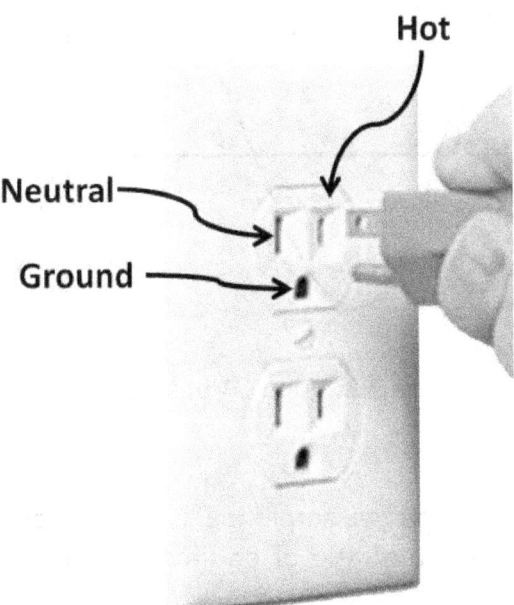

Figure 3.2a: Electrical outlet

Peak Voltage vs. Peak-to-Peak Voltage

From figure 3.1, the vertical axis is voltage. It "swings" up and down above and below the X-axis. The vertical amplitude is expressed in voltage (V). From the highest peak of the upper half waveform to 0 V on the X-axis, its amplitude is 120 V. This is the positive peak voltage (Vpeak), (see figure 3.3). The lower half of the waveform is peak voltage with a negative sign, i.e., − 120 V. Peak-to-peak voltage (Vpeak-to-peak) can be estimated from the highest peak voltage to the lowest peak voltage. In this example, **120 V − (− 120 V) = 240 V**. Vpeak-to-peak can also be viewed as the positive Vpeak multiplied by 2, i.e., **(120 V) X 2 = 240 V**. You can see that Vpeak is exactly half of Vpeak-to-peak.

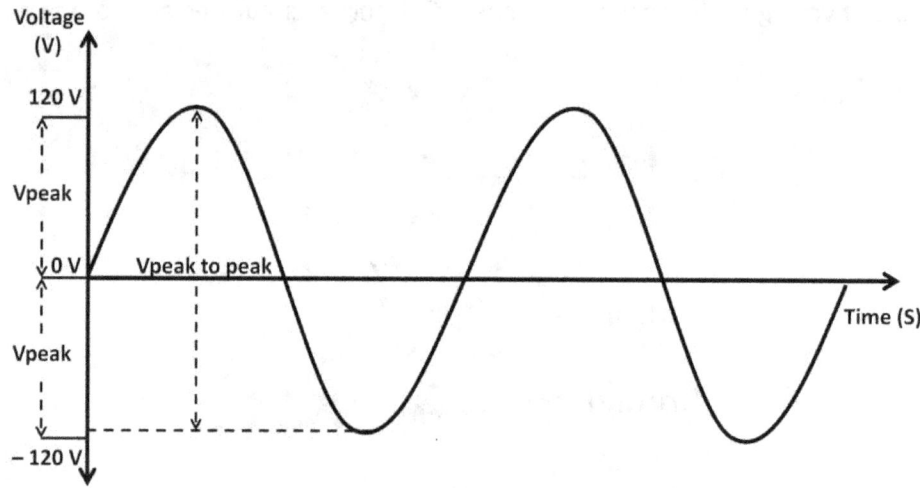

Figure 3.3: Vpeak, Vpeak-to-peak

Duty Cycle

So far, we have discussed peak voltage, amplitude, frequency, and period. Now, we will look at duty cycle. The duty cycle is the ratio of on-time over one period (on-time + off-time) expressed in percentage:

$$\text{Duty Cycle} = \frac{\text{On-Time}}{\text{On-Time} + \text{Off-Time}} \times 100\%$$

From figure 3.3, the time it takes for the upper half of one sine wave cycle to complete is on-time. The other half, off-time, is the time it takes for the lower half of the sine wave cycle. By definition, **On-time + Off-time = One period** (see figure 3.4). For a periodic waveform, on-time is exactly the same as off-time. Using the duty cycle equation, you can see that the duty cycle of a periodic 60 Hz sine wave is 50%:

$$\frac{\frac{1}{60\,\text{S}}}{\frac{1}{60\,\text{S}} + \frac{1}{60\,\text{S}}} \times 100\% = 50\%$$

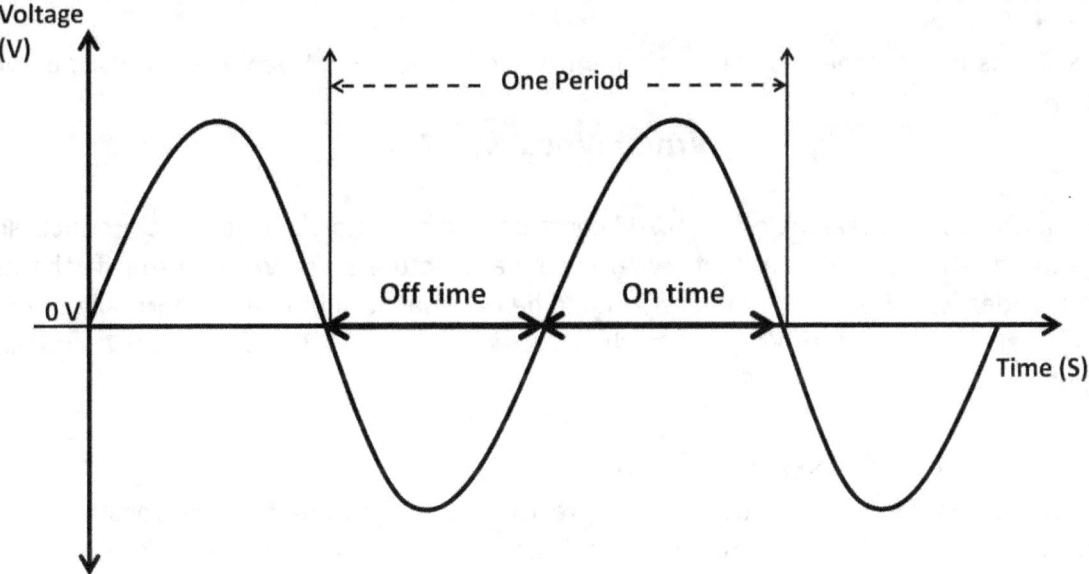

Figure 3.4: Period and duty cycle

In other words, a 50% duty cycle means the signal is "on" half of the time (one period) while the other half is "off." This concept applies to a signal in any frequencies, not just 60 Hz. As long as the waveform is periodic, the duty cycle is 50%. Not all AC signals are 50% duty cycle. Figure 3.5 shows a 10% duty cycle (a 0.1 GHz square wave).

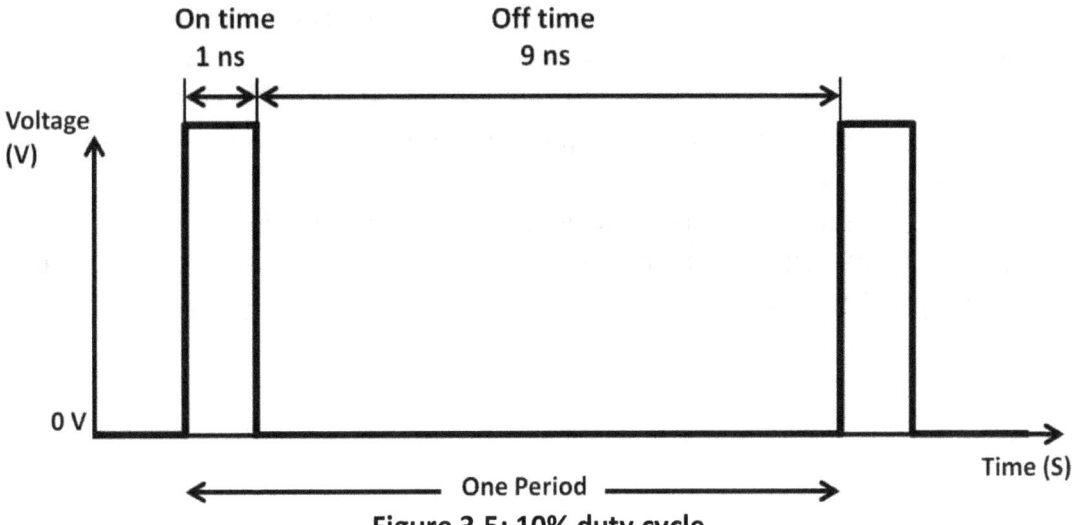

Figure 3.5: 10% duty cycle

Vrms

Vrms stands for root mean square voltage. It directly relates to Vpeak discussed in previous section:

$$\text{Vrms} = \text{Vpeak} \times 0.707$$

We will discuss the meaning of the 0.707 constant in detail shortly. Electronic products show Vrms information in the datasheets where the manufacturers use Vrms to specify the noise amount. Ideally, noise, as a parameter, should be minimal. Manufacturers normally use Vrms rather than Vpeak because Vrms is less than Vpeak by about 29.3% **(100% − 70.7% = 29.3%)**. If **Vpeak = 100 mV**, **Vrms = 70.7 mV**.

Impedance, Resistance, and Reactance

Until this point, we have strictly described resistance as a parameter that does not change with frequency. This is in fact true with ideal resistors. However, it is very different with AC. There are two electronic devices that are found in virtually all electronic systems that behave very distinctively when it comes to resistance and frequencies. These are capacitors and inductors. Before we discuss them, some essential resistance parameters are listed below:

$$\text{Impedance} = \text{Resistance} + \text{Reactance}$$

These three parameters all have units in Ohms. Impedance is the sum of resistance and reactance of an electronic component such as a resistor, capacitor, or inductor. For resistors, reactance is zero. Thus, resistance is equal to impedance. The resistor's value does not change with frequency:

$$\text{Impedance} = \text{Resistance} + 0 = \text{Impedance}$$

Reactance, however, changes with frequency. This causes the impedance to vary with frequency. This fundamental characteristic provides the framework for all AC electronic designs, circuits, and systems.

Capacitors

A capacitor is a passive electronic device that does not generate energy. However, it stores energy through an electric field. A capacitor is formed by two conductive plates separated by an insulator (dielectric). There are plenty of conductive plate materials as well as insulator types. The most common ones are tantalum, ceramic, polyester, and electrolytic. Figure 3.6 shows a capacitor graphical representation, capacitor schematic symbol, discrete tantalum, electrolytic capacitors, and film capacitors.

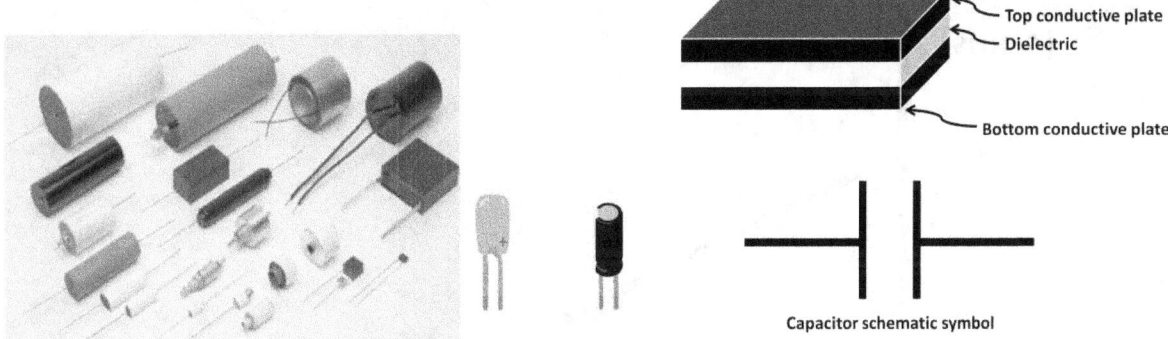

Figure 3.6: Capacitor structure, schematic symbol (top right), tantalum, electrolytic and film capacitors (bottom left); capacitor symbols (middle) created by Fritzing Software

Capacitor values (capacitance) are measured in farad (F), which exhibits reactance, called capacitive reactance (Xc). Capacitive reactance's units are in Ohms (Ω).

Capacitive Impedance = Resistance + Xc

The contribution of the capacitor resistance comes from capacitor packages, leads, and the intrinsic nature of capacitive materials.

Capacitive reactance (Xc) is defined as $\frac{1}{2\pi f C}$, where f is frequency of signal, π = 3.14, C is capacitive value in farad (F). If the signal frequency changes, Xc changes causing capacitor impedance to change. For example, if **signal frequency = 1 MHz, capacitance = 1 uF.**

$$Xc = \frac{1}{2\pi\,(1\text{ MHz})\,(1\text{ uF})} = 160\text{ m}\Omega$$

XC versus Frequency

If the frequency now increases to 2 MHz, then **Xc = 80 mΩ**. In short, Xc is inversely proportional to frequency. See bode plot in figure 3.7. A bode plot is a graph that shows AC (frequency) analysis. It includes X-Y axis where X-axis is frequency. Keep in mind that even though impedances changes with frequencies, the capacitance values remain the same. From above, capacitance remains 1 uF between the two frequencies.

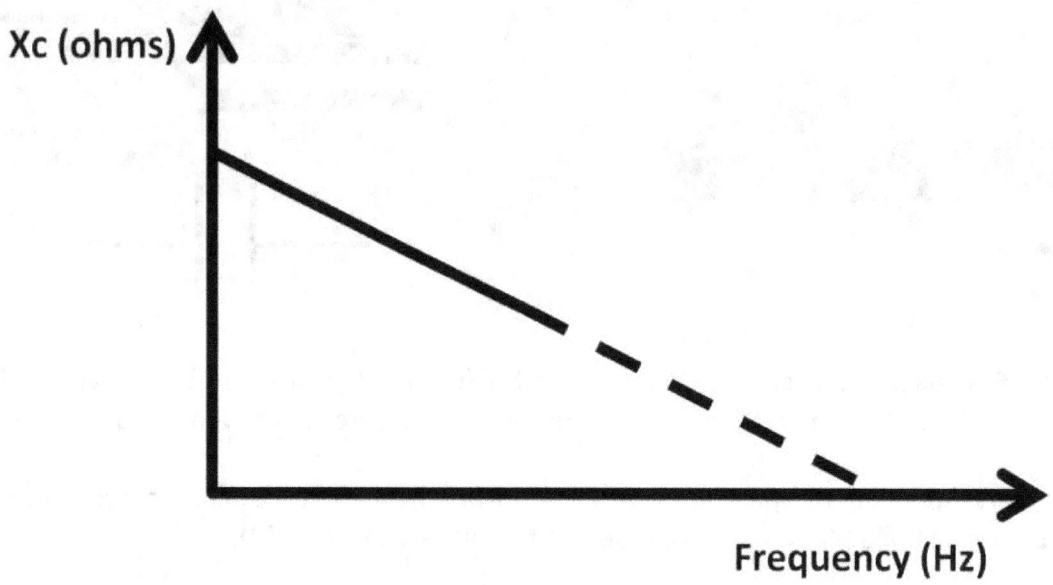

Figure 3.7: Xc vs. frequency

The above figure shows that Xc could reach zero ohms if frequency is extremely high. Using arithmetic rules, suppose frequency is infinite (∞); Xc would become zero (AC short).

$$\frac{1}{2\pi(\infty)(C)} = 0$$

Applying the same rule, if frequency is zero (DC), Xc would become infinitely large (DC block).

$$\frac{1}{2\pi(0)(C)} = \infty$$

Simple Capacitor Circuit

Let's use a simple capacitor circuit in figure 3.8 to further understand and apply this theory. Connecting a DC voltage source to a capacitor is equivalent to connecting the voltage source to an open circuit, i.e., infinite impedance. This implies that the capacitor is now "charged" to the positive voltage from the source. It's storing energy from the voltage source in the form of an electric field on the capacitor. Despite voltage drop across the capacitor, there is no current flowing through the capacitor (Xc is infinite, open circuit). According to Ohm's law, while impedance is infinite, there would be zero current flow. To simplify Xc, **1 / (2 π f C)**, can be converted to **1 / SC** or **1 / ωC**, where **S** or **ω (omega)** replaces **2 π f**.

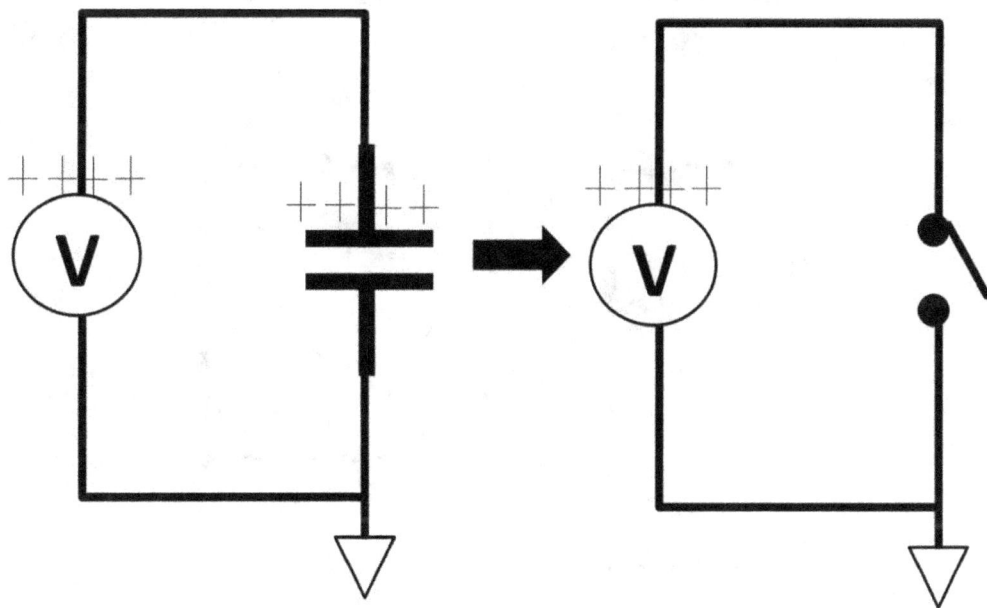

Figure 3.8: Simple capacitor circuit

A simple mathematical model can be used to represent the capacitor:

(Current) X (Time Change) = (Capacitance) X (Voltage Change)

I (Δt) = C (ΔV)

A simple four-step process circuit in figure 3.9 explains the I (Δt) = C (ΔV) equation.

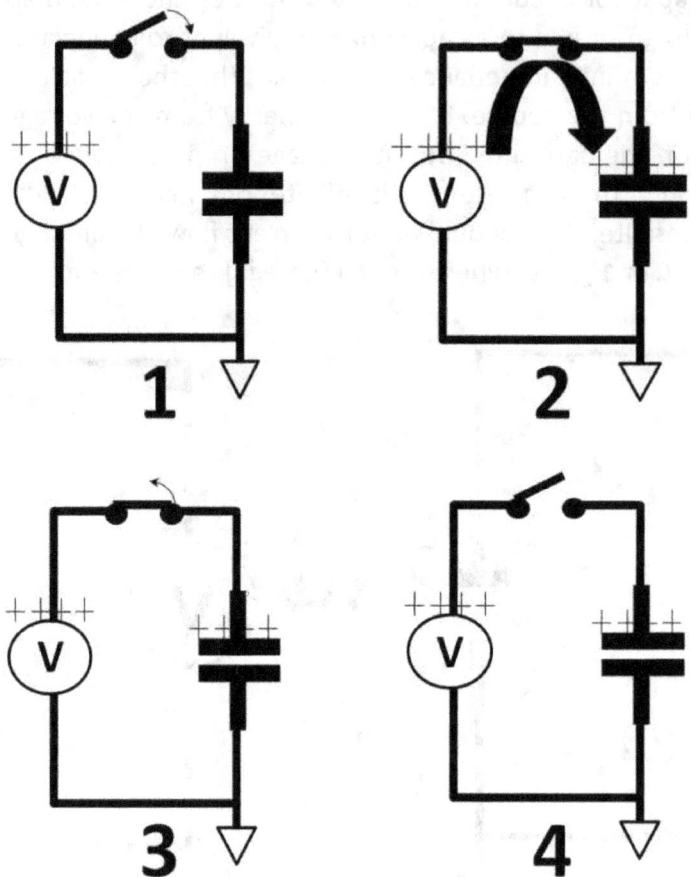

Figure 3.9: Four-step capacitor circuit

Step 1: Before the switch closes completely, there is no voltage across the capacitor, i.e., the voltage across top and bottom plates is 0 V.

Step 2: When the switch closes, electrons move towards the positive voltage source leaving the top capacitor plate positively charged. During this charge movement process, unless there is damage to the dielectric causing a short circuit, no electrical ions are able to pass through the capacitor due to the insulating dielectric. The time delay it takes for the voltage at the capacitor to be charged up to the voltage source amount is Δt. Dielectric damage can be caused by excessive voltage across the capacitor, thus breaking down the capacitor. The capacitor datasheet should spell out the maximum voltage.

Steps 3 and 4: When the switch opens, energy stored on the capacitor in the form of voltage has no other path to go (discharge) and therefore remains in the capacitor. The capacitor can now be viewed as a battery holding up the charge.

$$I (\Delta t) = C (\Delta V)$$

By knowing the voltage, capacitance, and Xc, Δt can be obtained:

$$\Delta t = C (\Delta V) / I$$

The charging behavior is further made clear in the waveform shown in figure 3.10. You can see that it takes time **(Δt)** to charge up the capacitor (dotted line) to full voltage.

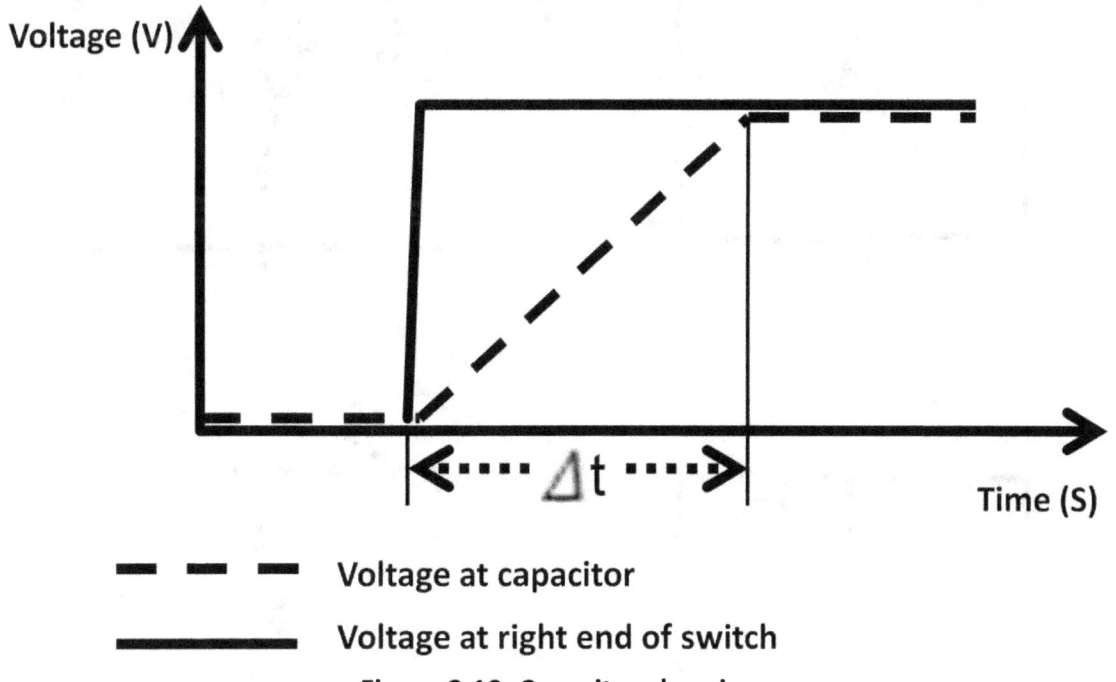

– – – Voltage at capacitor

⎯⎯ Voltage at right end of switch

Figure 3.10: Capacitor charging

Capacitor Charging and Discharging Circuit

Capacitors are used in applications where charging and discharging happens periodically. Flash lights applications found in cameras can be implemented using the circuit below (see figure 3.11).

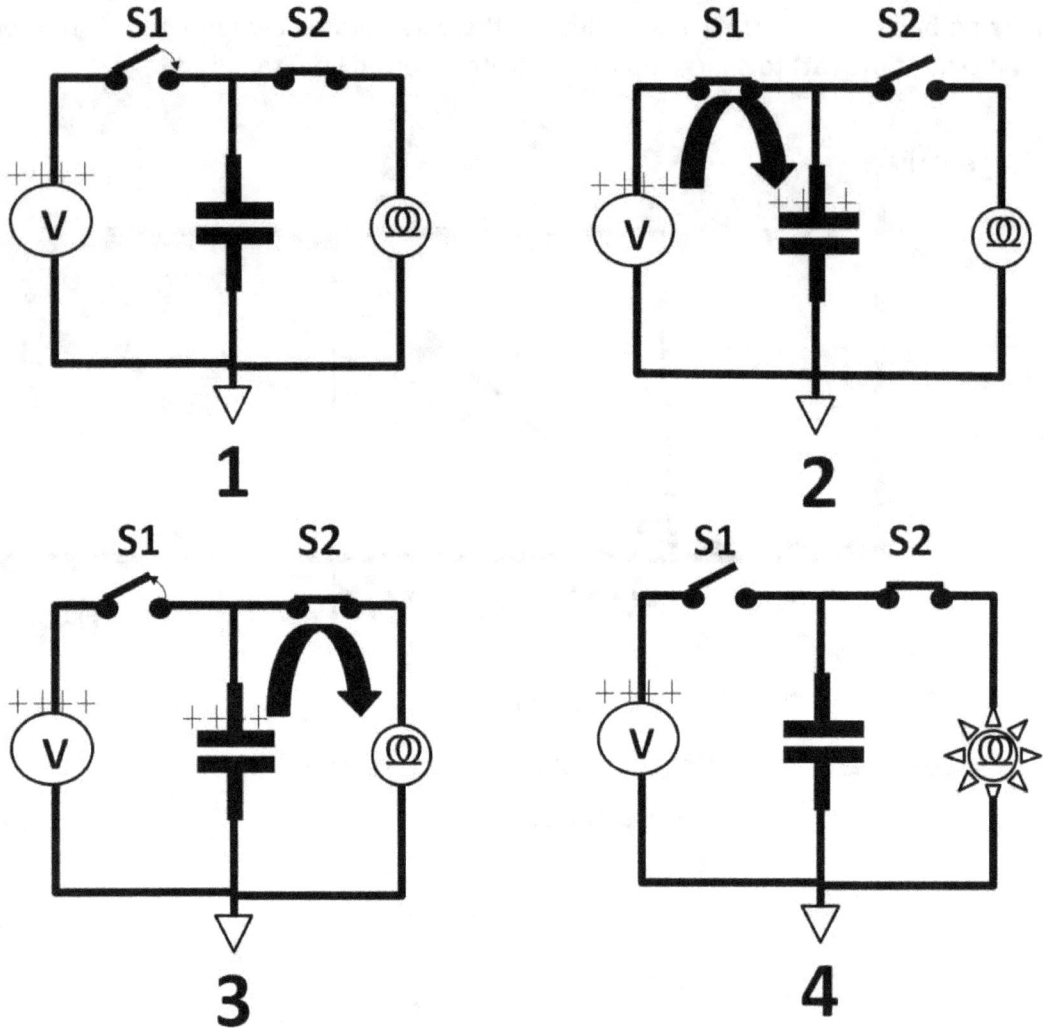

Figure 3.11: Capacitor flash light circuit

This circuit requires two switches, S1 and S2, operating in a complementary manner. When S1 opens, S2 closes and vice versa. The flash light acts as an electronic circuit load. The main difference between this flash light circuit and the capacitor charging circuit in figure 3.9 is that the flash light circuit dissipates charge to the load to light up the flash light (step 3). The electrical energy is transferred from the capacitor (battery) to the flash light (step 4). The capacitor voltage waveform is shown in figure 3.12.

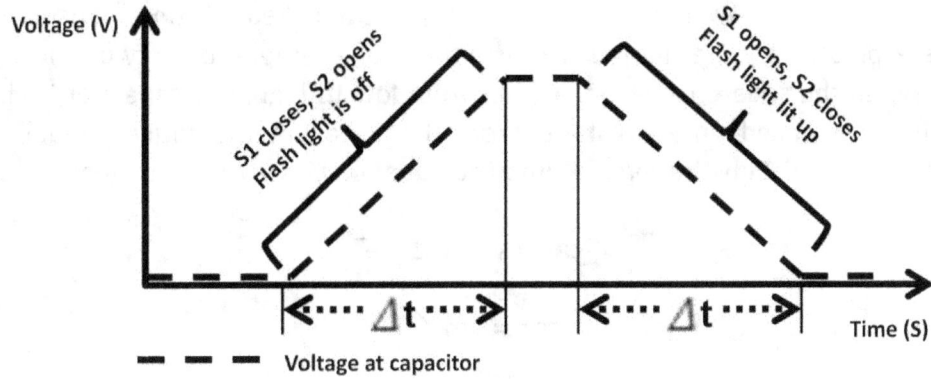

Figure 3.12: Capacitor charge and discharge

Combining resistors and capacitors creates several fundamental electronic circuits that are found in literally endless electronic systems (see figure 3.13). The circuit contains a square wave voltage source symbol (Vin). The circuit is analyzed using the following mathematical model,

$$V\_cap = Vin \times (1 - e^{-t/(rc)})$$

where V_cap is the voltage at the top capacitor plate. Vin represents input voltage; "e" is the exponential function in mathematics. R and C are resistance and capacitance. This is an exponential function. A V_cap waveform is shown in figure 3.14.

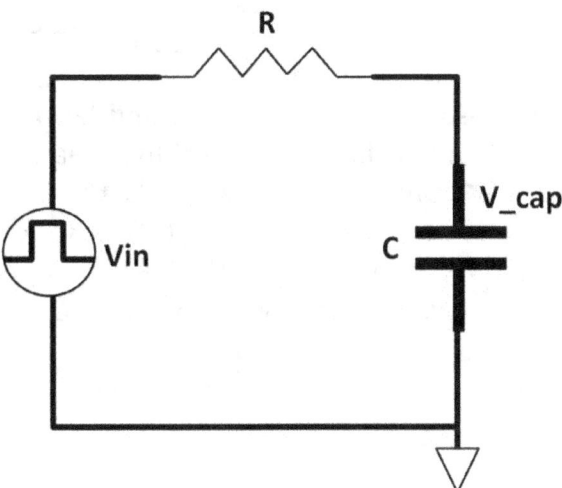

Figure 3.13: R, C series circuit

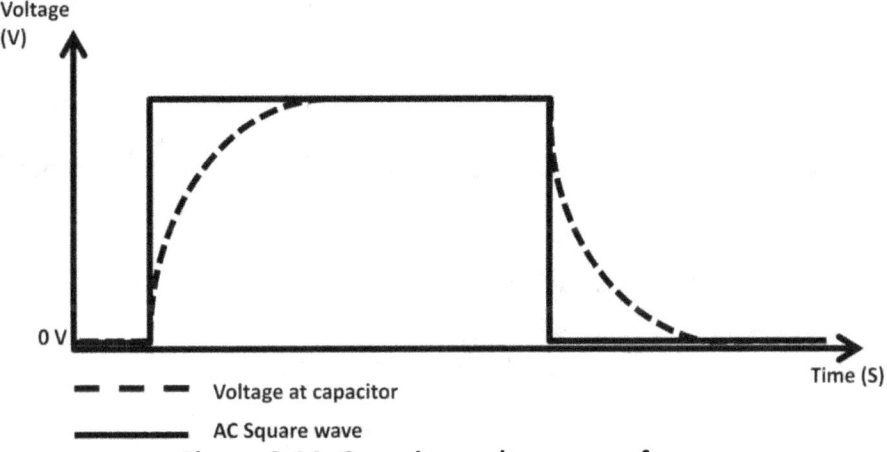

Figure 3.14: Capacitor voltage waveform

This circuit introduced a well-known electronic quantity called RC time constant. RC time constant is expressed by the number of instances, for example, one, two, and three time constants. When the square wave signal goes from low to high, the capacitor is charged up with a time delay called time constant. From the V_cap mathematical model, one time constant means **t = RC**. Substituting that into the equation yields the following:

$$V\_cap = Vin \times (1 - e^{-1})$$

$$V\_cap = Vin \times 0.64$$

From this result, at one time constant, voltage at the capacitor is 64% of the input voltage. For two time constants, time is equal to 2 X RC. V_cap now equates to:

$$V\_cap = Vin \times (1 - e^{-2}) = Vin \times 0.87$$

Plug in some realistic numbers and let us further demonstrate this concept. Suppose the lowest and the highest levels of the square wave are 0 V and 1 V respectively. **R = 10 kΩ** and **C = 1 uF**. One time constant yields **10 kΩ X 1 uF = 10 ms**. It means that it takes 10 ms for V_cap to reach 0.64 V (64% of 1 V). For two time constants, i.e., **2 X R C = 20 ms**, it takes 20 ms for V_cap to reach 0.87 V (87% of 1 V) (see figure 3.15). When the square wave goes from high to low, the capacitor was discharged (decay) using the same mathematical model.

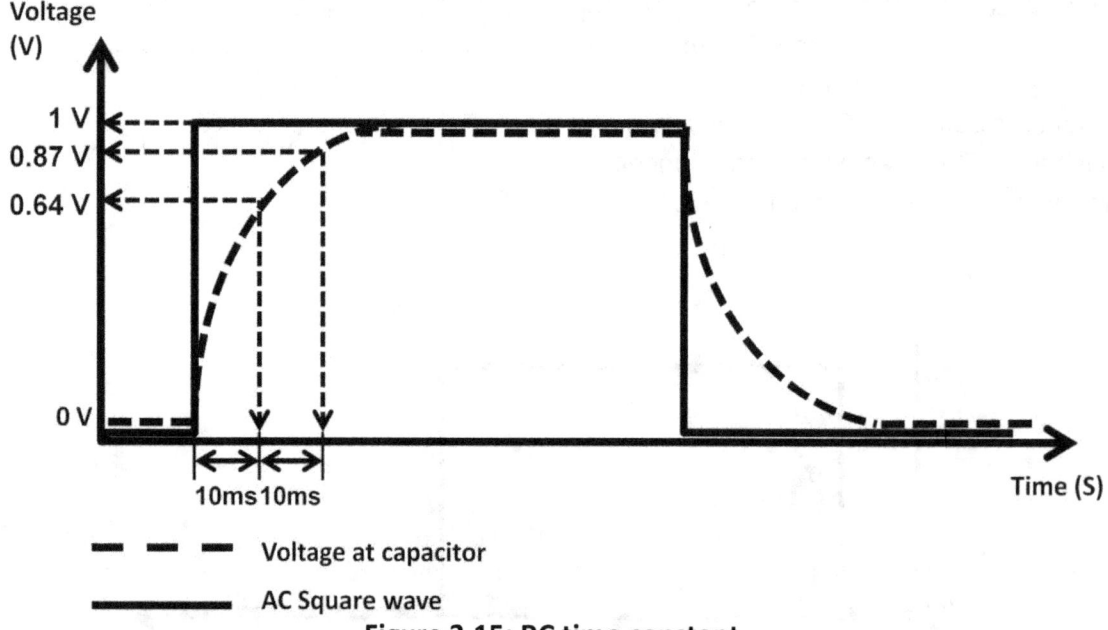

Figure 3.15: RC time constant

Parallel Capacitor Rule

Capacitors can be arranged in parallel with the following rules in figure 3.16. Two parallel capacitors' equivalence is the sum of two capacitances.

Figure 3.16: Parallel capacitor rule

Series Capacitor Rule

In figure 3.17, capacitors connected in series have equivalent capacitance (C_eq):

$$C\_eq = \frac{\text{Product of Capacitance}}{\text{Sum of Capacitance}}$$

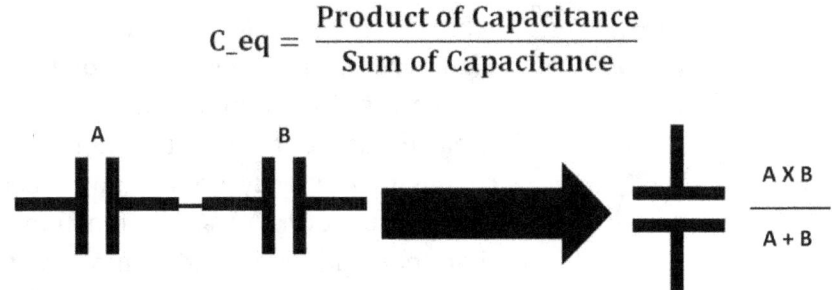

Figure 3.17: Series capacitor rule

You probably noticed that the capacitor rules are exactly opposite to the resistor rules. At this point, we focused on using time domain to illustrate capacitor circuits. It's more favorable to use frequency domain (bode plot, AC analysis) in electronic circuits. Frequency domain uses frequency on the X-axis and electrical quantities on the Y-axis. Before we take a deeper look at frequency domain, decibel or dB needs to be understood. dB is a ratio of two quantities. To calculate voltage ratio expressed in dB, logarithm (log) can be used.

For voltage:
$$\frac{V_{out}}{V_{in}} \text{ in dB} = 20 \log \left(\frac{V_{out}}{V_{in}}\right)$$

For current:
$$\frac{I_{out}}{I_{in}} \text{ in dB} = 20 \log \left(\frac{I_{out}}{I_{in}}\right)$$

For power:
$$\frac{\text{Power Out}}{\text{Power In}} \text{ in dB} = 10 \log \left(\frac{\text{Power Out}}{\text{Power In}}\right)$$

The difference between voltage, current, and power dB calculation is the constant 10 vs. 20.

Power Ratio in dB

For example, an audio power amplifier outputs 7.5 W, the input supply voltage, current is 10 V and 1 A. What is the power ratio in dB?

The input power is found by: **V X I = 10 V X 1 A = 10 W**:

$$\text{Power ratio in dB} = 10 \log \left(\frac{7.5 \text{ W}}{10 \text{ W}}\right)$$

$$\text{Power ratio in dB} = 10 \times (-0.125) \text{ dB}$$

$$\text{Power ratio in dB} = -1.25 \text{ dB}$$

R C Series Circuit

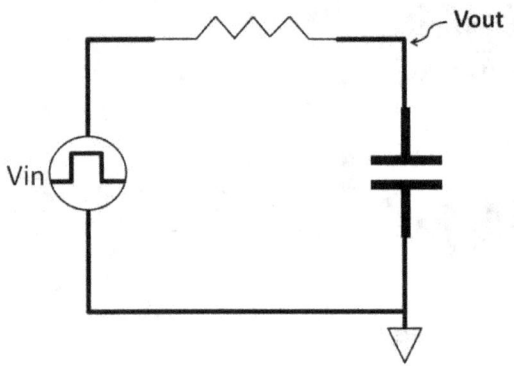

Figure 3.18: RC series circuit in frequency domain

We will now move onto using the same circuit from figure 3.13 to explain frequency domain in figure 3.18. We would define the square wave voltage source as Vin (input voltage), voltage at the capacitor is Vout (output voltage). To analyze this circuit in frequency domain, we need to derive a transfer function. A transfer function is an equation that spells out the relationship between input and output. If you look closely, it's nothing more than a voltage divider where the capacitor is an impedance varying resistor, (i.e., Xc). The transfer function thereby is:

$$\text{Vout} = \text{Vin} \times \left(\frac{-Xc}{-Xc + R}\right)$$

The "−" sign in Xc indicates C lags behind R by 90 degrees. This concept will be further explained shortly.

Recall **Xc =** or,

$$\text{Vout} = \text{Vin} \times \left(\frac{\frac{-1}{(S)(C)}}{\frac{-1}{(S)(C)} + R}\right) = \text{Vin} \times \left(\frac{\frac{1}{(S)(C)}}{\frac{1 + (R)(S)(C)}{(S)(C)}}\right) = \text{VIN} \times \left(\frac{1}{1 + (R)(S)(C)}\right)$$

- 20 dB per Decade

This transfer function shows that for given resistor and capacitor sizes, increasing frequency causes the Vout to decrease. The bode plot below elaborates this concept (see figure 3.19).

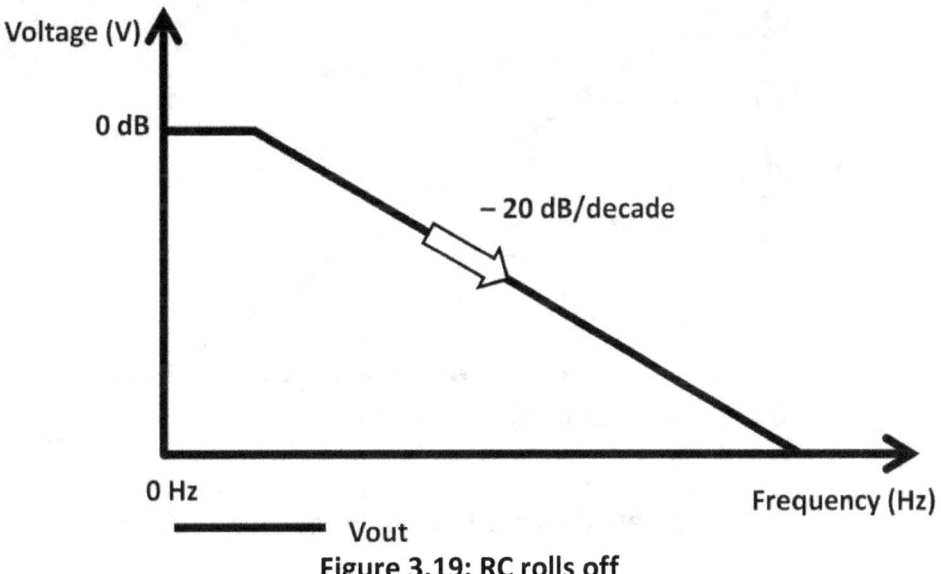

Figure 3.19: RC rolls off

At 0 Hz (DC), **Vout = Vin**,

$$Vin = Vout = (Vin) \times \frac{1}{1 + \cancel{(R)(C)(2\pi)}(0)} = Vin \times \left(\frac{1}{1}\right)$$

$$\frac{Vout}{Vin} \text{ in dB} = 10 \log 1 = 0 \text{ dB}$$

As frequency increases, voltage falls rolling off at – 20 dB per decade rate. A decade is 10 times change in frequency. Assume at 10 kHz, Vout starts to fall; Vin is at 10 V. Vout / Vin in dB; frequency and Vout are developed below in Table 3-2. There is a negative sign of dB after 0 dB. It's due to the – 20 dB per decade reduction rate.

Frequency	$\frac{Vout}{Vin}$ in dB	Vout
0 Hz (DC)	0 dB	10 V
10 Hz	−20 dB	1 V
100 kHz	−40 dB	0.1 V
1 MHz	−60 dB	0.01 V

Table 3-2: Frequency, Vout / Vin, Vout

The corresponding graph is shown in figure 3.20.

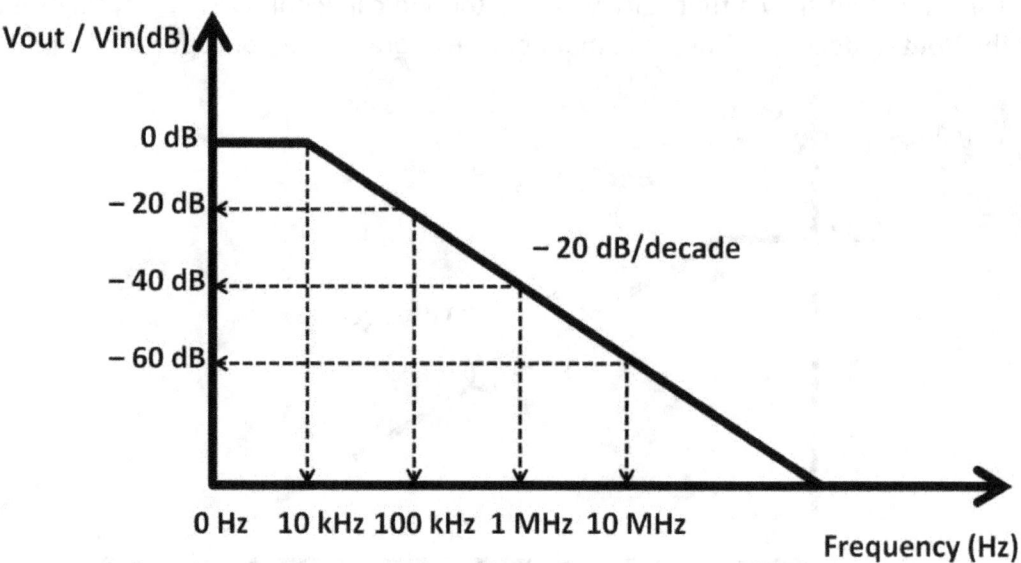

Figure 3.20: Vout / Vin vs. Frequency

Let's put some actual numbers to this RC circuit in figure 3.21. Vin = 0 V to 10 V at 10 kHz, 100 kHz, 1 MHz and 10 MHz, **R = 10 kΩ, C = 1 mF**

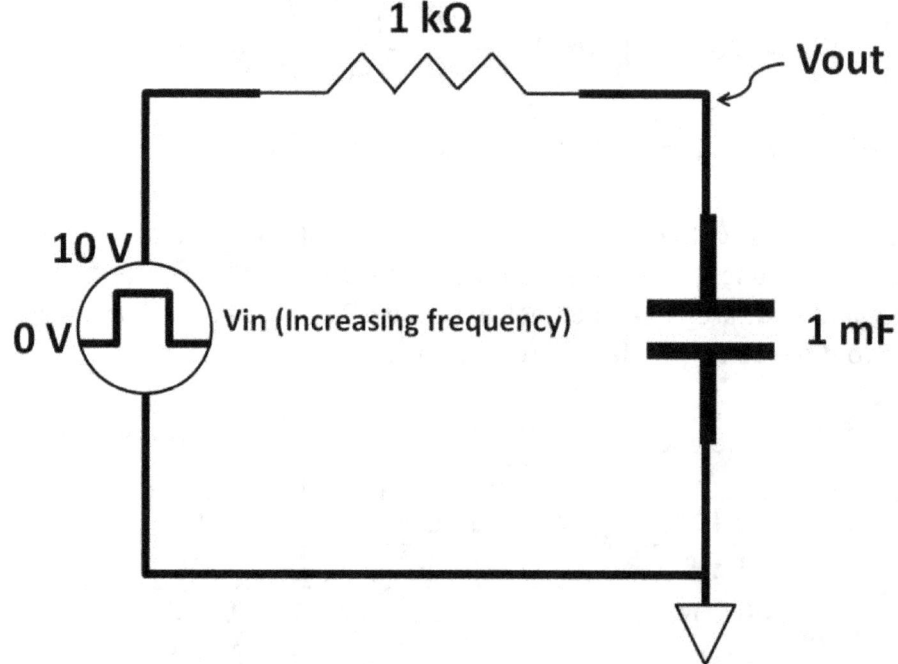

Figure 3.21: RC circuit with actual values

This circuit only contains passive devices, hence output cannot exceed input. The highest output can reach is input, denoted by 0 dB (Vin = Vout). As frequency increases; Vout decreases, the negative dB sign then follows. Every – 20 dB per decade roll-off signifies the 10 times increases in frequency. When **Vin = Vout = 10 V**, frequency is 0 Hz:

$$10 \text{ V} = (10 \text{ V}) \times \frac{1}{1 - (1 \text{ k}\Omega)(2\pi f)(1 \text{ mF})}$$

$$1 = 1 - 2\pi f$$

$$2\pi f = 0$$

$$f = 0 \text{ Hz}$$

At – 20 dB, Vout:

$$-20 \text{ dB} = 20 \log\left(\frac{V_{out}}{V_{in}}\right)$$

$$-1 = \log\left(\frac{V_{out}}{V_{in}}\right)$$

Vout = 0.1 V, where Vin = 10 V

Frequency is then found by:

$$(10 \text{ V}) \times \frac{1}{1 - (1 \text{ k}\Omega)(2\pi f)(1 \text{ mF})}$$

$$100 = 1 - 2\pi f$$

$$f = 16 \text{ Hz}$$

At – 40 dB, Vout:

$$-40 \text{ dB} = 20 \log\left(\frac{V_{out}}{V_{in}}\right)$$

$$-2 = \log\left(\frac{V_{out}}{V_{in}}\right)$$

Vout = 0.01 V, Vin = 10 V

Frequency is then found by:

$$0.01 = (10 \text{ V}) \times \frac{1}{1 - (1 \text{ k}\Omega)(2\pi f)(1 \text{ mF})}$$

$$1{,}000 = 1 - 2\pi f$$

$$f = 160 \text{ Hz}$$

Apply the same technique, it can be estimated that the frequency increases 10 times for every 10 times reduction in Vout. Table 3-3 summarizes these findings for **R = 1 kΩ, C = 1 mF**. The exercise above shows that you can design the circuit by changing the R and C values to fine tune a unique frequency in such a way that Vout starts to reduce. This concept extends to a very popular capacitor filter application: low-pass filter.

Frequency	$\frac{V_{out}}{V_{in}}$ in dB	Vout
0 Hz (DC)	0 dB	10 V
16 Hz	–20 dB	1 V
160 kHz	–40 dB	0.1 V
1.6 MHz	–60 dB	0.01 V
16 MHz	–80 dB	0.001 V

Table 3-3: For R = 1 kΩ, C = 1 mF

Low-Pass Filter

Figure 3.21 is a popular circuit called the low-pass filter. It allows signal to pass through only at low frequency filtering out high frequency signals. A low-pass filter is used to remove high frequency noise effectively improve electronic system performance. The capacitor used in this circuit is called decoupling or bypass capacitor. The down side to this noise reduction technique is the additional R and C components adding bill-of-materials (BOMs) costs and space on the printed-circuit board. BOMs are used to estimate overall system costs. They include all hardware components as well as printed circuit boards costs. One filter parameter often used is f –3dB. It denotes specific frequency value when the output starts to fall to 70.7 percent of the input. It's the point where the output just starting to roll-off. Using this characteristic, filter performance can be summarized (see figure 3.22).

$$-3\text{ dB} = 20 \log \left(\frac{V_{out}}{V_{in}}\right)$$

$$\frac{V_{out}}{V_{in}} = 0.707$$

Coincidently, the 0.707 is the same constant used in Vrms calculations.

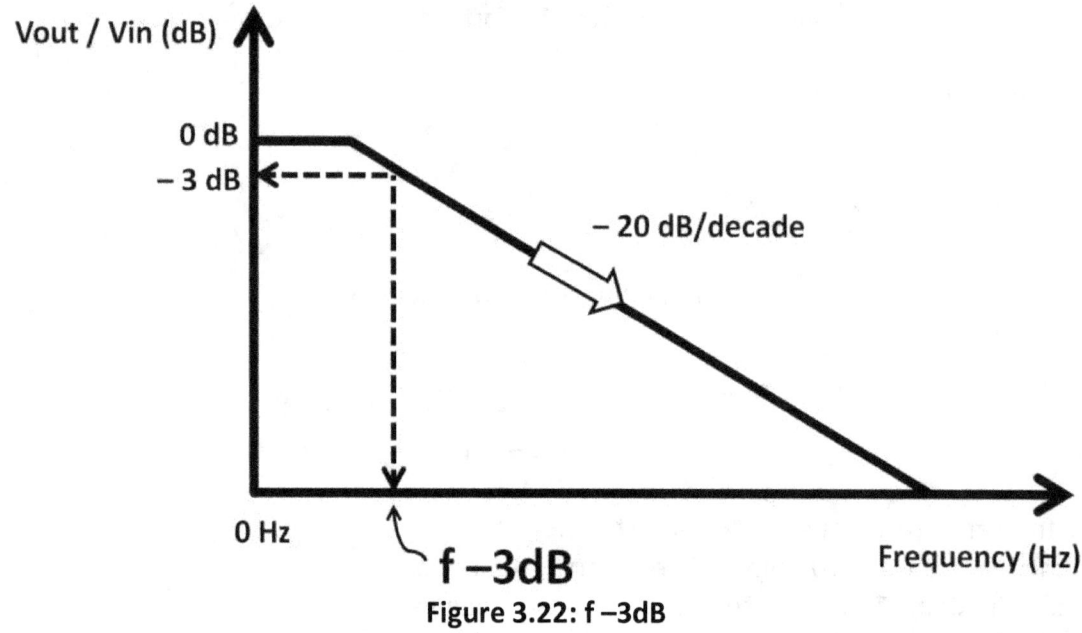

Figure 3.22: f –3dB

Phase Shift

In the RC low-pass filter circuit, there is a phase shift between the voltage at the resistor and capacitor. Phase shift is the time difference amount from the original timing position to a new one. A phase shift can be positive or negative. To understand phase shift in the low-pass filter, we use a 360-degree circle to interpret a full cycle sine wave in figure 3.23.

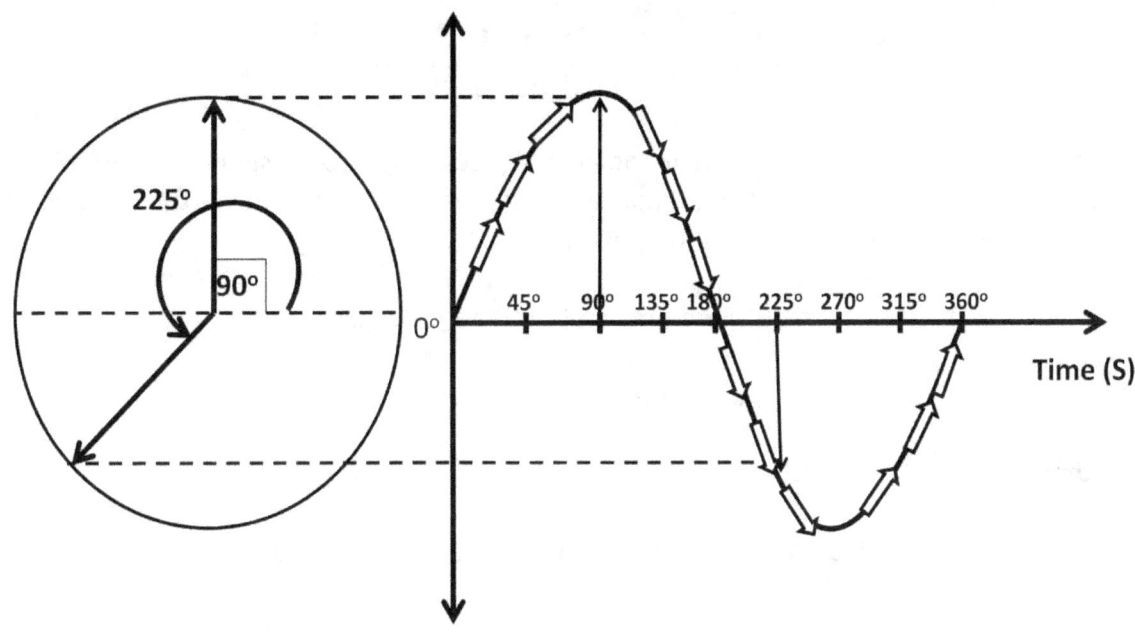

Figure 3.23: Sine wave with 360 degree circle

A periodic sine wave revolves continuously. It can be mapped to a 360 degree circle on the left-hand side of figure 3.23. A sine wave at a single point of time presents specific amplitude. It starts from 0 degree moving in anti-clockwise direction. In the figure 3.23, when the sine wave arrives at the positive peak, it represents 90 degree of the circle (top dotted line). When the sine wave continues to revolve towards the right-hand side of the timing waveform arriving at the lower half, it corresponds to 225 degrees. All these can be modeled as:

$$V(t) = V_{peak} \times \sin \Theta, \text{ where } \Theta \text{ is degree}$$
$$= V_{peak} \times \sin (\omega t)$$
$$= V_{peak} \times \sin (2 \pi f) \times (t)$$
$$= V_{peak} \times \sin (2 \pi)$$

For example, if **Vpeak= 5 V**, the rotation degree (Θ) = 45 degrees:

$$V(t) = 5 \text{ V} \times \sin (45 \text{ degrees}) = (5 \text{ V}) \times 0.707 = 3.53 \text{ V}$$

Radian

The equation on the previous page is a function where V(t) is a portion of Vpeak at any particular time. $\omega = 2\pi f$ or $(2\pi)/t$ is the angular velocity (distance divided by time). Figure 3.24 below shows the various radians around a full cycle sine wave and a table with degrees correlate with radian in π and decimal values. Radian can also be defined as:

$$\text{Radian} = \frac{\text{Arc of Circle}}{\text{Radius of Circle}}$$

If the arc distance and radius are equal, radian = 1. Because 2 π corresponds to the full 360-degrees of a circle, if given a radian, a particular degree is easily found via the 2 π and 360 degree ratio. For example, radian = 1, degree X:

(2 π / 360 degrees) = (1 radian / X degree)

X = 57.30 degrees, where π = 3.14

If **arc distance = 2, radius = 0.5**. Radian:

Radian = 2 / 0.5 = 4 radians = (1.27 X π) radians

The rotation degree can be calculated by the 2 π and 360-degree relationship. Let's assume the rotation degree is Y. Use 4 radians from the above example:

(2 π / 360) = (4 radians / Y)

(2 π / 360) = 1.27 π / Y

Y = 229.18 degrees

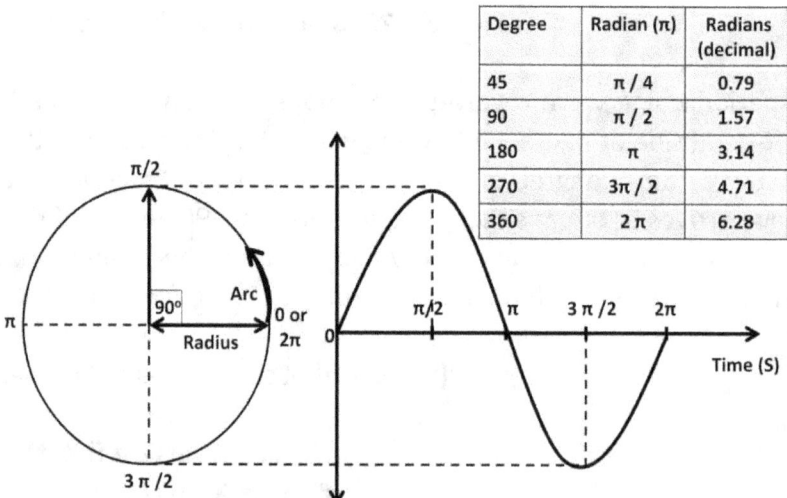

Degree	Radian (π)	Radians (decimal)
45	π / 4	0.79
90	π / 2	1.57
180	π	3.14
270	3π / 2	4.71
360	2 π	6.28

Figure 3.24: Radian vs. sine wave

ICE

In the RC filter circuit, the voltage at the capacitor lags current. Some use "I to C to E" (ICE) as a way to recognize this phenomenon. "I" corresponds to current, "C" is the capacitor, and "E" is voltage potential across the capacitor. In figure 3.25, voltage at the capacitor appears on the right while both capacitor and resistor currents (on the left) lead capacitor voltage by 90 degrees. Because the resistor and capacitor are connected in series with only one current branch, the capacitor current has the same phase as resistor current and voltage. In other words, the capacitor voltage is lagging capacitor and resistor currents by 90 degrees. This explains why Xc has a negative sign in the R C circuit calculation in figure 3.18. It designates the 90-degree phase shift.

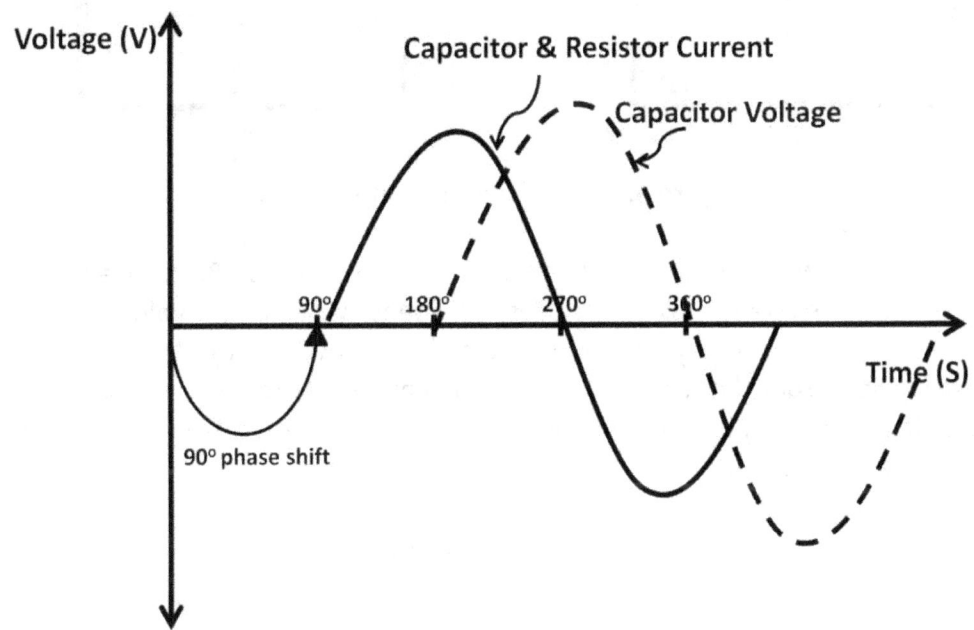

Figure 3.25: ICE

We derived in chapter 1, DC: **Current = I = ΔQ / t**

The capacitor is represented by this model,

$$I (\Delta t) = C (\Delta V)$$

Substituting I with **ΔQ / t** in this model results in

$$(\Delta Q / t)\Delta t = C (\Delta V)$$

$$\Delta Q = C (\Delta V)$$

This equation can be realized by the circuit below in figure 3.26.

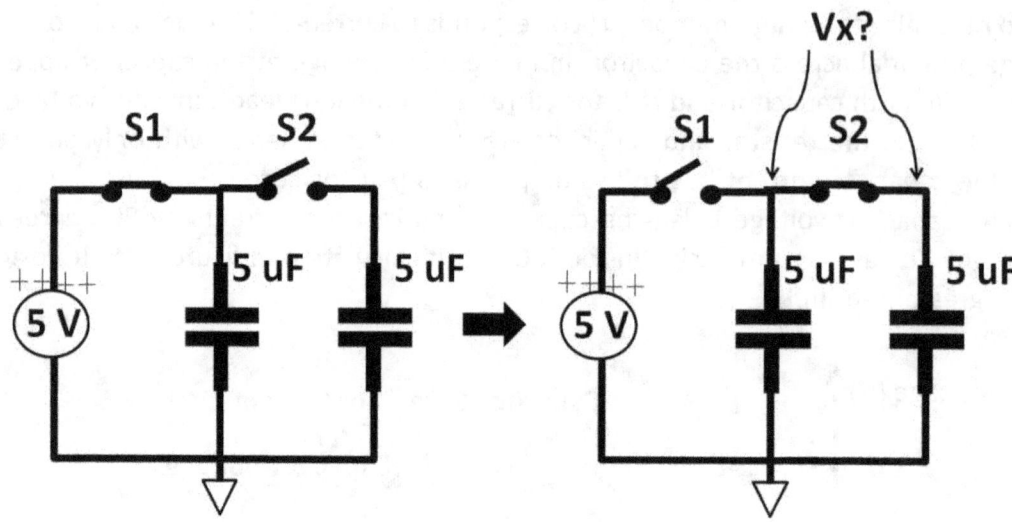

Figure 3.26: Capacitor circuit question

With a 5 V source, two 5 uF capacitors are connected in parallel. Both S1 and S2 are ideal switches with zero resistance. S1 first closes, S2 opens. What is the voltage at Vx after S1 opens while S2 closes? You may be tempted to say Vx is 5 V, but it isn't. We can use **ΔQ = C (ΔV)** to prove that. From the formula, the electric charge (Q) remains the same before and after the switch opens and closes. Applying energy conservation from the law of physics, when S1 is closed, S2 opens:

$$Vx = 5 \text{ V}, C = 5 \text{ uF}$$

Q = C V:

$$Q = (C) \times (Vx) = (5 \text{ uF}) \times (5 \text{ V}) = 25 \text{ u coulombs}$$

After that, S1 opens, S2 is closed, and Q remains the same. We are now looking at two capacitors in parallel **(5 uF || 5 uF = 5 uF + 5 uF = 10 uF)**:

$$Q = (C) \times (V) = 25 \text{ u coulombs} = (10 \text{ uF}) \times (Vx)$$

$$Vx = 2.5 \text{ V}$$

We will go over more practical capacitor circuits at the end of the chapter after the inductor section.

Inductors

An inductor is an electronic passive device that does not generate energy but rather stores energy in a magnetic field. Inductors are typically made of wounded coil in multiple forms and sizes. Common inductor materials are iron, copper, and ferrite. Many characteristics are exactly opposite that of a capacitor. Figure 3.27 shows several inductor types and its schematic symbol.

Figure 3.27: Assorted inductors (top) and inductor schematic symbol (bottom)

Inductor value (inductance) is measured in unit Henry (H). They exhibit reactance, called inductive reactance (XL) measured in Ohms (Ω). The inductor symbol is L. Some inductors are constructed in the microelectronic scale housed in small semiconductor packages. The smaller sizes save area, however, small sized inductors offer much less inductance.

Inductor Impedance = Resistance + XL

The contribution of the inductance resistance comes from inductor packages and leads and the intrinsic nature of inductive materials. **XL = 2 π f L**, where f is signal frequency and L is inductance with units in Henry (H). When the signal frequency change, XL change causing change in inductor impedances. If **signal frequency = 1 MHz, inductance = 1 uH**:

$$XL = (2\pi) \times (1\ MHz) \times (1\ uH) = 6.28\ \Omega$$

XL versus Frequency

If the frequency now increases to **2 MHz**, then **XL = 12.56 Ω**. In other words, XL is proportional to frequency (see figure 3.28). Keep in mind that even with changes in impedances with frequencies, the inductive values remain the same. Inductance remains 1 uF in both frequencies.

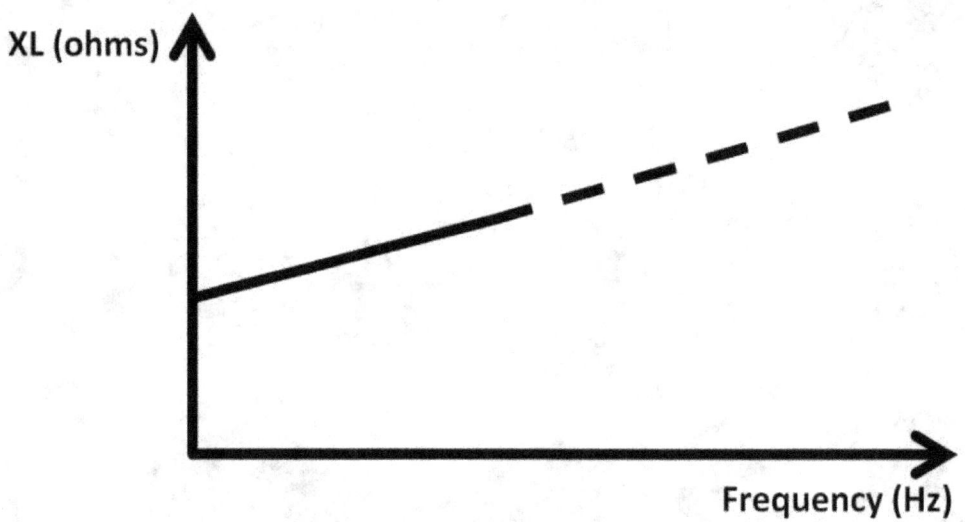

Figure 3.28: XL vs. frequency

The above diagram shows that XL could reach infinity if frequency is extremely high. This only applies to ideal inductors. You will see several non-ideal inductor characteristics later on. Using arithmetic rules, assuming frequency is infinite, XL would become infinite (AC choke):

$$XL = 2\pi\ (\infty)\ (L)$$

∞ >> 2 π, L,

$$XL = \infty$$

Applying the same rule, if frequency is zero (DC), Xc would become zero.

$$XL = 2\pi\ (0)\ (L) = 0$$

Let's use a simple inductor circuit to explain this in figure 3.29. Connecting a DC voltage source to an inductor is equivalent to connecting the voltage source to a zero Ω resistor. This implies that the inductor practically is non-existent (DC short).

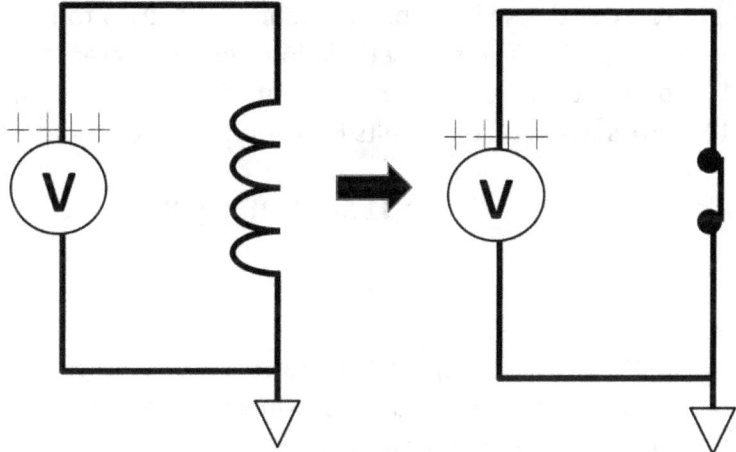

Figure 3.29: Simple inductor circuit

V (Δt) = L (ΔI)
A simple mathematical model can be used to represent a capacitor.

$$(Voltage) \times (Time\ Change) = (Inductance) \times (Current\ Change)$$

$$V\ (\Delta t) = L\ (\Delta I)$$

A simple inductor circuit in figure 3.30 explains the above model.

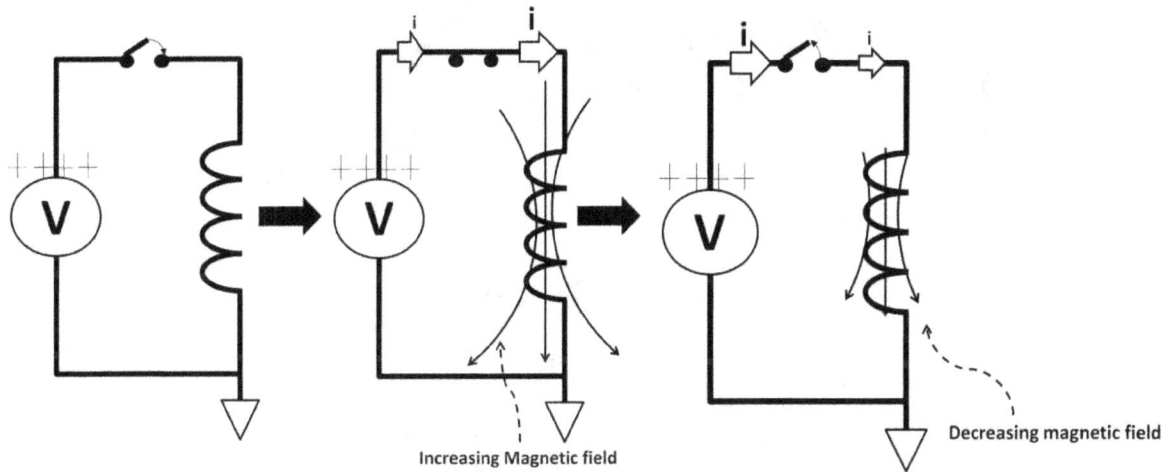

Figure 3.30: Inductor circuit explained

After the switch is closed, current starts to ramp up and the magnetic field starts to increase. It takes Δt for the inductor to build up the magnetic field and current to its maximum level. The field strength and current depend on the inductance amount, which relates strongly to the inductor materials and proportionally to the coil number. Immediately after current ramps up to the highest level, the switch opens. The inductor tries to maintain current flow and the inductor will flip polarity. The magnetic field strength decreases and current ramps down. Figure 3.31 shows the current ramp (current ripple) waveform. If, for example, inductance is 1 H, 1 V across the inductor results in current ramp of 1 A in 1 S.

$$(1 \text{ V}) \times (1 \text{ S}) = (1 \text{ H}) \times (\Delta I)$$

$$(\Delta I) = 1 \text{ A}$$

Power management applications step up and/or down input voltage to provide higher or lower output voltage, current, or power. Some power management designs operate in DC such as those (diodes, zener diodes) mentioned in chapter 1, DC. Many power management applications operate in AC and at much higher frequency. For example, a 400 kHz (2.5 ms period) switching regulator takes 12 V input voltage and regulates to 1.2 V output. It specifies that 10% duty cycle (0.25 ms) is required with maximum 2 A ripple current at the output. The inductor size can be calculated:

$$(12 \text{ V}) \times (0.25 \text{ ms}) = L (2 \text{ A})$$

$$L = 15 \text{ mH}$$

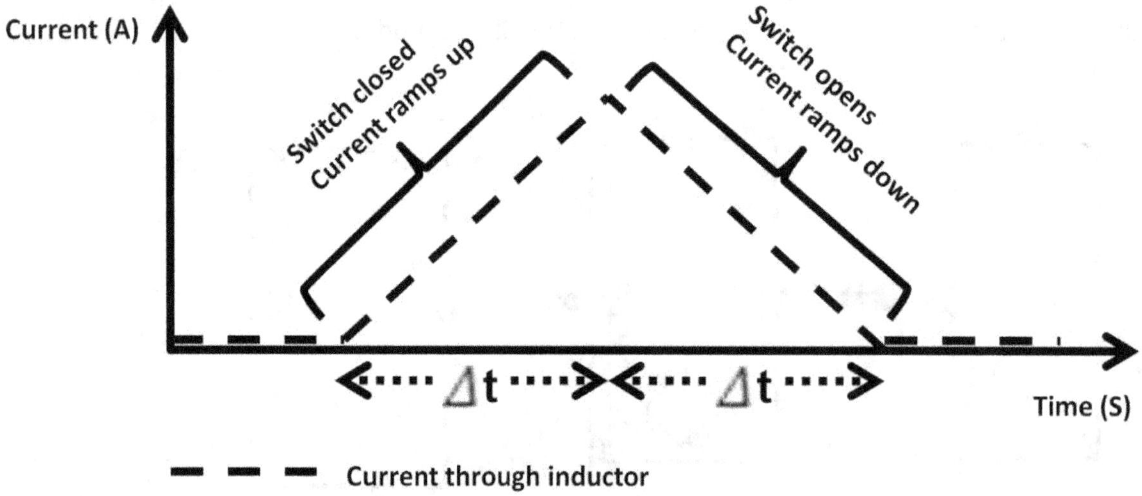

Figure 3.31: Inductor current ramps

ELI

As for voltage leads and lags, inductors behave exactly the opposite of capacitors. We use "E to L to I" ("ELI") where "E" is potential difference, "L" represents the inductor, and "I" corresponds to current. Inductor voltage is leading the current by 90 degrees, shown in figure 3.32.

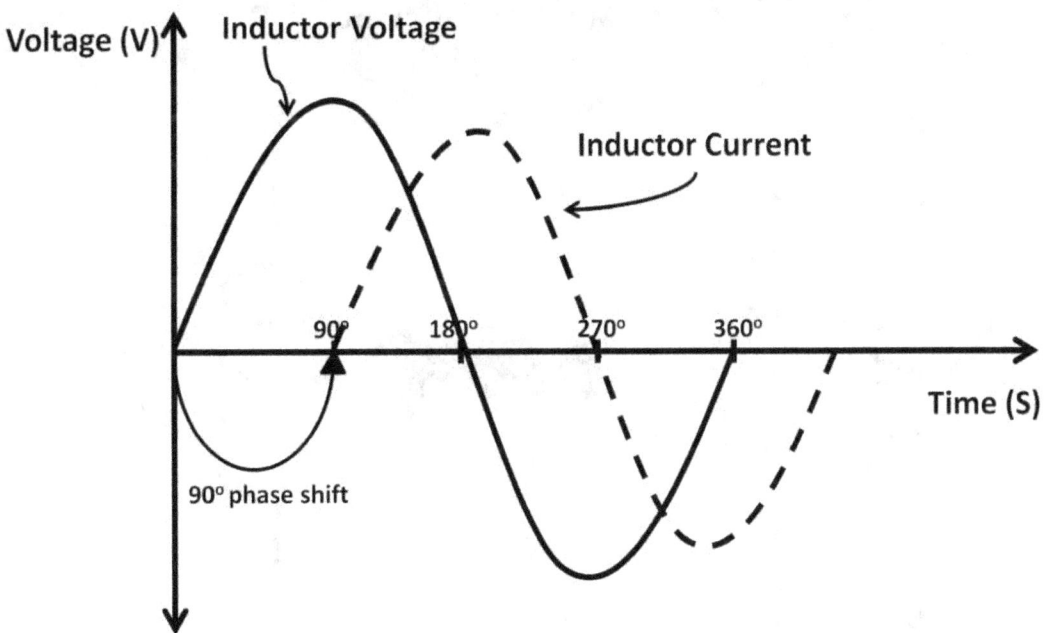

Figure 3.32: ELI

Q Factor

The inductor quality factor (Q) dictates how good an inductor is. This factor determines how much loss the inductor incurs in terms of heat and magnetic field losses. Q factor is modeled by 2 π f, inductance (L), and the inductor's internal electrical resistance (R).

$$Q \text{ factor} = \frac{2\pi f \ (L)}{R}$$

With this simple model, an ideal inductor (lossless) Q factor is infinite **(R = 0)**:

$$Q = \frac{2\pi F \ (L)}{0} = \infty$$

A surface-mount 10 nH inductor with 1 mm by 0.5 mm in dimension can have a Q factor as high 20 at 500 MHz.

Parallel Inductor Rule

Recall capacitor parallel rules. Inductors can be arranged in parallel with the following rules in figure 3.33. It's again opposite to the capacitor rule. The parallel inductor rule is the same as the resistor rule. The equivalent of two parallel inductors, L_eq:

$$L\_eq = \frac{\text{Product of Inductance}}{\text{Sum of Inductance}}$$

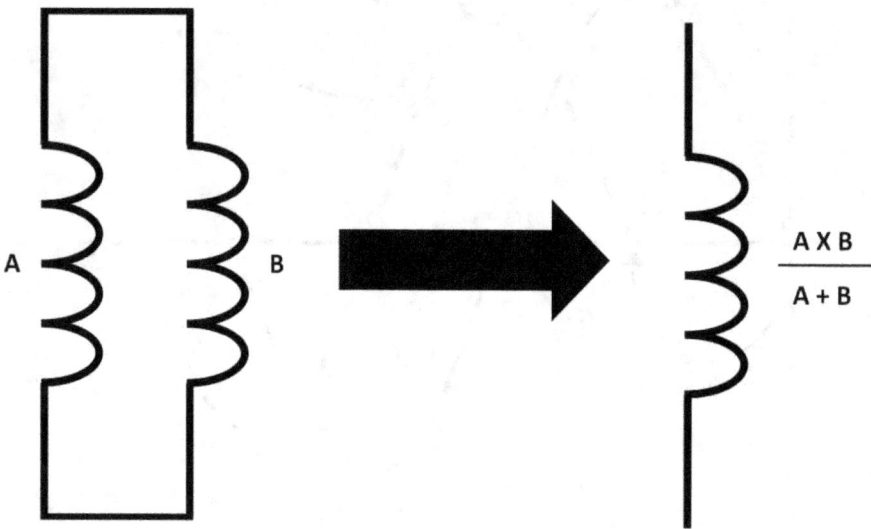

Figure 3.33: Parallel inductor rule

With two wire-wound inductors connected in parallel, each has 100 nH. Equivalent inductance, L_eq is:

$$L\_eq = \frac{100 \text{ nH} \times 100 \text{ nH}}{100 \text{ nH} + 100 \text{ nH}} = 50 \text{ nH}$$

The same as the resistor rule, if the parallel inductors are the same sizes, the equivalent inductance is the inductance divided by the number of inductors, (i.e., 100 nH / 2 = 50 nH). If 3 inductors connected in parallel are all equal in inductances, the equivalent inductance:

$$L\_eq = \frac{\text{Individial Inductance}}{3}$$

Series Inductor Rule

The equivalence of two inductors in series yields the sum of two inductances shown in figure 3.34.

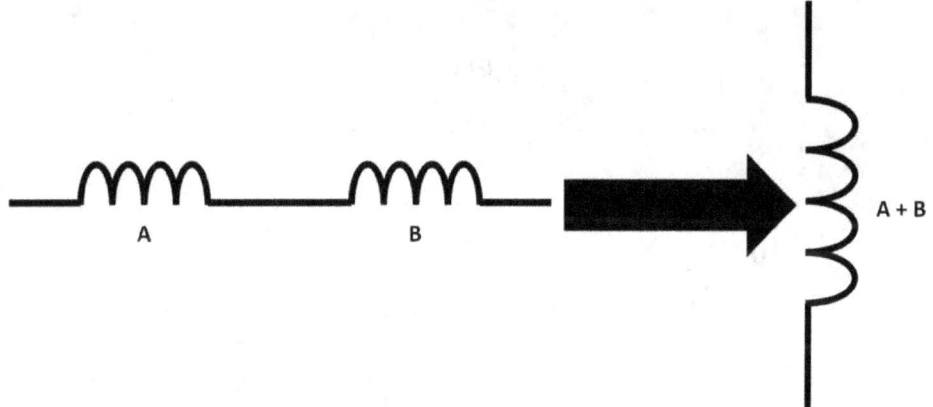

Figure 3.34: Series inductor rule

Two inductors, each having 5 uH connected in series, yields **5 uH + 5 uH = 10 uH**.
A combination of series and parallel inductor can be evaluated using these rules to yield equivalent inductance. In figure 3.35, L_eq:

Step 1: First combine B, E, and D, (B + E + D)
Step 2: Parallel (B + E + D) || F
Step 3: Add A, results from step 2, and C together

$$A + (F \;||\; (B + E + D)) + C$$

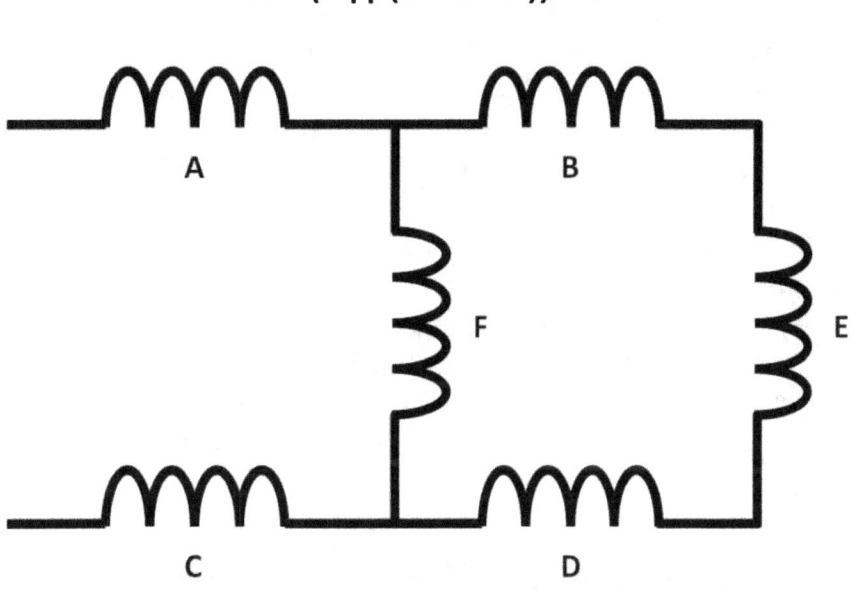

Figure 3.35: Inductor combinations

High-Pass Filter

Let's use a simple inductor circuit in frequency domain (AC analysis), illustrated in figure 3.36, to help our inductor knowledge sink in further.

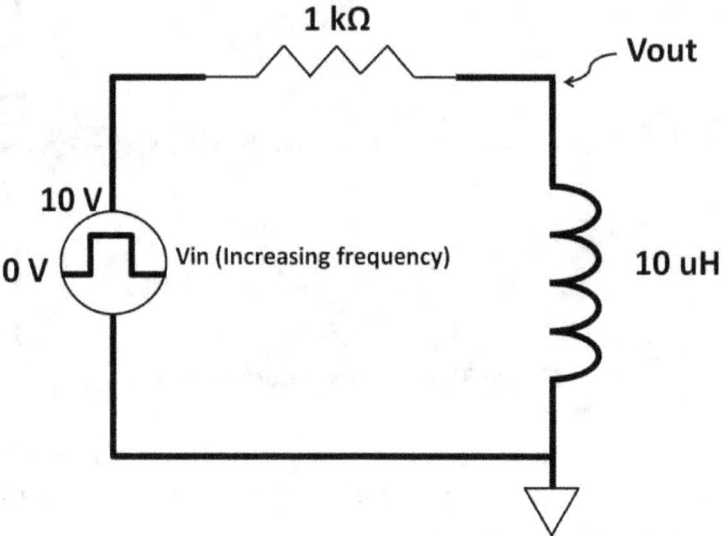

Figure 3.36: R L circuit

This circuit is called high-pass filter. It contains passive devices only; a 1 kΩ and 10 uH inductor connected in series. AC square wave source frequency increases. Output cannot exceed input. It can only go as high as input, denoted by 0 dB, **Vin = Vout**. To derive Vout, we cannot simply use a standard voltage divider because of the ELI behavior, (i.e., inductor current leads inductor voltage by 90 degrees). A creative way to analyze an AC circuit is to use the vector diagram shown in figure 3.37. Instead of using time on the X-axis, the vector diagram uses voltage or current on both the X- and Y-axis along with degree rotations. Because this circuit is a series circuit, current in this series R L circuit is the same, (i.e., the same phase). This makes the voltage and impedances 90 degrees out of phase. In order to figure out the voltage across resistor and inductor, total impedance Z first needs to be found. Z is the resultant impedance between XL and a 1 kΩ resistor. For a given frequency, if XL and R are the same, the resultant Z would be angled at 45 degrees (half of 90 degrees). If XL is higher than R, (i.e., Vin frequency is higher), XL pulls Z upward with more degree rotation. We can use standard trigonometry or the Pythagorean Theorem to calculate the resultant Z and angle. Once Z is found, the divider rule is used to evaluate the voltages across inductor and resistor. In figure 3.37, to achieve a 45 degree angle, **XL = R = 1 kΩ = 2 π f L, and Vin frequency:**

$$\text{Vin frequency} = \frac{1\ k\Omega}{2\pi L} = \frac{1\ k\Omega}{2\pi(10\ uH)} = 159\ \text{MHz}$$

Use the Pythagorean Theorem:

$$Z = (R^2 + XL^2)^{0.5}$$

$$Z = (1\ k\Omega^2 + 1\ k\Omega^2)^{0.5}$$

$$Z = 1.414\ k\Omega$$

$$V_{out} = V_{in} \times \frac{1\ k\Omega}{1.414\ k\Omega}$$

When Vin reaches peak at 10 V,

$$V_{out} = (10\ V) \times \frac{1\ k\Omega}{1.414\ k\Omega}$$

$$V_{out} = 7.07\ V$$

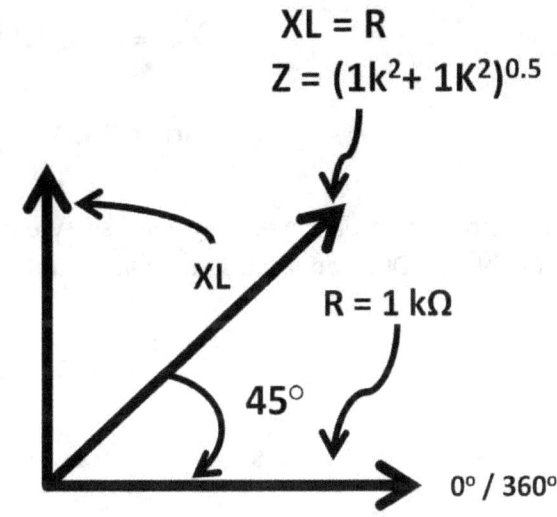

Figure 3.37: R L vector diagram

Figure 3.38 shows the timing waveform between inductor and resistor voltage. They differ by 45 degrees. If the input frequency changes, the resultant Z changes as well as the rotation angle (phase shift).

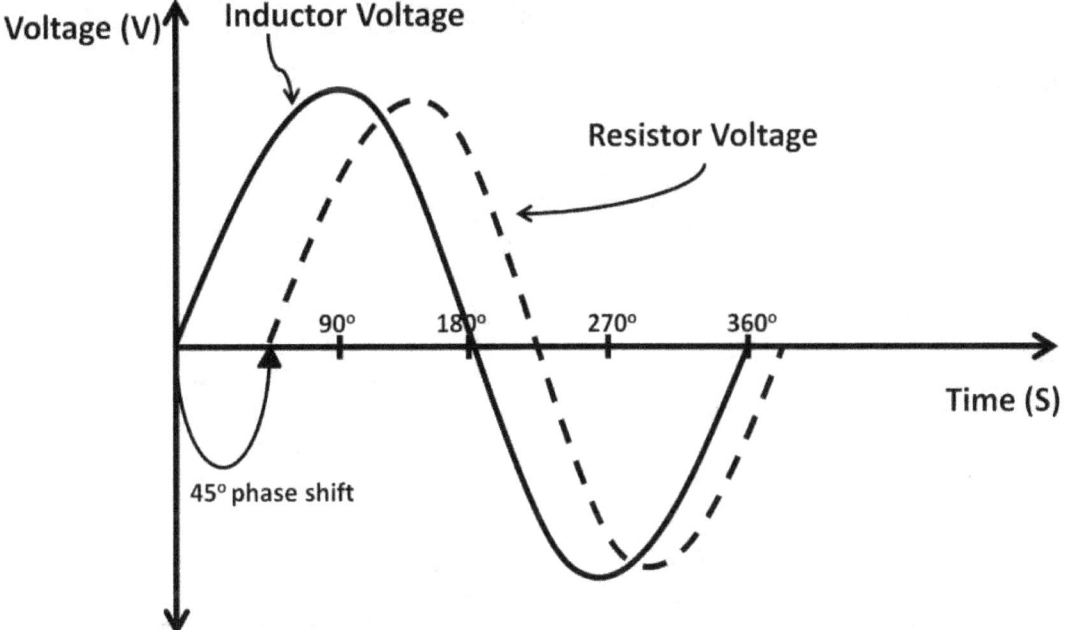

Figure 3.38: RL timing waveform, XL = R, 45-degree phase shift

At DC, XL is zero yielding **Vout = ground**.

$$XL = 2\pi \times 0 \times 10\, uH = 0\, \Omega$$

$$Vout = (Vin) \times \frac{0}{0 + 1\, k\Omega} = 0\, V$$

Using the dB equation covered previously, we can figure out the Vout/ Vin versus frequency relationship. At DC, frequency is zero and Vout / Vin in dB:

$$\frac{Vout}{Vin} \text{ in dB} = 20 \log \left(\frac{Vout}{Vin}\right)$$

$$= 20 \log \left(\frac{0}{Vin}\right) = -\infty\, dB$$

Assuming frequencies step up gradually to infinitely high, **Vout / Vin** in dB:

$$Vout = Vin \times \frac{\infty}{\infty + 1\, k\Omega} = Vin$$

∞ >> 1 kΩ,

$$Vout = Vin \times \frac{\infty}{\infty + \cancel{1\, k\Omega}} = Vin$$

dB when Vout / Vin = 1:

$$\frac{Vout}{Vin} \text{ in dB} = 20 \log (1)$$

$$= 20 \times 0\, dB$$
$$= 0\, dB$$

See figure 3.39 for Vout / Vin in the dB bode plot. At DC, Vout is at ground (0 V). With − ∞ dB, as frequencies increase, negative dB goes up with Vout going up to 0 dB (Vin). This is why the circuit is called a high-pass filter. At DC or low frequency, Vout is close to ground or no signal at the output. The signal only passes through at higher signal frequency.

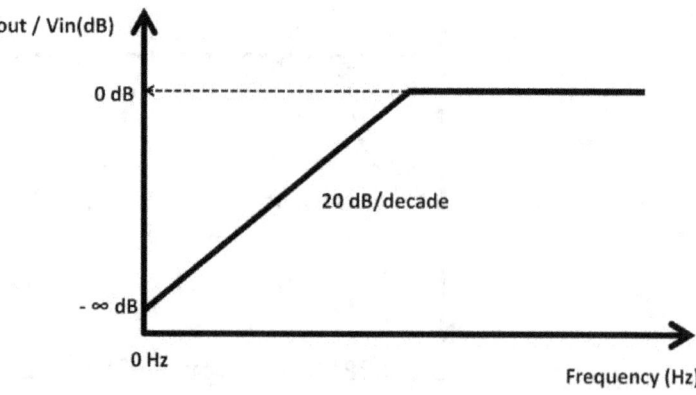

Figure 3.39: Vout / Vin vs. Frequency

To determine −3db bandwidth of a high-pass filter:

$$-3\text{ dB} = 20 \log\left(\frac{V_{out}}{V_{in}}\right)$$

$$-3\text{dB} = 20 \log(0.707)$$

$$0.707 = \frac{V_{out}}{V_{in}} = \frac{X_L}{X_L + 1\text{ k}\Omega}$$

$$0.707 \times (X_L + 1\text{ k}\Omega) = X_L$$

$$0.707 \times (X_L) + 707\text{ }\Omega = X_L$$

$$0.29 \times (X_L) = 707\text{ }\Omega$$

$$0.29 \times (2\pi) \times (f - 3\text{dB})(10\text{ uH}) = 707\text{ }\Omega$$

$$f - 3\text{dB} = 38.82\text{ MHz}$$

Real L and C

Before we step into real world circuits, it's beneficial to know capacitors and inductors device models. Device models include additional components (R, L, and C) that are called parasitic. Although these components are small in quantity, they could have major effects on circuit performance. A non-ideal capacitor model is shown in figure 3.40. It includes the capacitor itself, leakage resistor, equivalent series resistor (ESR), and equivalent series inductance (ESL). ESR contributes to heat loss. ESL contains XL. 10 mΩ is considered good performance for a 500 uF aluminum capacitor. Recall that Xc decreases with increasing frequency. In reality, if the frequency is high enough, XL from ESL would eventually kick in, tilting the overall impedance upward (see figure 3.41). The uptick in Xc occurs at extremely high frequency. Some datasheets will not show them because it's above the normal operating frequency for a specific capacitor.

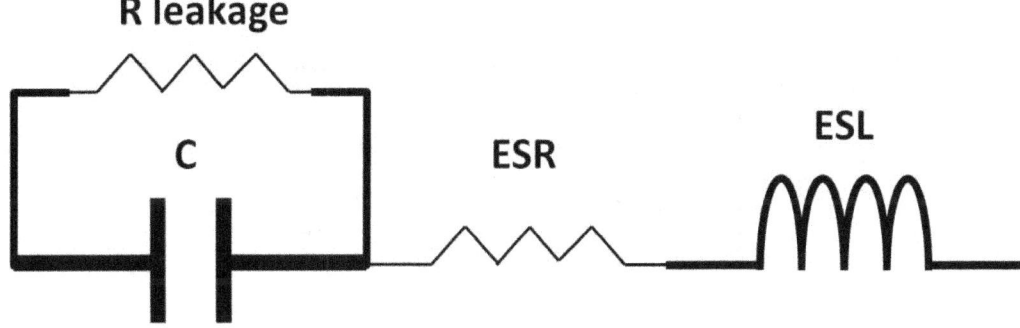

Figure 3.40: Capacitor model

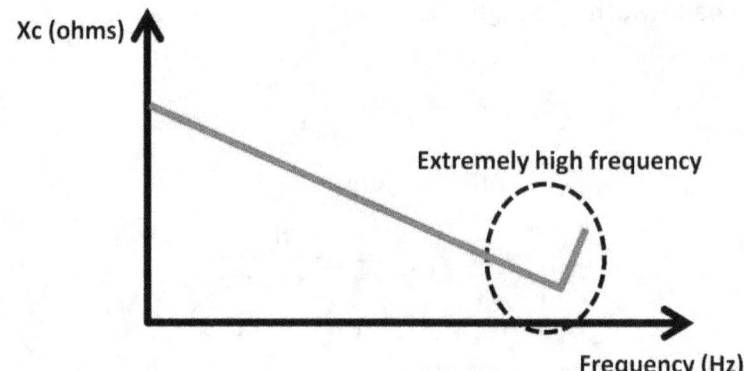

Figure 3.41: Xc vs. frequency

Inductor contains parasitic devices as well. Figure 3.42 showcases a real inductor model. The ESR comes from leads and package resistance. Parasitic capacitance comes from tiny gaps between coils.

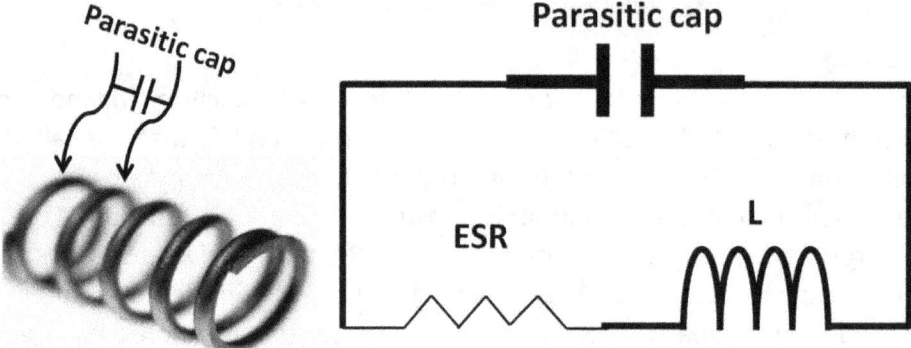

Figure 3.42: Inductor parasitic

At high signal frequency, XL ultimately decreases (see figure 3.43). For a high frequency surface-mount inductor with 100 nH, ESR can be as low as 500 mΩ.

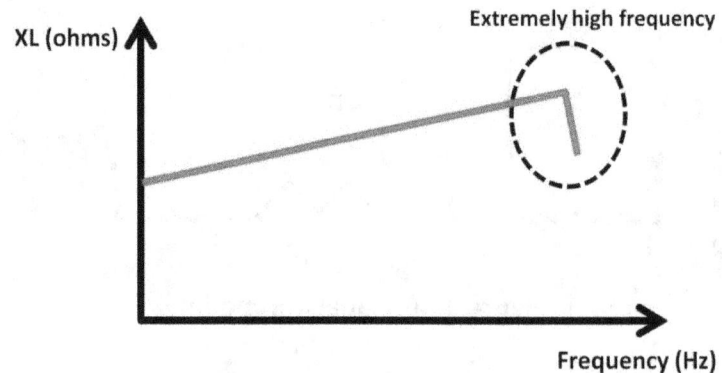

Figure 3.43: XL vs. frequency

Practical AC Circuits

Figure 3.44a is a typical circuit used in a Frequency Modulation (FM) radio circuit. It eliminates noise by "clipping" out signal above the upper and below lower diode voltage. More AC circuits will be presented in chapter 4, Analog Electronics. In this circuit, D1 conducts in the positive half AC cycle leading Vout at +2 V, during the negative half AC cycle, the D2 forward-bias resulting in – 2 V at Vout while the D2 anode stands at ground. See Vin and Vout waveforms in figure 3.44b.

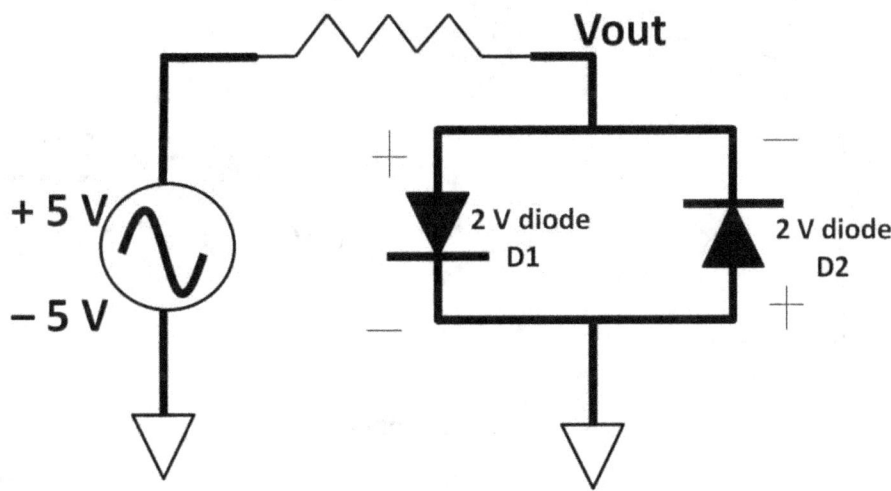

Figure 3.44a: FM noise clipper

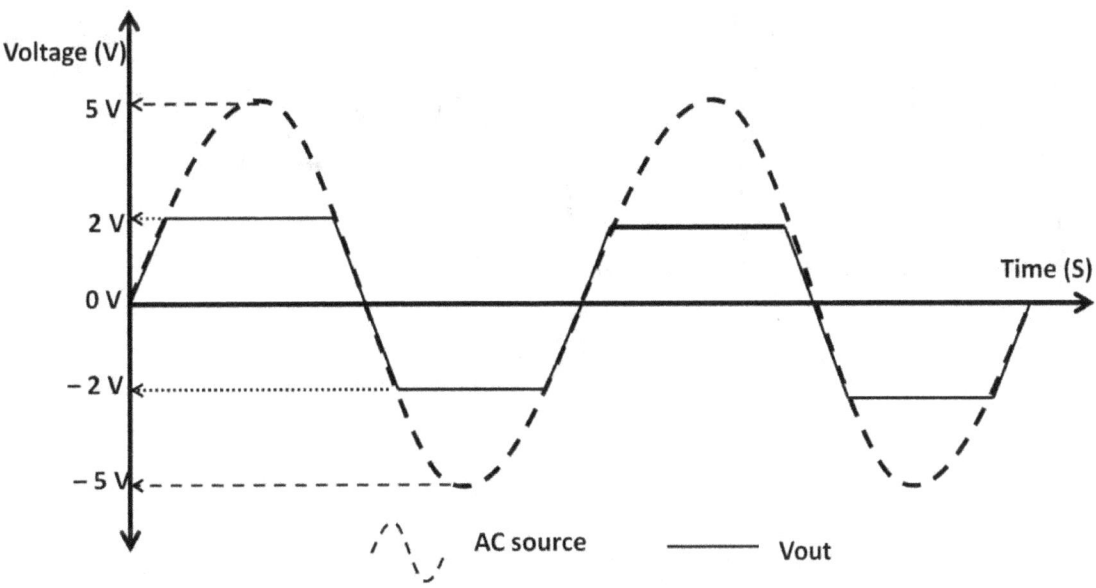

Figure 3.44b: FM noise clipper, Vout and Vin waveforms

Ringing and Bounce

The RC circuit was mentioned previously as a filter. A practical scenario below in figure 3.45 has a "real" switch connecting to an electronic load. When a "real" switch closes (step response), the transition takes finite delay time with physical bounce back and forth causing ringing, which is a form of oscillation and undesirable. Ringing is also referred to as undershoot and overshoot of the signal. Using an RC filter in figure 3.46, this bounce can be eliminated or high frequency noise can be filtered out (dotted trace). The trade-off is a longer time to reach peak value, completely closing the switch.

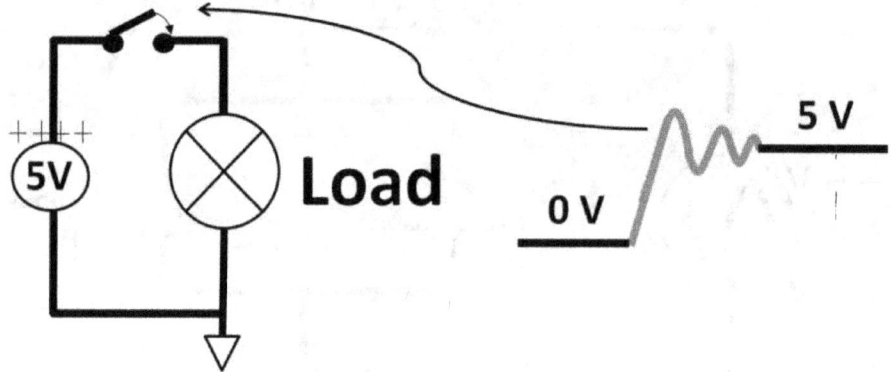

Figure 3.45: Switch connects to load

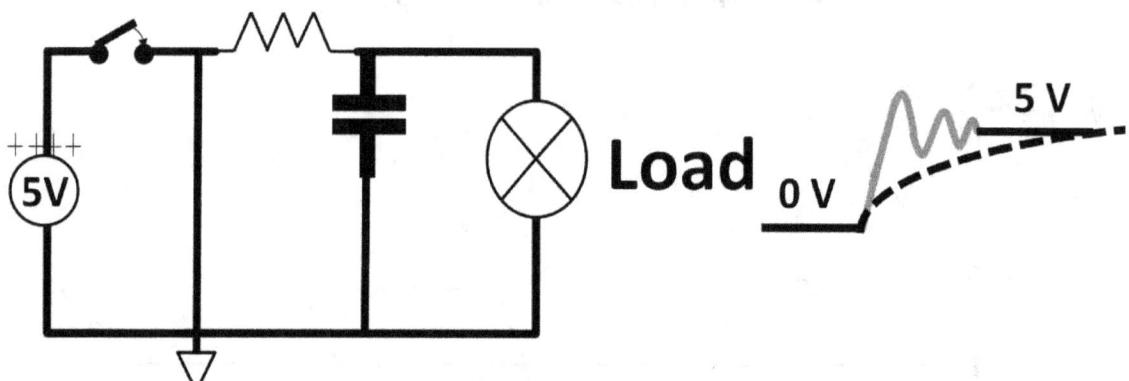

Figure 3.46: R C eliminates bounce

Inductive Load

You need to be mindful about an electronic load that is inductive during switching. In figure 3.47 below, when the switch is fully closed, the inductive load turns on.

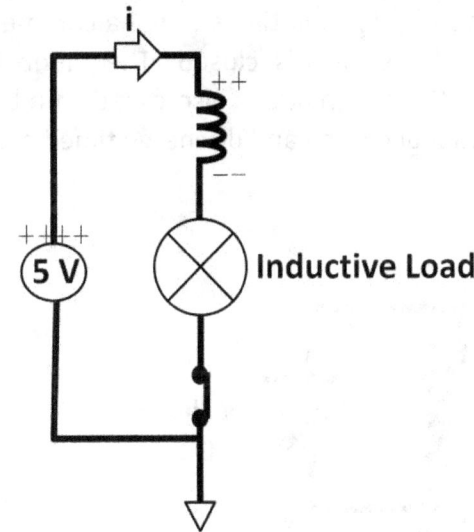

Figure 3.47: Switch close with inductive load

When the switch opens in figure 3.48, the inductor flips polarity, attempting to maintain current flow. The bottom end of the inductor is open, and voltage is unknown. There is no limit to how high this voltage can be. There is no mechanism to define the bottom end inductor voltage. The electronic load could be damaged by exposing it to large voltage amount. We call this phenomenon inductive kick.

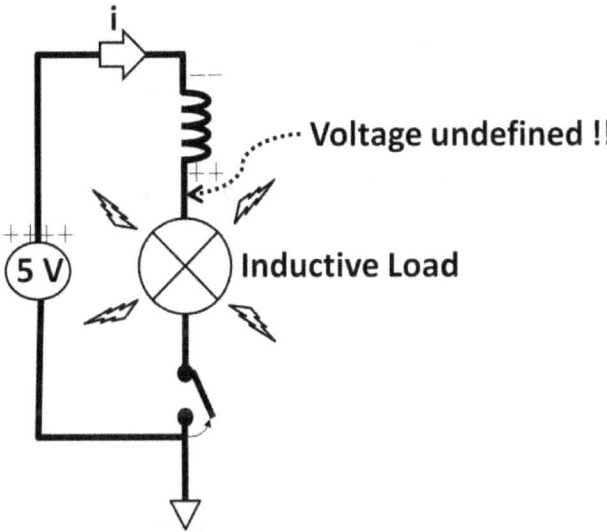

Figure 3.48: Switch opens, voltage undefined

Diode Clamp

To solve this problem, a diode can be added (see figure 3.49) in parallel with the inductor. When the switch opens, the diode now conducts and holds (clamps) the diode anode at 2 V above the 7 V source. In other words, the electronic load voltage is safely "clamped" at no more than 2 V above 7 V. This diode, sometimes called a commutating diode, has no impact on normal operation when the switch is closed. The diode in fact is reversed-biased, appearing as an open circuit. This technique is also called snubber circuit. The disadvantage of using the diode is the additional charge and discharge time because of the diode resistance and parasitic.

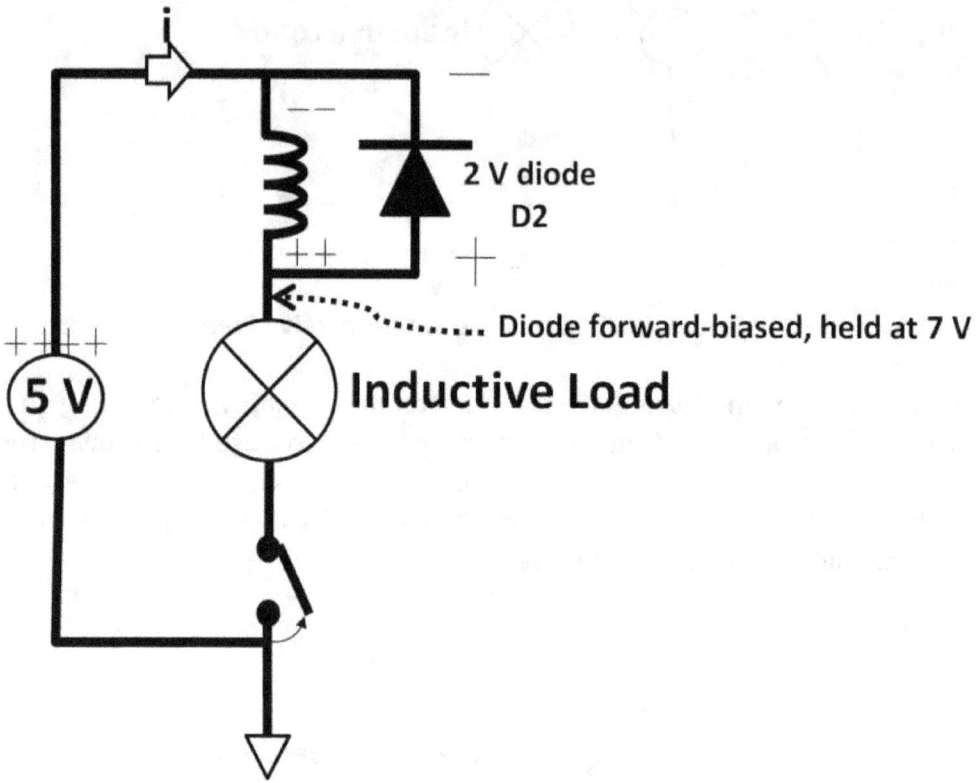

Figure 3.49: Diode clamp (snubber)

Series R L C Circuit

Recall the voltage, current lead, and lag differently among R, L, and C. This interesting feature creates many useful circuits like the R L C series circuit in figure 3.50.

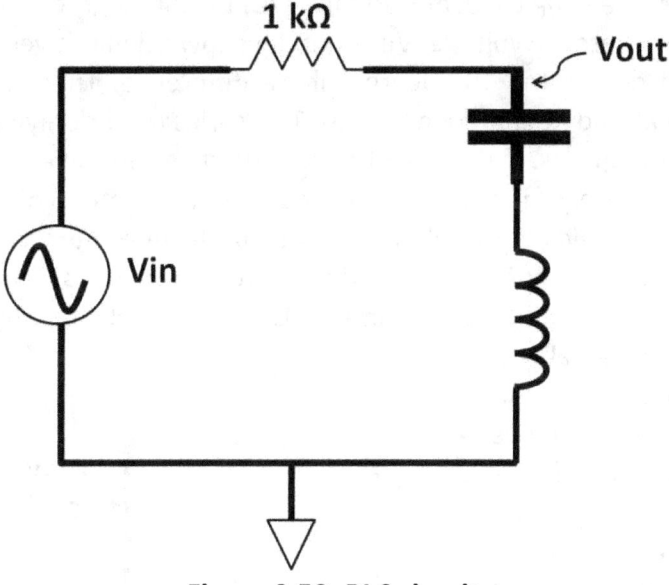

Figure 3.50: RLC circuit

Applying voltage, current lead, and lag rules, the following waveform in figure 3.51 can be obtained.

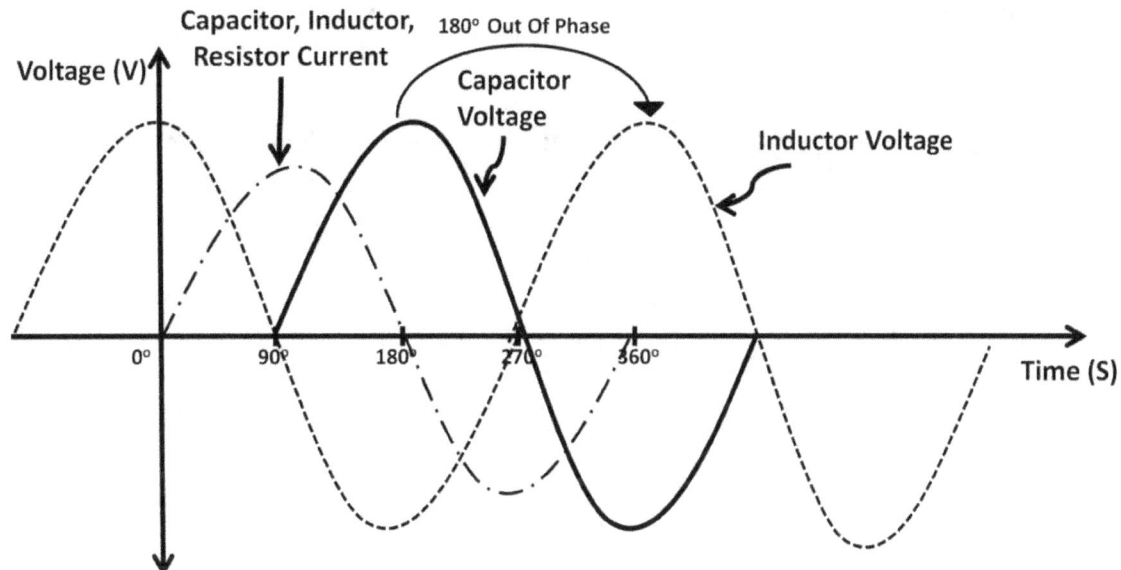

Figure 3.51: Capacitor and inductor voltage lag and lead

All currents in a series circuit are in phase. Capacitor voltage lags its current by 90 degrees (ICE) while inductor voltage leads its current by 90 degrees (ELI). This results in inductor voltage leading capacitor voltage by 180 degrees. Using the same AC principles, voltage, current, and phase information can be extracted. In figure 3.52, the vector diagram shows inductor voltage (VL) is leading capacitor voltage (VC) by 180 degrees. The resistor voltage is at zero degree as the reference voltage. VL is standing upward in the vector diagram while VC is pulling downward due to the 180-degree phase difference. There is a net voltage sum depending upon the XL and XC impedance sizes. This series circuit only has one current going through all three components. If the C and L were designed to have the same impedances, the resulting circuit is purely resistive, i.e., no phase shift between voltage and current. The net VL and VC voltage yields zero voltage. This leads to maximum current flowing in the circuit with minimum impedance. This particular frequency is called resonant frequency. Frequency affects both XL and Xc. L C reactance is heavily controlled by frequency. Keep in mind, however, the non-ideal R, L, and C nature could become factor in the RLC circuit. We can derive resonant frequency using figure 3.52: **Vout = 0** when XL and XC cancel each other out: **XL = Xc** or **XL − Xc = 0**. Maximum current occurs when **XL − Xc = 0**, i.e., minimum impedances. To look for the resonant frequency, we simply apply XL = XC then solve for resonant frequency, f:

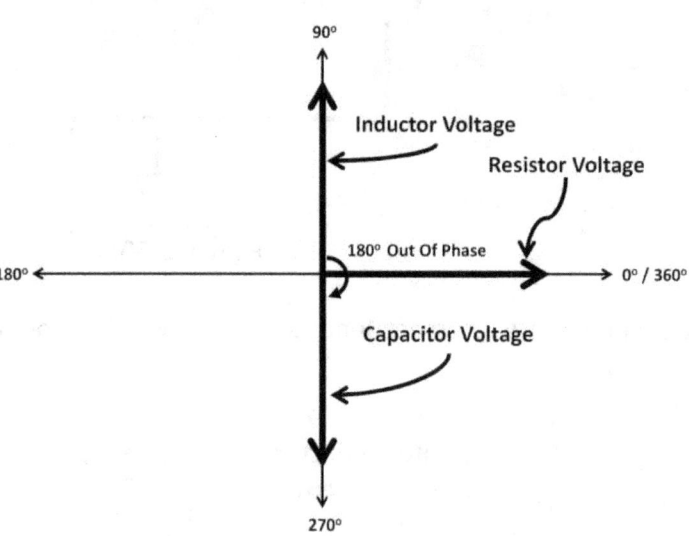

Figure 3.52: Inductor voltage leads capacitor voltage

$$XL = XC$$

$$2\pi f L = \frac{1}{2\pi f C}$$

$$(2\pi f)^2 = \frac{1}{LC}$$

$$\text{Resonant Frequency} = \frac{1}{2\pi\sqrt{LC}}$$

LRC Parallel (Tank) Circuit

The popular LRC parallel circuit is called a tank circuit (see figure 3.53). It includes the inductor and capacitor connected in parallel. The voltage across L and C is the same. The current throws through the inductor and capacitor are 180 degree out of phase. The vector diagram in figure 3.54 shows the vector diagram. Varying (tuning) LC component values allows us to determine and adjust resonant frequency. Similar to series an LC circuit, a tank circuit's resonant frequency is:

$$\text{Resonant Frequency} = \frac{1}{2\pi\sqrt{LC}}$$

At resonant, **XL = Xc**, the total reactance is at maximum while circuit current is minimum. Positive peak inductor current cancels out the negative peak capacitor current (see figure 3.54). Resonant frequency can easily be tuned by varying inductor and capacitor sizes for a given frequency. For example, to achieve 1 MHz resonant frequency using a 10 mH inductor, the capacitor value can be evaluated:

$$\text{Resonant Frequency} = 1\,\text{MHz} = \frac{1}{2\pi\sqrt{10\,\text{mH} \times C}}$$

$$2\pi \times 1\,\text{MHz} = \frac{1}{\sqrt{10\,\text{mH} \times C}}$$

$$\sqrt{10\,\text{mH} \times C} = \frac{1}{2\pi\sqrt{1\,\text{MHz}}}$$

$$C = 2.53\,\text{pF}$$

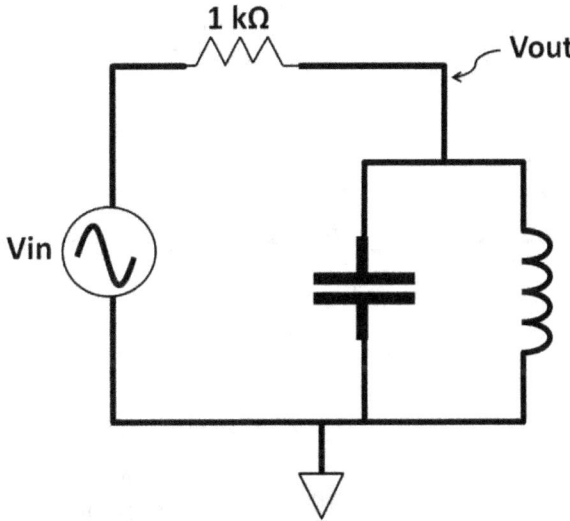

Figure 3.53: LRC parallel tank circuit

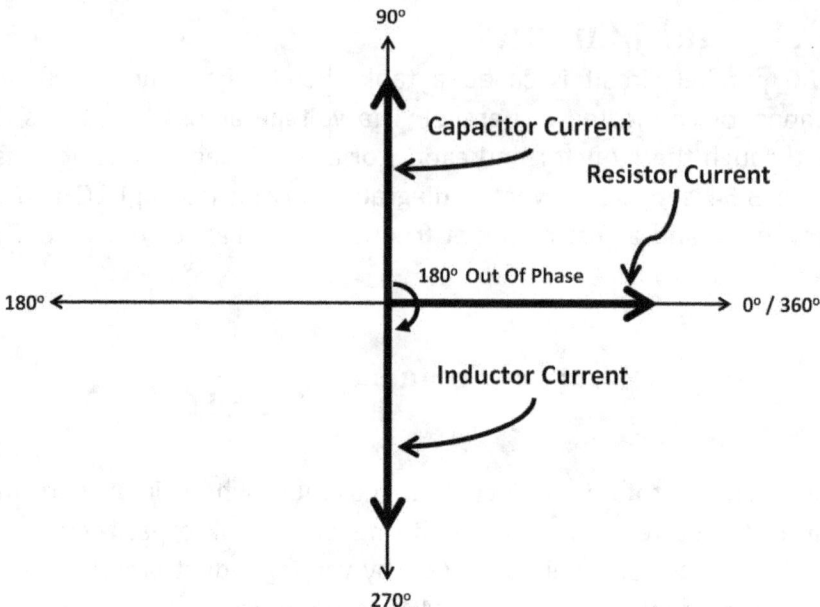

Figure 3.54: Parallel LC vector diagram

Figure 3.55 demonstrates the transient waveform among capacitor and inductor voltage and current. Inductor and capacitor voltage are in phase. Inductor current lags inductor voltage (ELI) by 90 degrees while capacitor current leads inductor voltage also by 90 degrees. This results in a 180-degree phase shift between capacitor and inductor currents. The applications of a tank circuit include oscillators and wireless transmitter and receivers. These applications will be further explored in chapter 6, Communications.

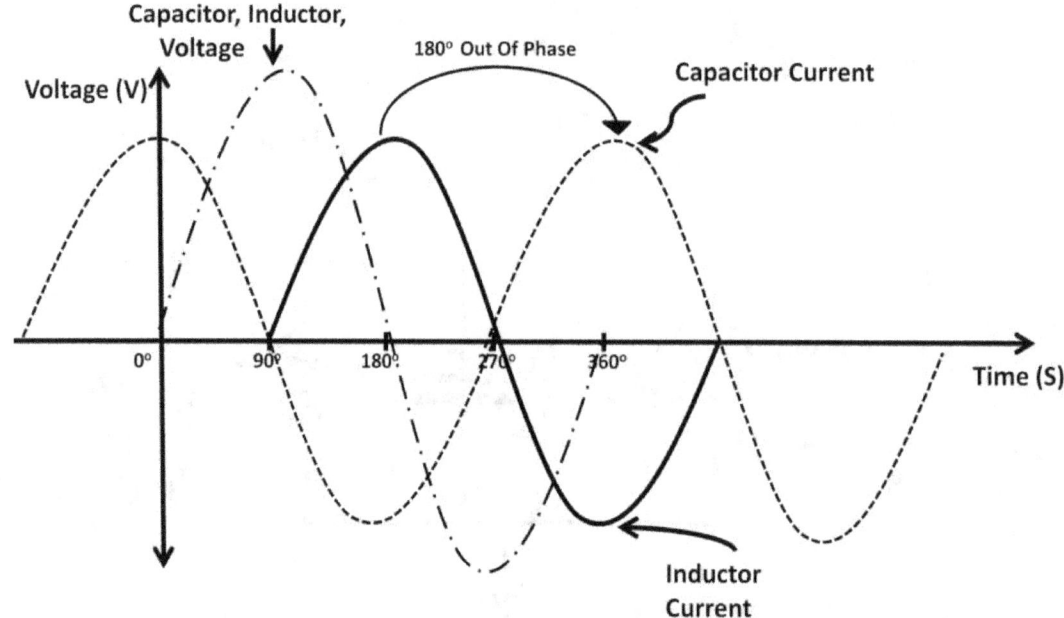

Figure 3.55: Tank LC current waveforms

Transformers

A transformer is an AC circuit that steps up or down AC voltages. The operation of a transformer can be explained by electromagnetic theory. Transformers are used in many applications such as electric power generation and electronic device charging, (e.g., laptop and cell phone battery chargers). A transformer requires at least two sides to operate: primary and secondary sides. Multiple secondary sides are often found in complex transformer designs. The key to transformer operation is electromagnetic theory where changing voltage and current "induce" voltage and current on the other side of the circuit through electric and magnetic fields generated on both sides. In figure 3.56, the primary side on the left is powered by an AC voltage source which connects to wires. The wires are rounded with many turns (turn numbers), called N1. These turns are tightly wrapped around the core, which is made of conductive materials. The wires on the secondary side wrap around the core also with fixed turn numbers (N2). For step-down applications, from a household electrical outlet (120 V AC) to DC, the secondary turn number is less than the primary one. The input AC voltage (Vin) is generating magnetic and electric fields from the wire carrying AC current. These fields are directed to the secondary side via the core, inducing changing voltage and current on the secondary side. The current amount and voltage at the output (Vout) are determined by the turn number ratio between primary and secondary sides. Let's use some real numbers to further elaborate it.

N1 = 100, N2 = 10, Vin = 120 V, Vout =?

$$\frac{N1}{N2} = \frac{Vin}{Vout}$$

$$\frac{100}{10} = \frac{120 \text{ V}}{Vout}$$

$$Vout = 12 \text{ V}$$

12 V is an AC voltage. To achieve a 5 V DC, usually found in portable electronic devices such as iPod and smartphone car chargers, additional voltage reduction is required. The benefit of using a transformer is electrical isolation. It offers safety in addition to the ability to increase or decrease voltages at the output. You may ask how it could provide isolation if the coils are tightly wrapped around the conductive core. The answer is that the wire surface is coated with non-conductive materials. Despite tight wrapping with the core, there isn't any direct electrical connection from primary to the core and the secondary side. All actions are purely relied reliant on electromagnetic theory where voltage and current are created by changing electric and magnetic fields.

In electrical plant operations, power plants step up the voltage to tens of thousands of volts. Then, travelling through cable before arriving at the household, these high voltages are eventually stepped down through multiple substations before they get to 120 V AC (household rating). It's very typical that few thousand volts are present at the sub transformer located right outside residential homes. The motivation for this high step-up voltage is power loss reduction. For power conservation rule,

(Input Voltage) X (Input Current) = (Output Voltage) X (Output Current)

Suppose all components are ideally lossless. This means if the input voltage is extremely high, current would be lowered for the same input and output power. Less input current means less $I^2 R$ power loss. These losses are mainly due to heat and electromagnetic field losses when current flows through the utility cables. For example, to provide 12,000 W of power, input voltage is at 1,200 V drawing 10 A of current. On the output side, power remains the same assuming all components are ideal. Transformer steps down 120 V output. This equates to 100 A of output current.

(1,200 V) X (10 A) = (120 V) X (100 A)

Many countries have their own standards. 120 V AC is the United States standard. Asia, Europe, and other parts of the world have different ratings numbers ultimately affecting transformer designs.

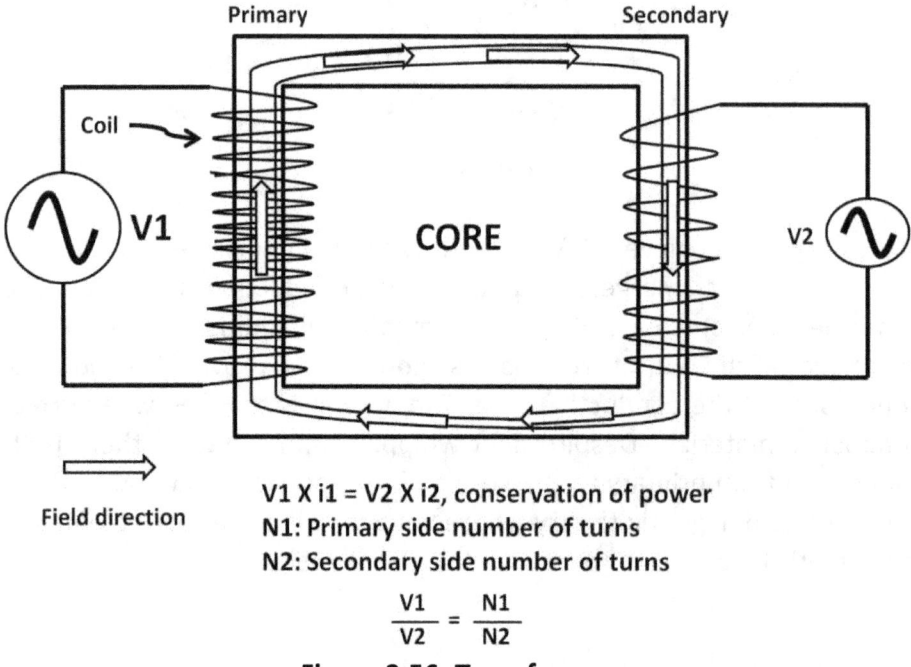

V1 X i1 = V2 X i2, conservation of power
N1: Primary side number of turns
N2: Secondary side number of turns

$$\frac{V1}{V2} = \frac{N1}{N2}$$

Figure 3.56: Transformer

Half-Wave Rectifier

Using a diode rectifier, a zener linear regulator could further transform AC voltage to DC. A half-wave rectifier is a classic example shown in figure 3.57. The diode only conducts during the positive Vin half cycle. Vout is at 0 V during the negative half (Diode reverse-biased) cycle. By adding a capacitor in the circuit, a "DC-like" output is acquired similar to figure 3.58. During the positive half cycle, the capacitor is charged up to the Vin peak. During the negative half cycle, the diode turns off, and charges accumulated on the top capacitor plate slowly discharge to the resistor

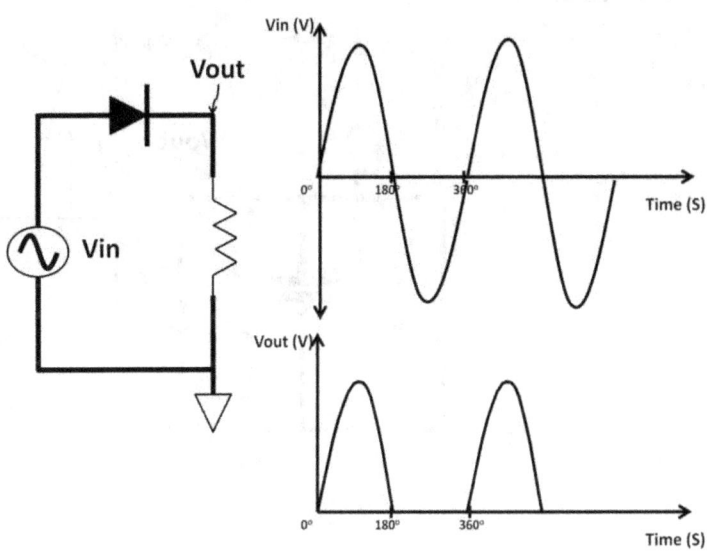

Figure 3.57: Half-wave rectifier

delayed by the RC time constant. This output is not a stable DC voltage due to the fact that the voltage is being charged and discharged. The amplitude of this charge and discharge voltage is called ripple voltage. Ripple voltage defines how well the Vout is compared to a stable DC voltage.

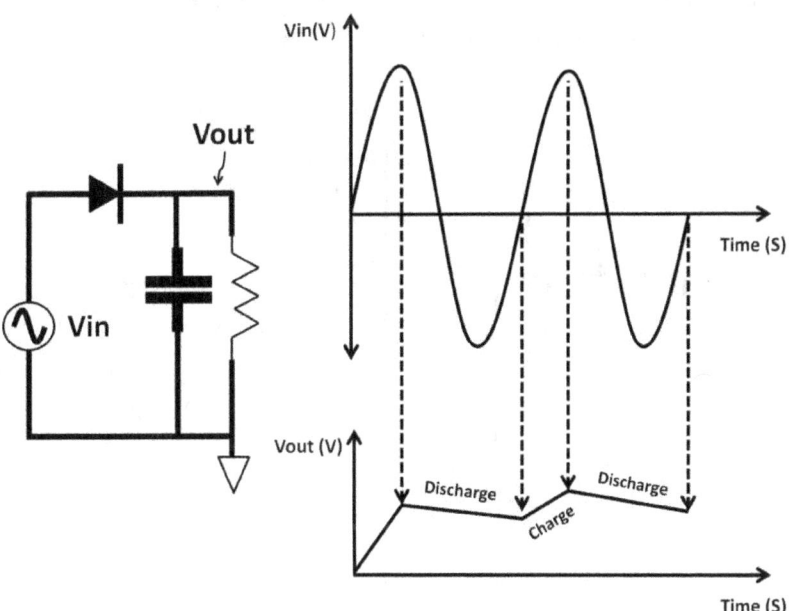

Figure 3.58: DC voltage with capacitor

A zener regulator or DC-to-DC regulators may be used to achieve more stable Vout as shown in figure 3.59.

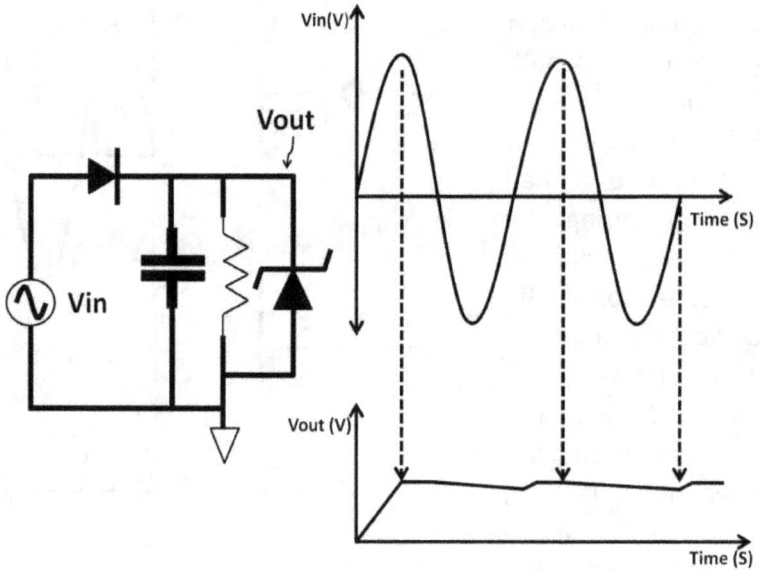

Figure 3.59: Diode, RC with zener diode

The AC signal frequency and RC sizes are important design considerations to produce stable Vout. For example, in figure 3.60, Vin peak-to-peak is 10 V running at 10 kHz (0.1 ms period). R C is initially designed to be 10 kΩ and 1 uF (10 ms time constant). The problem with this design is that the RC time constant is too long. Recall time constant definition: it takes 2 time constants to reach 87% of the input. The Vout in this design never had enough time to reach noticeable output. Sizing RC accordingly is the key to designing this type of regulator successfully.

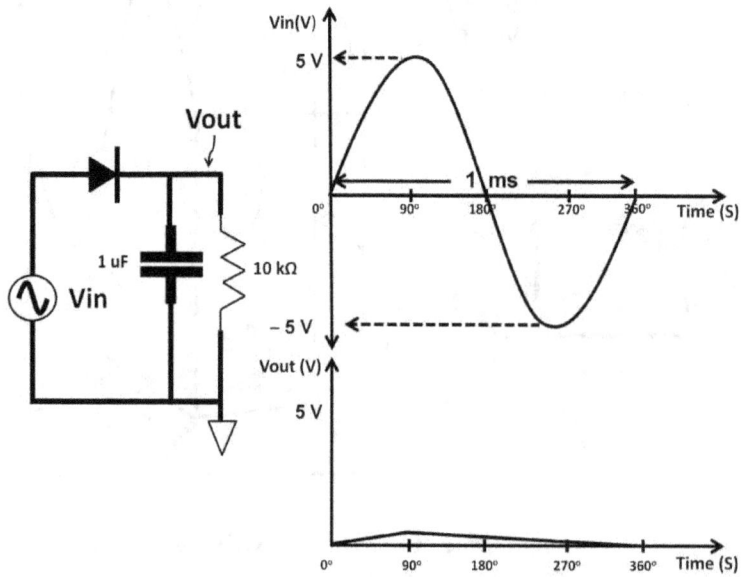

Figure 3.60: Large RC time constant

Switching versus Linear Regulators

By definition, voltage regulators provide constant DC output voltage to a load. A switching regulator's output is an AC signal with minimum ripple voltage behaving like a DC signal. Most switching regulators require a controller circuit and a switch toggling on and off. This increases circuit complexity. It's because of this reason switching regulators are more efficient because devices are only on only part of the time. Some switching regulators can run with as high as 90% efficiency. This is extremely beneficial in portable applications when longer battery life is required. Conversely, a linear regulator does not have any switching actions, is easy to use, makes less noise, and costs less, but suffers from lower power efficiency because the active device remains on (heat sink may be required) the entire time during voltage regulation. Typical linear regulator efficiency is less than 50%.

	Switching regulator	Linear regulator
Complexity	High	Low
Power consumption	Low	High
Power efficiency	High	Low
Cost	High	Low
Load driving capability	Low	High

Table 3-4: Switching vs. linear regulators

Both types come in many topologies and are found in plenty of portable applications such as smartphones, digital cameras, robots, computers, etc. Popular topologies of switching regulators are step-up (boost) and step-down (buck). The zener diode and low drop out regulator (LDO) are common linear regulators. Both switching regulators and LDO use feedback control circuitry to regulate the output. A summary of the major differences between switching and linear regulators is shown in table 3-4.

Buck Regulator

Lastly, a switching voltage regulator is shown in figure 3.61. This is a buck regulator circuit invented in the 1970s. It continues to be popular in power management systems. It merely consists of three devices: a switch, a diode, and an inductor. This circuit steps down a higher voltage to a lower one. One application is the 5 V DC outlets in automobiles. They run off a 12 V lead-acid car battery, which is stepped down to lower voltages for portable electronics used inside automobiles. **V(t) = L (ΔI)** can be used to explain this circuit. When the switch is closed, the diode is reverse-biased and inductor current starts to ramp up with a fixed voltage across it. The time it takes for the current to ramp to its peak is ton (on-time). During this time, the switch is closed. The voltage across the inductor is **(Vin − Vout)**. The switch then opens, and the inductor flips polarity trying to maintain current flow (see figure 3.62).

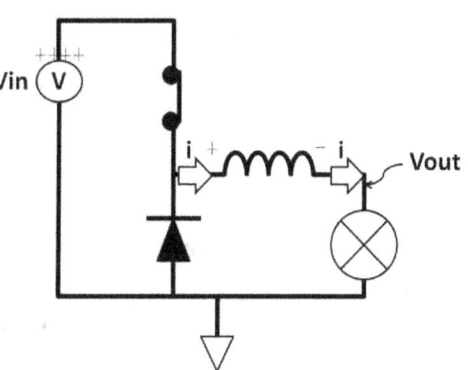

Figure 3.61: Switching regulator

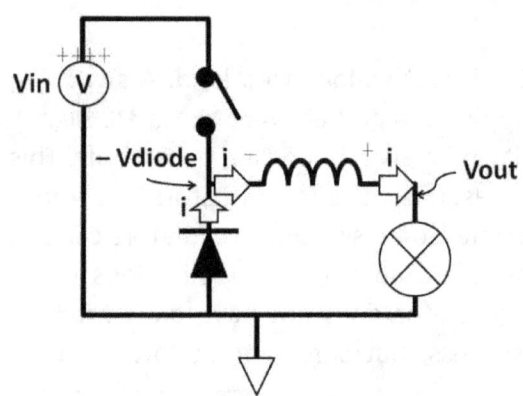

Figure 3.62: Switch opens

The only current path is through the diode, which is now forward-biased (see figure 3.62). This causes the diode's cathode to be one diode below ground (– Vdiode). This diode sometimes is called a "catch diode." It's intended to be switching fast to keep up with the on-off switching action. To achieve just that, it's quite typical for a catch diode threshold to be as low as 200 mV. Using KVL, the voltage across the inductor is now

(Vout – (– Vdiode)) = Vout + Vdiode.

The switch-open time duration is toff. For a given inductor size, current change (either ramping up or down) has the same amplitude (see figure 3.63). The inductance and ΔI literally are constants.

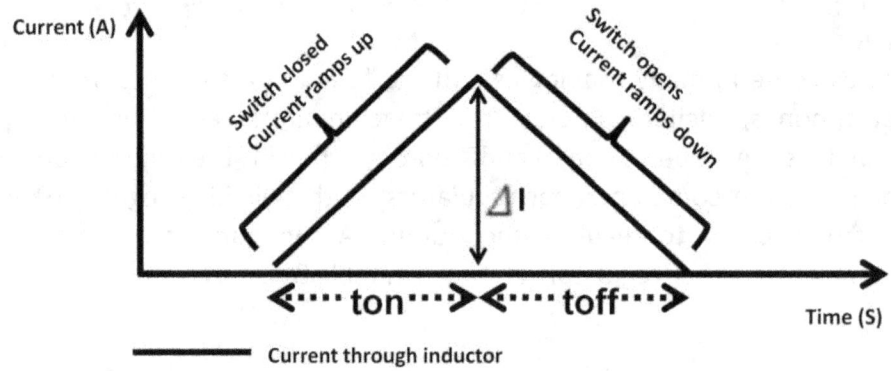

Figure 3.63: Inductor current ramp

L (ΔI) = (Vin – Vout) X ton = (Vout + Vdiode) X toff

Vdiode is designed to be as low as possible. Assume Vdiode is much smaller than Vout and becomes negligible.

(Vin – Vout) X ton = Vout X toff

Solve for Vout:

Vin X ton – Vout X ton = Vout X toff

Vout (ton + toff) = Vin X ton

$$\text{Vout} = (\text{Vin}) \times \frac{\text{ton}}{\text{ton} + \text{toff}} = \text{Duty Cycle}$$

Duty cycle needs to be less than or at least equal to one (duty cycle ≤ 1) in order for Vout to be lower than Vin. For example, if Vin is 10 V, regulated output voltage is 5 V. 50% duty cycle is needed **(ton = toff)**. If the switching frequency is 400 kHz:

duty cycle = 0.5 = ton / (ton + toff)

(ton + toff) = Period = 1 / 400 kHz
   = 2.5 us

0.5 = ton / 2.5 us

ton = toff = 0.5 X 2.5 us = 1.25 us

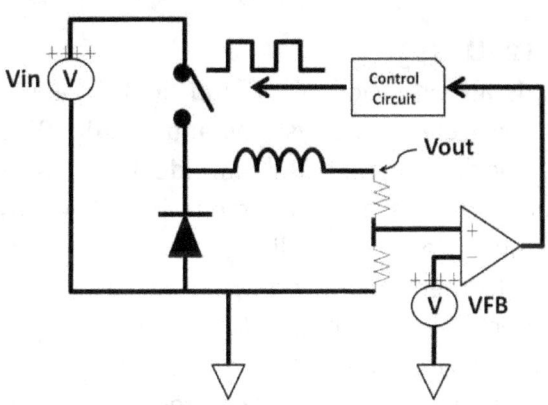

Figure 3.64: Buck regulator controlled circuit

To change Vout, you need to control the switch's duty cycle using control circuits, a voltage divider, and a comparator. In figure 3.64, the output is always in AC, i.e., the output voltage is toggling back and forth. The amplitude of AC output waveform is quantified as ripple voltage. The peak-to-peak value of the ripple voltage determines how well the output "looks like" a DC signal. The smaller the ripple, more stable the output would be. Due to the load attached to the Vout causing uneven current flow, noise in the system, and possible intermittent Vout disconnection, Vout could change erratically. The triangular symbol with the + and − signs inside is the operational amplifier (op-amp). The op-amp and control circuit are part of the feedback mechanism. It takes a voltage sample and compares it to a fixed value (VFB). The result of the difference feeds back to the control circuit. The control circuit then alters the switch's duty cycle. The whole concept is to maintain constant output by adjusting the switch's duty cycle according to the feedback from the Vout. If Vout goes too high, the switch turns on less to bring the Vout back down, and vice versa. The op-amp in this circuit is a comparator that compares voltage between VFB and Vout. The positive op-amp terminal changes if Vout changes (voltage divider) causing the op-amp output changes. The control circuit takes this change then adjusts the duty cycle. If the Vout drops, the switch turns on longer (increasing the duty cycle), bringing the Vout back to its original value. Feedback techniques in the op-amp are used in countless electronics products. They will be further examined in chapter 4, Analog Electronics. Using a voltage divider can control Vout level easily by changing the resistors ratio. For example, target **Vout = 2.5 V, VFB = 1.25 V. If resistors are the same size, then:**

$$(2.5\ V) \times \frac{R}{R + R} = VFB$$

$$(2.5\ V) \times \left(\frac{R}{2\ R}\right) = (2.5\ V) \times \frac{1}{2} = 1.25\ V$$

This feature allows you to "program" the output voltage by using different resistor sizes. The buck regulator is a simple, yet powerful architecture demonstrating the simplicity of basic AC theories in creating useful and practical electronic circuits.

Summary

AC is an extension of DC and diode theories. AC characteristics empower large number of modern electronic systems and circuits. We covered basic AC parameters, definitions, and components. Ideal and non-ideal capacitors and inductor characteristics were reviewed followed by simple LC circuits including low- and high-pass filters. Series and parallel LRC circuits were then discussed with several other practical AC applications (rectifiers, transformers, diode clamps, and snubber circuits). We also explored resonant frequency, vector diagrams, bode plots, and switching and linear regulators towards the end of the chapter. Only with a solid foundation in DC, diodes, and AC, can more complicated electronic circuits be understood, designed, tested, and analyzed. Table 3-5 is a summary of inductor and capacitor characteristics.

	Inductor	Capacitor
Schematic symbol	⏛	⊣⊢
Symbol	L	C
Unit	Henry (H)	Farad (F)
Impedance	$X_L = 2\pi f L$	$X_C = \dfrac{1}{2\pi f C}$
Energy stored in	Magnetic field	Electric field
Impedance over frequency	Proportional to frequency	Inversely proportional to frequency
Δt rule	$V(\Delta t) = L(\Delta I)$	$I(\Delta t) = C(\Delta V)$
Phase lead, lag	ELI (Voltage leads current)	ICE (Current leads voltage)
Parallel rule	$\left(\dfrac{1}{L_1} + \dfrac{1}{L_2}\right)^{-1}$	$C_1 + C_2$
Series rule	$L_1 + L_2$	$\left(\dfrac{1}{C_1} + \dfrac{1}{C_2}\right)^{-1}$

Table 3-5: Inductor and capacitor summaries

Quiz

1) The signal is 5 sin (2 π 1000 t + 20 degrees). What are the signal frequency and Vpeak?

2) The peak of an AC voltage (Vpeak) may be calculated as: Vrms X Constant. What is the constant value?

3) The ideal inductor stores energy in _____ field.

4) The ideal capacitor stores energy in _____ field.

5) If an AC signal is running at a 25% duty cycle, and on-time is 250 ns, what is the frequency?

6) Derive the Vout to Vin transfer function of the boost-switching regulator (see figure 3.65). Hint: Assume diode forward voltage drop is 1 V.

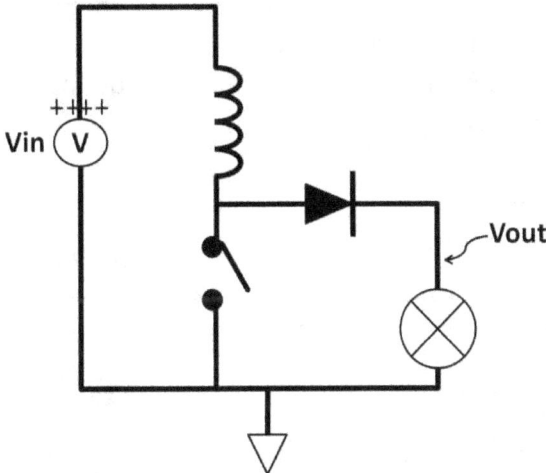

Figure 3.65: Boost-switching regulator

7) Design a high-pass filter using an inductor and resistor. This circuit allows a signal to pass through at the output starting at 10 MHz assuming resistor value is 1 kΩ.

8) Ceramic capacitors are often used in filtering noise due to their small size and low cost. Figure 3.66 shows a simple application using a ceramic capacitor to filter out high frequency noise to the IC. In actual implementation, the location of the capacitor needs to be as close as possible to the chip to minimize any noise pickup along the board traces. What is the purpose of the diode from VCC to the external pin? If the VCC contains AC noise running as high as 100 kHz, what is the size of the ceramic capacitor in order to reduce noise starting at f −3dB, assuming the output impedance of the external pin is 100 Ω?

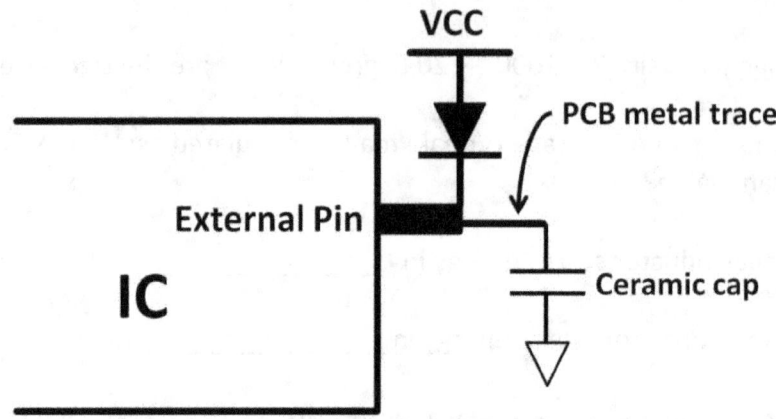

Figure 3.66: Ceramic capacitor noise filter

9) A full-wave rectifier in a power supply generates a rectified AC voltage (DC) signal. As opposed to a half-wave rectifier in figure 3.57, a full-wave rectifier converts both first and second halves of an AC input (secondary side of the transformer) to the output (see figure 3.67). In this circuit, the dotted lines show the current directions during the positive half of the transformer output. Only D2 and D4 are conducting. According to this design, if Vin's peak voltage is 10 V and frequency is 100 kHz, what is the voltage waveform at the Vout?

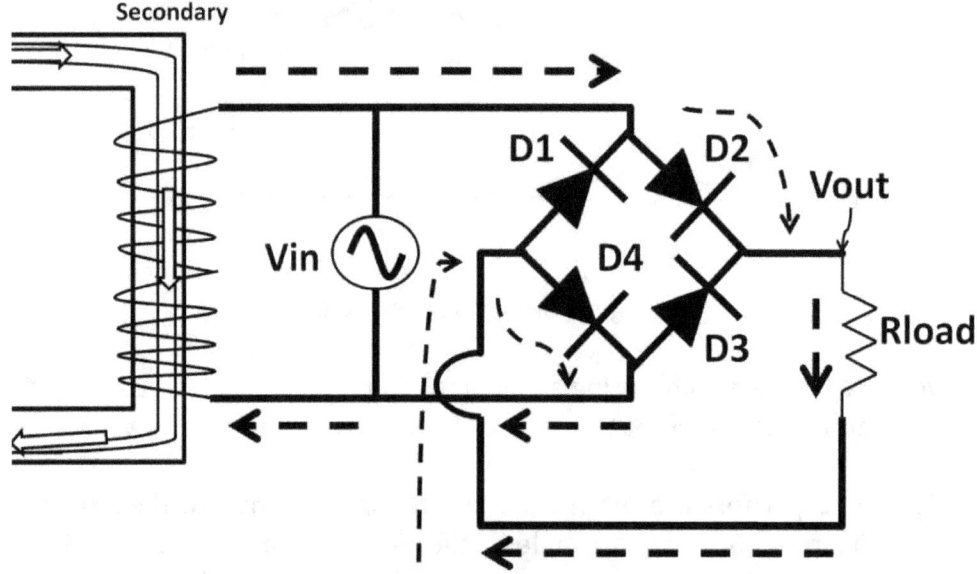

Figure 3.67: Full-wave rectifier

10) Voltage can be easily doubled by using switches and capacitors. Figure 3.68 shows a voltage-doubler circuit called a charge pump. While switches 1 and 4 are closed, switches 2 and 3 are open, and vice versa, charge pumping the capacitor. If Vin is 10 V, examine the circuit and draw the transient response waveform of Vin and Vout, assuming Vout connects to a resistive load and the RC time constant is negligible.

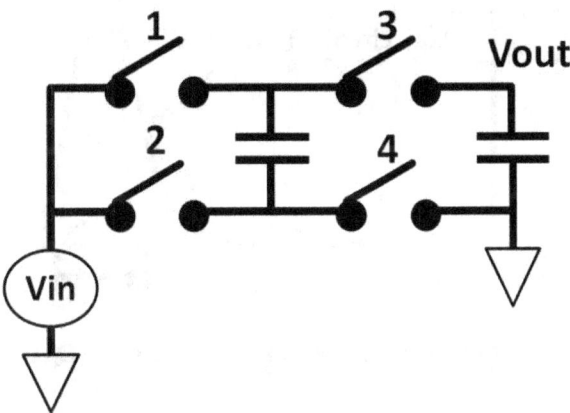

Figure 3.68: Charge pump circuit

11) A tank circuit shown in figure 3.69 consists of 10 mH and 100 pF capacitors. What is the resonant frequency of this tank circuit? The Q factor of a resonant circuit can be used as a figure of merit to describe how good the tank circuit is. The higher the Q, the smaller the bandwidth. This results in sharper AC response, as shown in figure 3.70. Bandwidth is measured from the peak reactance to both rising and falling at 70.7%. What is the bandwidth of this tank circuit if Q is 100?

$$\text{Bandwidth} = f_{resonant} / Q$$

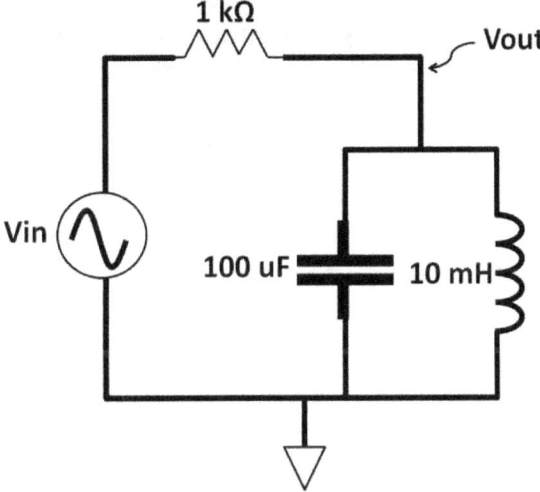

Figure 3.69: LC tank circuit

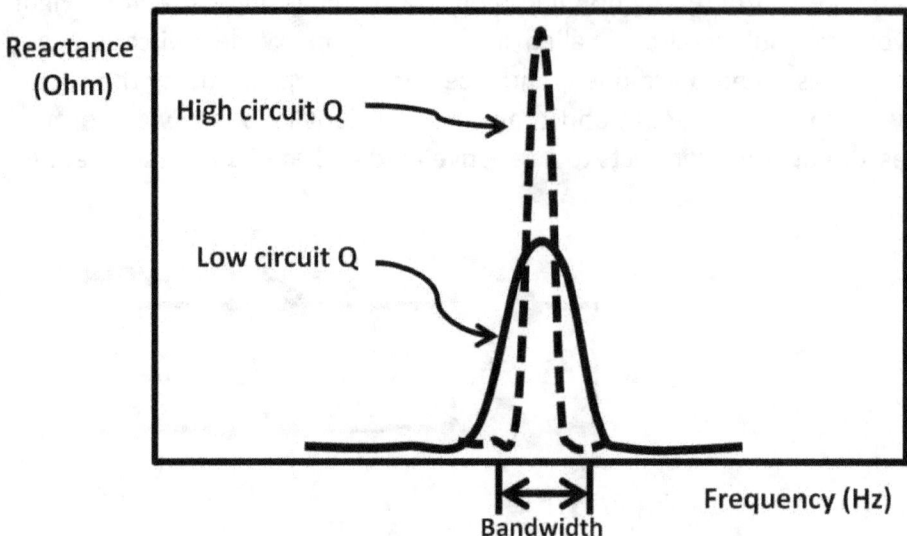

Figure 3.70: High and low circuit Q AC response

12) The vector diagram of a RC filter is shown in figure 3.71. If Vin is 25 V at 500 kHz, calculate the total impedance of the circuit, and calculate the voltage at the output and phase shift. Hint: Use the Pythagorean Theorem to calculate Z, then use the voltage divider rule to calculate Vout, and trigonometry to calculate the phase angle.

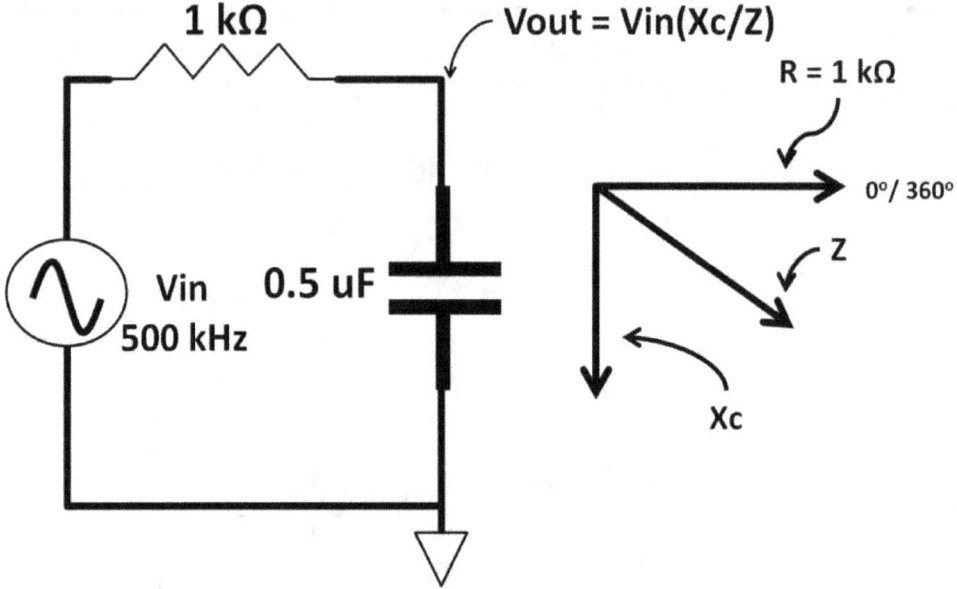

Figure 3.71: R C circuit

Chapter 4: Analog Electronics
What Is Analog?

Let's first define and clarify what an analog signal is. We experience analog signals daily. Sound, light intensity, speed, temperature, pressure, humidity, weight, height, voltage, current, and power are examples of analog quantities. Analog signals consist of infinite combinations of levels or numbers. In chapter 3, AC, sine waves were presented. They are analog waveforms having an infinite number of combinations between two points. There are no discrete levels at a single point of time (see figure 4.1).

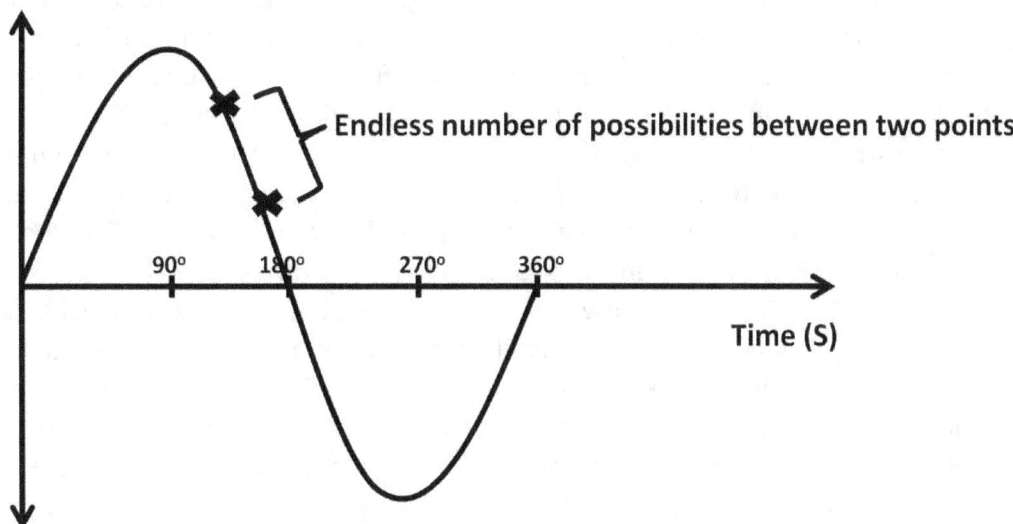

Figure 4.1: Analog signal

An analog signal can also come in irregular patterns. Figure 4.2 shows a temperature pattern as a function of time depicted in a timing waveform. At any particular point in time, the temperature reading can have infinite digit numbers after the decimal points.

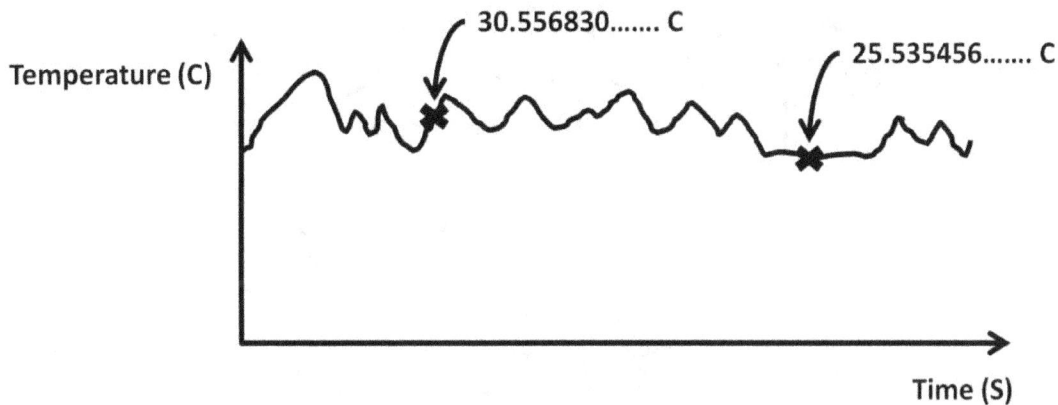

Figure 4.2: Temperature in time domain

Analog IC Market

Before we dive into the world of analog electronics, let's take a look at how big the analog market really it. The analog IC market size is about US $17 billion according to 2011 market data from research firm Databeans. Plenty of electronic products deal with analog signals. When you speak into your smartphones, your voice is an analog signal that is processed and digitized before being transmitted through the air. The top five analog IC vendors account for almost 40% of total market share. They are Texas Instruments, STMicroelectronics, Infineon Technologies, Analog Devices and Qualcomm. Low costs and technological advancement in electronics technology made electronics an ideal choice for processing analog signals. Electronics take analog signals as input; then the signals are filtered and amplified before passing to the next processing phase. Such a process is called signal conditioning. Major analog electronics products include standard amplifiers, comparators, analog-to-digital converters (ADCs), digital-to-analog converters (DACs), radio frequency (RF) chips, power management systems, and more. Many analog electronics are now combined with digital electronics. The industry terminology of such framework is mixed-signal design. These systems contain a mix of analog and digital design combined into one semiconductor chip. Some refer these chips as "system-on-a-chip" (SOC). Products containing mixed-signal electronics are plentiful. Figure 4.3 shows several industries that use mixed-signal electronics and the market and product categories within. In each market, numerous electronic functions and applications are employed: audio, video, automotive, LED lighting, Ethernet network, wireless network, telecommunication applications, medical equipment, motor control applications, renewable energy, aerospace, military, defense applications, touch screen, smartphone applications, industrial testing, manufacturing equipments, and the list goes on.

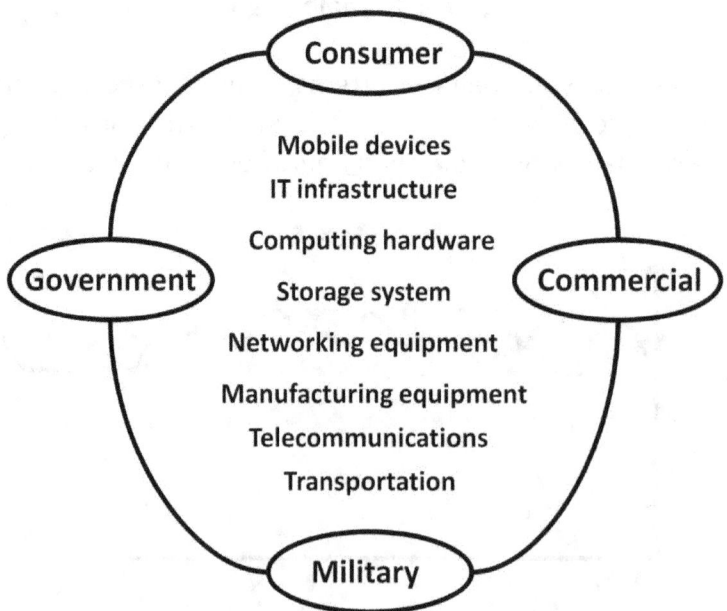

Figure 4.3: Mixed-signal electronics industries and markets

What Are Transistors Made Of?

Transistors are the building blocks of analog electronic systems. A great deal of understanding is required before attempting to understand more complex analog circuits. The transistor was invented in 1947 at Bell Labs. It has since gone through tremendous developments. Today's computing microprocessors easily hold several hundred million transistors on a single chip measured in 10 mm X 10 mm. Transistors come in different types with either a discrete or in an IC package. Popular transistor types are manufactured via bipolar and Complementary-Meta-Oxide-Semiconductor (CMOS) processes (see chart 4-1). BiCMOS process has been popular in recent years combining both bipolar and CMOS into one single manufacturing process. BiCMOS process offers the best of both bipolar and CMOS technologies undermined by its high cost. The materials used to manufacture transistors are mainly silicon based. For high-speed application, germanium and gallium arsenide are considered alternatives. There are two types of bipolar transistors: NPN and PNP. CMOS transistors are not made of diodes although both bipolar and CMOS transistors operate similarly. For CMOS transistors (MOSFETs: Metal Oxide Semiconductor Field Effect Transistor), there are two types—NFET (N-type field-effect transistor) and PFET (P-type field-effect transistor). NFET and PFET can also be called NMOS and PMOS transistors respectively. Chart 4-1 below shows all transistor and process types. We will focus on NPN, PNP, and enhancement mode NFET and PFET in this book.

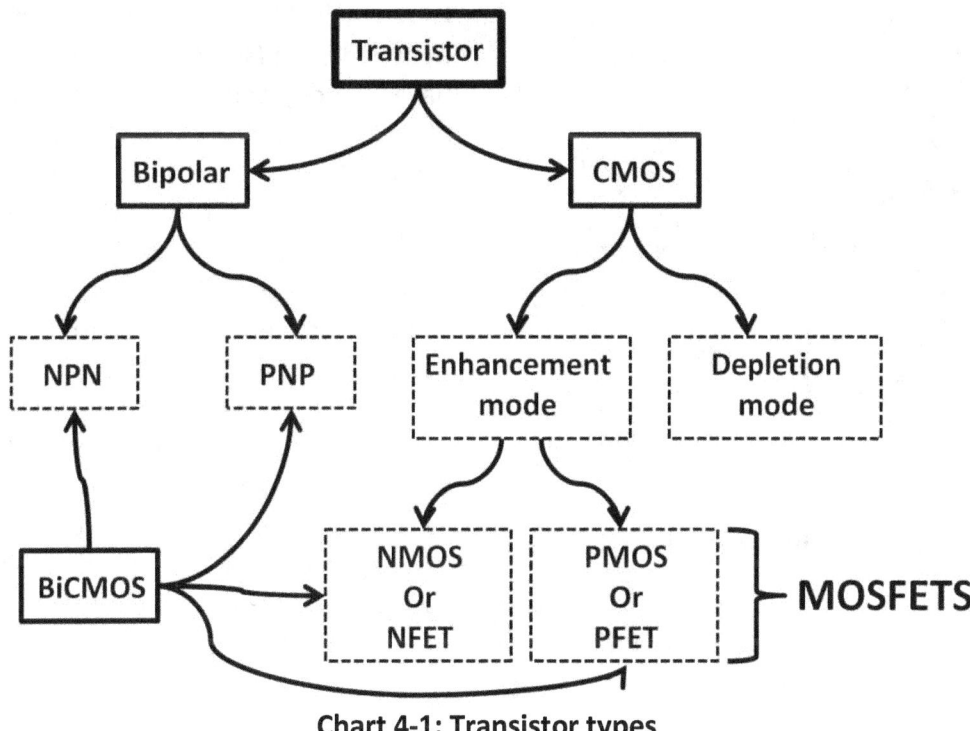

Chart 4-1: Transistor types

NPN and PNP

We will first go over bipolar transistors. NPN and PNP each have three terminals. Each is made of a triple-layer sandwich of N, P, and N-type materials for NPN; P, N, and P-type materials for PNP (see figure 4.4). On NPN, the P-type junction (base) is sandwiched by two N-type junctions (collector and emitter). For PNP, the N-type junction (base) is sandwiched between two P-type junctions (emitter and collector). The terminal names—collector, base, and emitter have special meanings and are not randomly assigned. Each

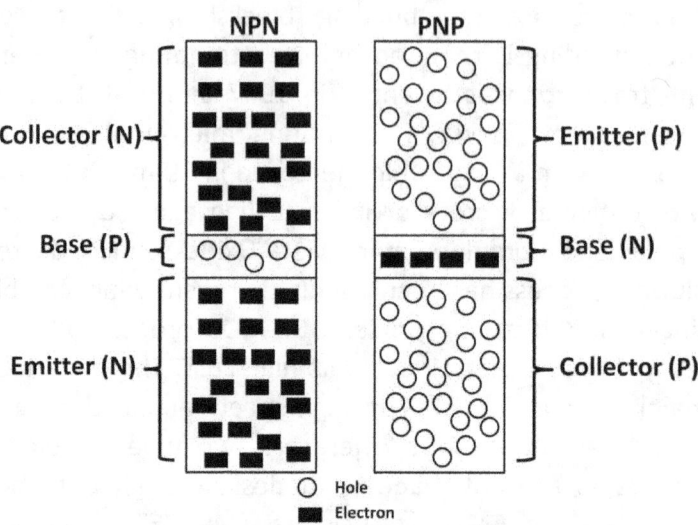

Figure 4.4: NPN and PNP structures and terminal names

terminal name represents a specific transistor action. Their meanings will become obvious in the next section.

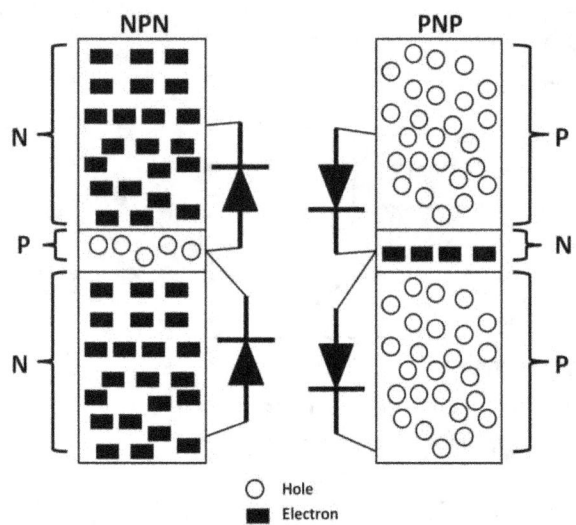

Figure 4.5: Two diodes on NPN and PNP

NPN and PNP transistors are each formed by two diodes (see figure 4.5). For NPN, the base (anode) and emitter (cathode) is one of the two diodes. The second diode is formed by the base (anode) and collector (cathode). You can see that the base is shared between the two diodes in NPN. Both the NPN collector and emitter are highly concentrated with electrons. Relatively speaking, there are more electrons in the emitter than in the collector. The base, on the other hand has higher holes concentrations. PNP is also formed by two diodes. The terminal names are: P (emitter), N (base), and P (collector).

From a performance point of view, NPN switches faster than PNP due to electrons moving at a higher speed than holes. Even with this performance difference, both NPN and PNP are used frequently together.

NPN and PNP Symbols

The NPN and PNP schematic symbols and discrete NPN and PNP transistors are shown in figure 4.6.

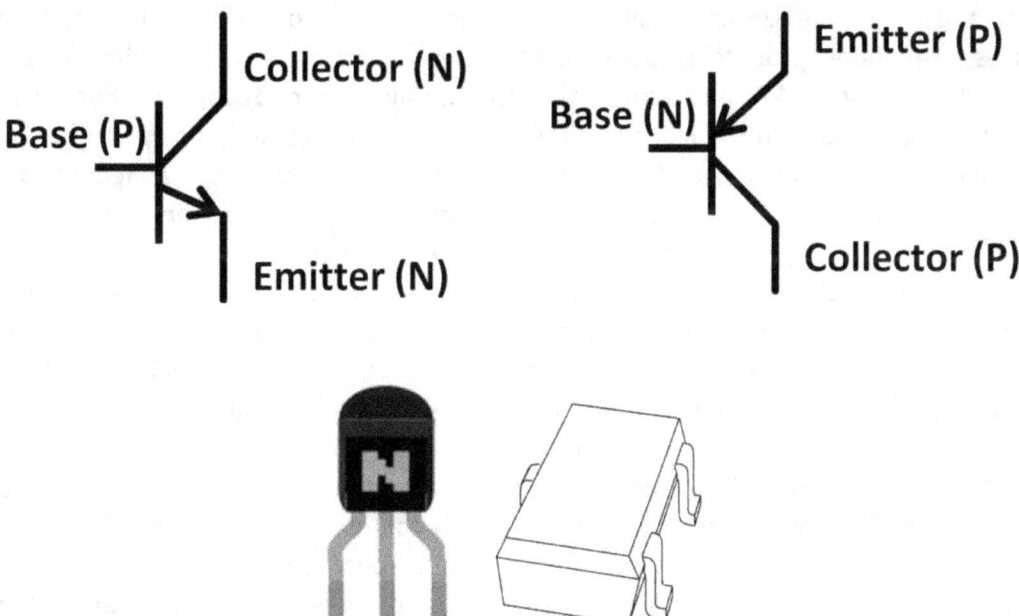

Figure 4.6: NPN and PNP schematic symbols (top); discrete NPNs (bottom left, created by Fritzing software); and PNP in plastic package, 2.2mm in length (bottom right)

The NPN schematic symbol has two diodes (see figure 4.7), base-collector and base-emitter diodes. The base-emitter diode is a part of the NPN symbol and looks a like an arrow.

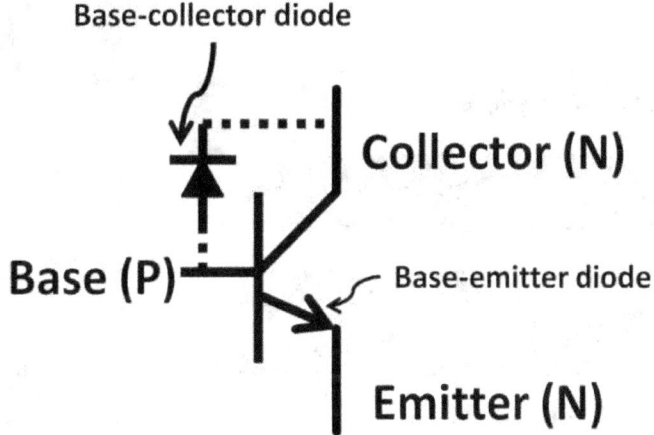

Figure 4.7 NPN diodes in schematic symbol

Transistor Cross-Section

A conceptual NPN device and a realistic Silicon Germanium (SiGe) transistor cross sections are shown in figure 4.8. Transistor regions are created one layer at a time starting from the bottom, involving hundreds of steps by complex semiconductor manufacturing equipment. Many of these steps utilize chemicals in gas or liquid form. Ion implantation and diffusion processes are major process steps in creating junctions, also called diffusions. In the conceptual view (figure 4.8), the substrate is an area filled (doped) with positive ions (holes) by chemical reactions with the silicon wafer used as the device-supporting structure. The N pocket (junction or diffusion) is doped with electrons supporting the collector. Base and emitter junctions are built on top of the collector. The dimensions and thickness of the junction vary from one process to another. Nonetheless, they are measured in the order of micrometer (um). Because germanium requires less energy than silicon to excite electrons from one energy band to the next, transistors made by germanium are faster, consume less power, and generate less electrical noise. Its disadvantage is lower reliability compared to silicon, especially in higher temperatures and the high cost of manufacturing. However, by combining both silicon and germanium in one process, we can leverage the low cost silicon manufacturing capability while gaining the performance benefits of germanium. In 1989, IBM Microelectronics first introduced a mainstream, high volume SiGe IC process. Since then, IBM has pioneered SiGe with other major semiconductor companies following suit with their proprietary SiGe processes. The latest development of SiGe has demonstrated that CPUs successfully operate at more than 100 GHz clock speed (conventional desktop computers' CPU clock speeds are less than 10 GHz). This is ideal for wireless and high-speed applications. On the silicon germanium cross-section diagram below, the base emitter region is the critical area defining transistor performance in terms of switching speed, noise, and power consumptions. In the SiGe process, germanium is doped in the base region, improving operating frequency, reducing noise, and increasing power capabilities.

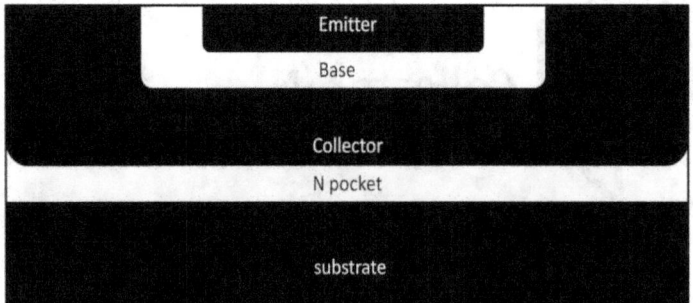
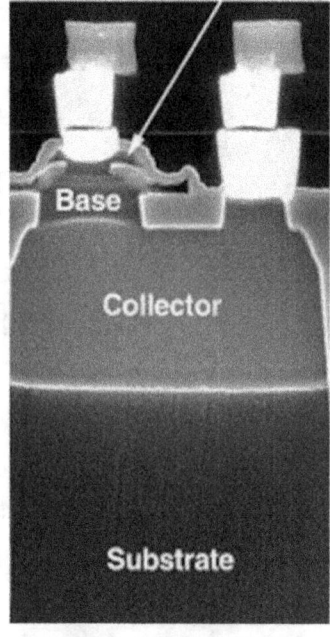

Figure 4.8: NPN, SiGe transistor side view, cross sections (Courtesy of Dr. Steve Voldman)

Bipolar Transistor Terminal Impedance

Before we go into transistor circuit design, let's get familiar with device characteristics. First, we will take a look at bipolar transistor impedances. Table 4-1 below shows the base has the highest impedance, and then the collector, followed by the emitter. Base's high impedance is due to narrow base width and low carrier concentration. Emitter's high doping level contributes to low impedance, and collector's impedance is moderately higher than emitter's but less than that of the base. Real world circuits will be discussed later to echo back to this impedance concept.

	Impedance
Base	Extremely High
Collector	High
Emitter	Low

Table 4-1: Base, collector, and emitter Impedances

IC, IB, IE, and Beta (β)

Let's now use a simple circuit in figure 4.9 to examine how a transistor functions. The NPN base connects to a variable DC input voltage. The output is at the collector that connects to a resistor. The top end of resistor ties to a DC voltage source. The emitter is grounded. In this example, DC input voltage sweeps from 0 V to 5 V. Assume the base-emitter diode threshold (minimum voltage required to forward-bias the base emitter diode) is 1 V. At 0 V, it's reverse-biased. Consider the transistor as a switch. At this input voltage level, the switch is inactive (off, open). No current flows through the transistor. As the input voltage continues to sweep higher to a point where it reaches the 1 V threshold, it starts to conduct forming the base current (Ib). A nice feature of the transistor is that there is a much bigger current (IC) now starting to flow through the collector down to the emitter (IE). This is why transistors are active devices. Current and/or voltage are larger at the output with gain. So, how could a small Ib generate a bigger IC and IE?

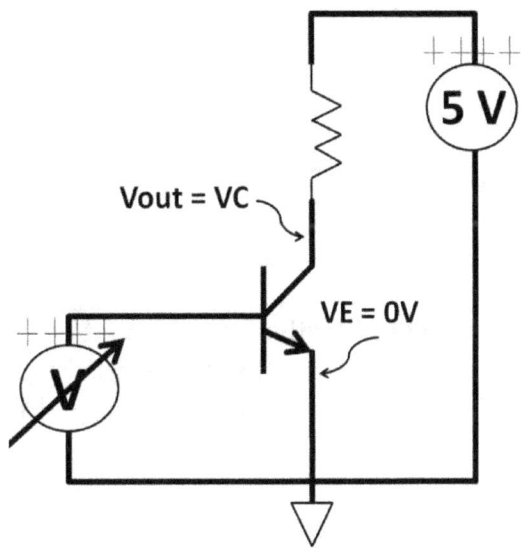

Figure 4.9: Simple transistor circuit

The graphical representation of the NPN circuit operation explains the reason (see figure 4.10). The base junction is filled with positive ions (holes). The base size (width) is relatively smaller than the collector and emitter. Only small numbers of electrons can "emit" from the emitter towards the base forming a small base current (Ib). The majority of electrons are swept across the base junction, "collected" by the collector as long as the collector is tied to a positive terminal (a positive 5 V attracts electrons). Recall from chapter 1, DC, that electrons and current flow in reverse directions. This collector current (IC) combines with the base current (Ib) flowing downwards to form emitter

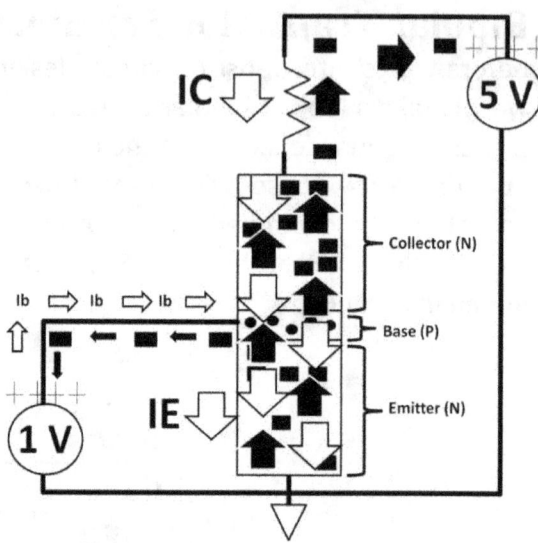

Figure 4.10: NPN operations

the current (IE). To turn on the transistor, base-emitter voltage (VBE) needs to be at least equal to or larger than the diode threshold (VBE ≥ forward-bias threshold). The second condition is that the collector has to be more positive relative to the base. The current transfer function:

$$IE = IC + Ib$$

Beta (β) is used to specify current gain:

$$\beta = \frac{IC}{IB}$$

Many academic textbooks claim that the base current is zero for simplicity reason. This is a false assumption. In the real world, Ib is non-zero and it could adversely affect circuit performance. Typically Beta (β) is around 100 to 200. If **Ib = 1 uA, beta = 100**:

$$\beta = \frac{IC}{IB}$$

$$100 = \frac{IC}{100 \text{ uA}}$$

$$IC = 100 \times (1 \text{ uA}) = 100 \text{ uA}$$

Beta in the real world doesn't stay constant and change over temperature. This imperfect characteristic could become a major design challenge. Many design tasks are to compensate for these changes, keeping the circuit running in stable conditions over wide temperature range. We use Alpha (α) to specify the IC to IE (IC / IE) ratio. For a non-ideal transistor, where Ib is non-zero, α is always less than 1.

VBE

As input voltage (VBE) continues to go up, this causes IC and IE to increase as well. The VBE transfer function is:

$$VBE = VT \times \ln\left(\frac{IC}{(A)(IS)}\right)$$

VT = Thermal voltage (KT / Q), K is the Boltzmann's constant, and T is temperature. The ln symbol is the natural log math function, IS = Saturation current, and IC = collector current. A = Transistor Area, measured in width and length.

Within VT, K is a constant that is fixed for a specific transistor manufacturing process. T is absolute temperature measured in Kelvin (K). Q is electron charge (1.6 X 10^{-19} C). VT is approximately 26 mV at room temperature (27°C). Saturation current (IS) is a complex function that is inversely proportional to temperature. Recall from chapter 2, Diode, that the temperature coefficient of a diode is negative. The VBE equation is a reflection of that. Despite the fact that VT goes up with temperature, with strong temperature dependence of IS in the denominator, VBE actually decreases with temperature. Applying VBE function, IC increases with VBE exponentially as follows:

$$IC = A \times IS \times e^{(VBE/VT)}$$

IE = IC + IB

Up to this point, the VBE diode is fully on. The transistor is operating in the active region. On the other hand, the base collector diode is kept intentionally off. You will see in the next section that it is critical to keep this diode off for optimal transistor operation. Furthermore, you will see in the next section that IC will eventually stop increasing even with increasing VBE. PNP, in contrast, works in an opposite manner in a sense that the current flows from emitter to base and collector. Figure 4.11 shows the NPP and PNP current directions using schematic symbols. Knowing how to connect the terminals to appropriate voltage levels or bias the transistors the right way gives you great control over a transistor's operations.

IE = IC + IB

Figure 4.11: NPN and PNP current flow directions

IC versus VCE Curve

The following IC versus VCE curves in figure 4.12 reveals more information about transistor. These curves are great tools to examine transistor operations.

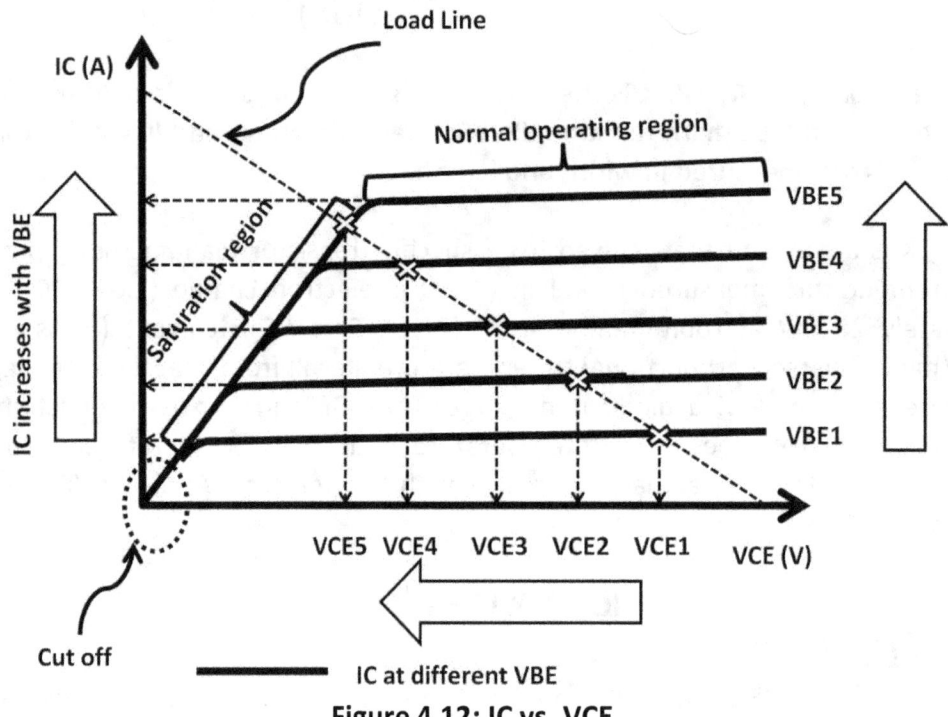

Figure 4.12: IC vs. VCE

The graph shows collector current (IC) on the Y-axis, voltage across the collector, and the emitter (VCE) on the X-axis. The emitter is connected to ground (0 V), thus,

$$VCE = VC - VE = VC - 0\ V = VC$$

This graph shows 5 IC vs. VCE curves. Each curve's VBE is different. VBE1 > VBE2 > VB3...etc. As mentioned in previous section, as VBE increases, IC increases accordingly, as shown by the vertical up arrows on the left of figure 4.12. The dotted line intercepts (cross) each curve to form a load line. For each VBE increase, (e.g., VBE1), the load line intersects the IC curve and extrapolates down to VCE1. Increase VBE1 to VBE2, and the load line intersects with the IC curve leading to VCE2. This process continues as VBEs increase. The load line shows that as input voltage (VBE) increases, IC increases and VCE decreases, shown in the horizontal arrow right below the X-axis. From VCE1 to VCE4, the transistor current is constant at each VBE. In other words, the IC has little effect on decreasing VCE. This is explained by a device physic effect called emitter current crowding. This effect reduces base-emitter area reducing the current gain significantly. This is why the IC starts to bend down at higher VCE.

From VCE 1 to 4, the transistor is said to be running in the normal operating region (almost constant current). Noticed I mentioned "almost" constant. Within this region, the collector current actually goes up slightly with increasing VCE instead of remaining absolutely constant. The Early effect explains this phenomenon. The Early effect was discovered by Mr. James Early in 1952. This effect states that base width is modulated as the VCE changes in the operating region. As the VCE continues to rise, the effective base width gets reduced further due to the spreading of depletion region into the base, increasing current gain slightly and therefore the uptick in IC (ΔIC) (see figure 4.13). VBE continues to increase to VBE5 and VCE goes down further to VCE5. At this point, the IC starts to fall. Continued VCE reduction leads to more IC decreases. This region is called saturation. The small region where IC just started to fall (bend) is called the "knee" region. It defines the point where the transistor starts going into saturation. During saturation, the current is changing, modulating with changing VCE. This collector current change could cause unstable system operation if a constant current is expected. If VCE goes down even more, IC would eventually reach zero current and the transistor is now in cut-off region (Zero IC). Ideally, you would want to operate the transistor in the normal operating region where the IC is relatively constant. In addition to a stable collector current, the normal operating region offers the highest voltage output swing. This is the optimal operating region often referenced as the transistor Q point.

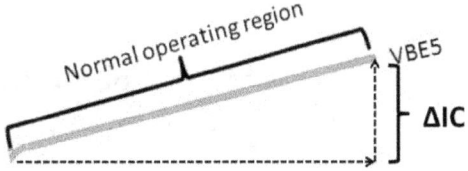

Figure 4.13: Early effect

Common Emitter Amplifier

Let's apply individual transistor understanding into a simple circuit: an amplifier. An amplifier by definition provides voltage, current, and/or power gain. Amplifiers are regularly used to amplify input signals and produce a larger output signal. The circuit in figure 4.9 is categorized as a single-ended amplifier. There is one single input and output. This is a common emitter amplifier, which means the emitter is common (DC) to a fixed potential where the output is located at the collector. There are many ways to build amplifiers using transistors. We will look at several in this section. Reusing the circuit back in figure 4.9, we revised it to replace the input with a sinusoidal source shown in figure 4.14. When Vin is zero, NPN stays off (Zero VBE). No IB, IC, and IE are zeros. There is no voltage drop across the resistor. Vout is 5 V according to Ohm's law:

Top end of resistor = 5 V, I = 0, voltage across resistor = I X R = 0 X R = 0 V

5 V – (Voltage at resistor bottom end) = 0 V

(Voltage at resistor bottom end) = 5 V – 0 V = 5 V

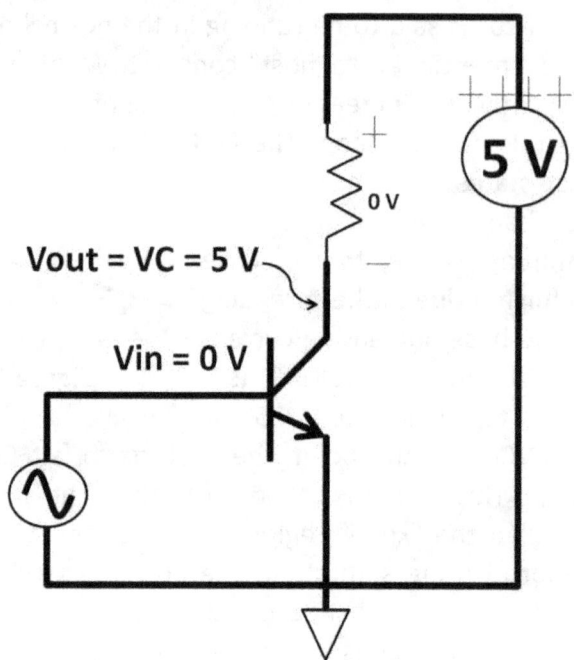

Figure 4.14: Revised NPN circuit

As Vin rises above the diode forward-biased threshold in figure 4.15, IB, IC, and IE start to flow. Suppose **IC = 1 mA** at 1 V input. **Transistor Beta = 100 (these numbers are device-specific), R = 1 kΩ, VC = 4 V, Ib = 10 uA, and IE = 1.01 mA**

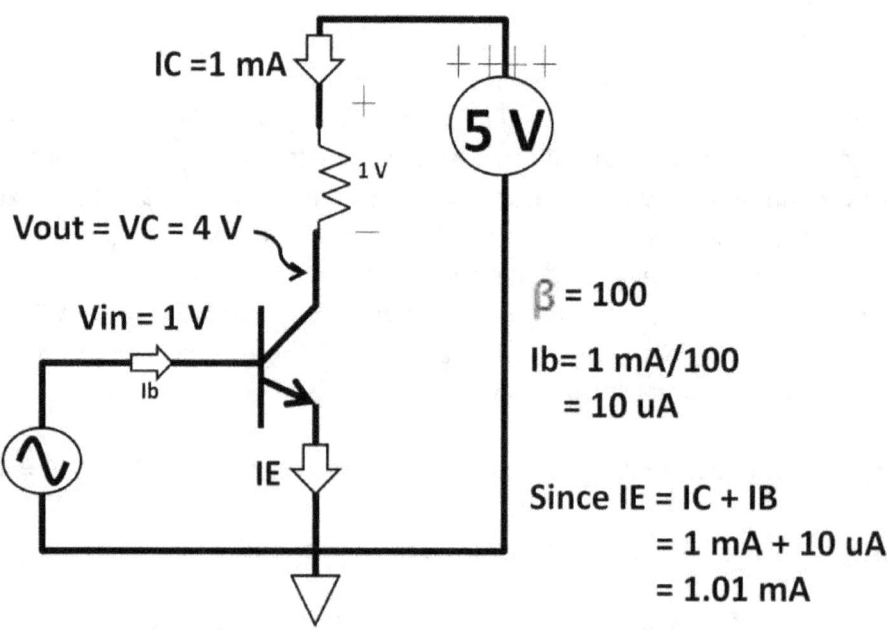

Figure 4.15: NPN "on"

Repeating the steps shows that increasing Vin leads to decreasing VC and vice versa. This agrees with the IC vs. VCE graphs in figure 4.12, revised in figure 4.16. Vin is a sinusoidal wave, and the Vout will have the waveform as shown in figure 4.17. You can see that Vin and Vout are 180 degrees out of phase where Vout is larger than Vin. This circuit offers an inverter function meaning when input is low, output is high and vice versa. More importantly, it's an amplifier that provides voltage gain (hfe). By definition, hfe:

$$hfe = \frac{\Delta Vout}{\Delta Vin}$$

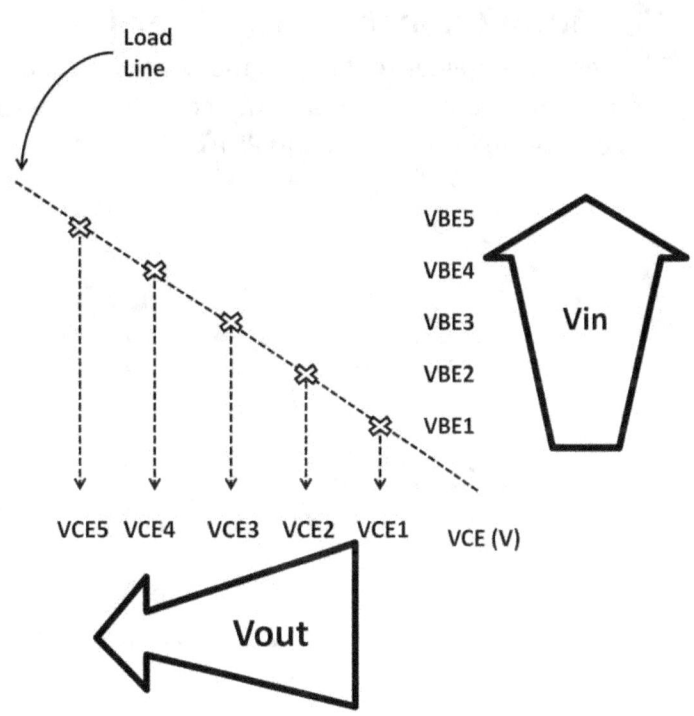

Figure 4.16: IC vs. VCE revised

Voltage gain (hfe) is unit-less because this is a division. An amplifier, by definition, is to create a larger output signal from a lower one. The larger output signal can be in the form of voltage, current, and/or power.

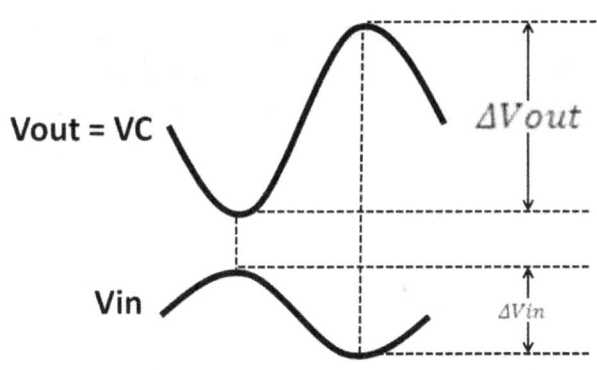

Figure 4.17: Sinusoidal Vin

All circuits discussed so far have ground (0 V) being the lowest circuit potential. Many amplifiers were designed to accept negative voltage on the bottom supply (rail). For personal safety and to minimize the chance of damaging the parts, do pay close attentions to maximum positive and negative voltages the device could withstand from the datasheets. In fact, all components in the circuits can be used in conjunction with diodes, inductors, and capacitors for an unlimited number of circuits. It depends on the application's requirements when making a part-selection decision.

Common Collector Amplifier (Emitter Follower)

The second amplifier topology is the common collector amplifier. Its output is at the emitter. Input remains at the base, and the collector connects to the positive rail. Figure 4.18 shows two emitter followers (NPN and PNP).

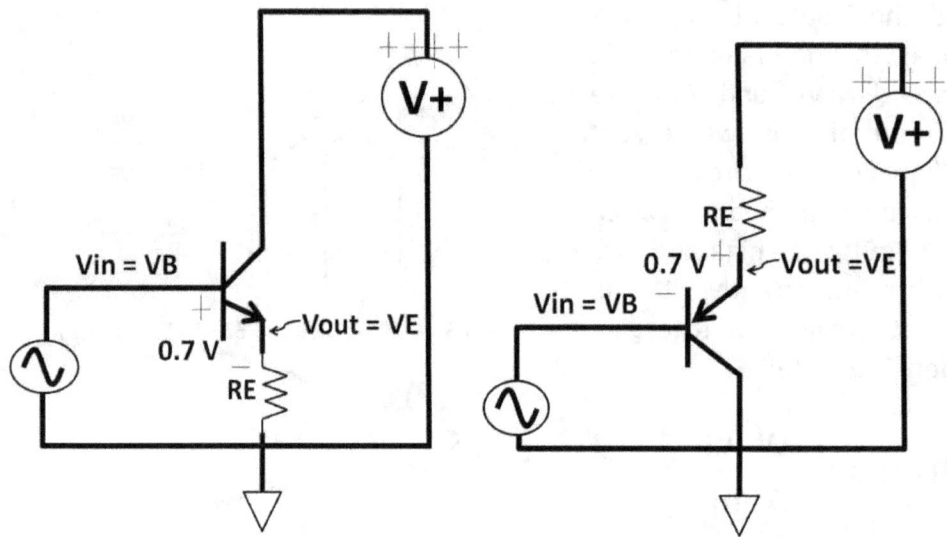

Figure 4.18: Common collector amplifier (Emitter follower)

NPN and PNP common collector amplifiers works the opposite way. For PNP, the device was flipped upside down (emitter to rail, collector to ground). In both NPN and PNP, an emitter resistor (RE) is added in the circuit. The collector resistor needs not be there as long as it's connected to a positive source. Alternatively, this circuit is named emitter follower because the emitter output (VE) "follows" the base input (VB). The emitter follower's voltage gain is one. To prove that, assume a 0.7 V VBE threshold, and Vin swings from 1 V to 2 V. This leaves the output swinging from 0.7 V to 1.3 V (0.7 V drop across VBE). Voltage gain (hfe):

$$hfe = \frac{\Delta Vout}{\Delta Vin}$$

hfe = (1.3 V – 0.3 V) / (2 V – 1 V) = 1

Because voltage gain has no unit, to describe it in some form of measured unit, dB is used:

$$\frac{Vout}{Vin} \text{ in dB} = 20 \log (1)$$

$$= 0 \text{ dB}$$

On phase shift, the output follows the input. Both are in phase. Figure 4.19 shows the phase relationship between Vout, Vin, and the 0 dB gain. The emitter follower's gain in reality is slightly less than 1, which will be discussed in the next section. If you question why use the emitter follower if the ΔVout = ΔVin, voltage gain sometimes may not be your primary goal. First, the emitter follower has current gain from beta. Secondly, the strong appeal of the high input, low output impedances makes the common collector (emitter follower) an ideal choice as a buffer (more on buffer in chapter 4, Analog Electronics). The following model in figure 4.20 illustrates this buffer idea with a multi-staged amplifier design.

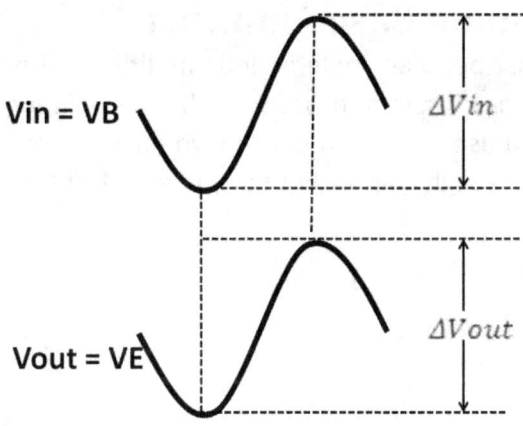

Figure 4.19: Vin vs. Vout phase

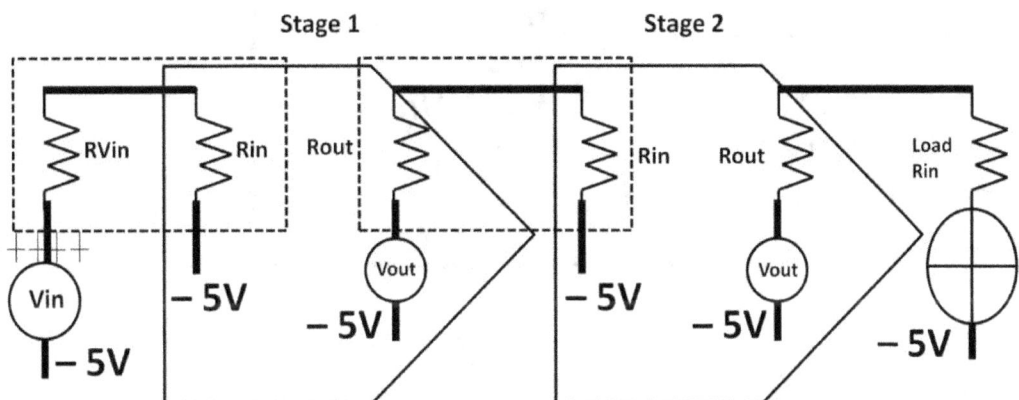

Figure 4.20: Multi-staged amplifier block diagram

Starting from the left, Vin has finite impedance, and RVin connects to a Stage 1 amplifier that presents finite input impedance, Rin. From chapter 1, DC, we know that this forms a voltage divider denoted by the dotted rectangle. To achieve the closest voltage possible at Stage 1 input from Vin, RVin ideally would be zero while Rin would be infinite. These situations, however, are not practical in the real world (RVin > 0, Rin < infinite). Instead, we choose the right amplifier topology to give us the highest possible impedance (highest output voltage level) as possible. At Stage 1 output, we need to "condition" the output to have the lowest possible output impedance. It faces the same issue where the output ties to Stage 2's input forming a voltage divider. Finally, Stage 2 output connects to a load. Ensure Stage 2 Rout's low impedance is critical especially when Load Rin may not be easily changed due to system requirements. Using an emitter follower (high input, low output impedance) is a good choice for Stage 2 to drive the load.

Common Base Amplifier

The last popular single-ended amplifier is the common base amplifier. Figure 4.21 shows an NPN-based common base amplifier. Its input is at the emitter, its output at the collector, while base ties to the fixed voltage source (common DC). The common base amplifier provides high gain without any phase shift (see figure 4.21).

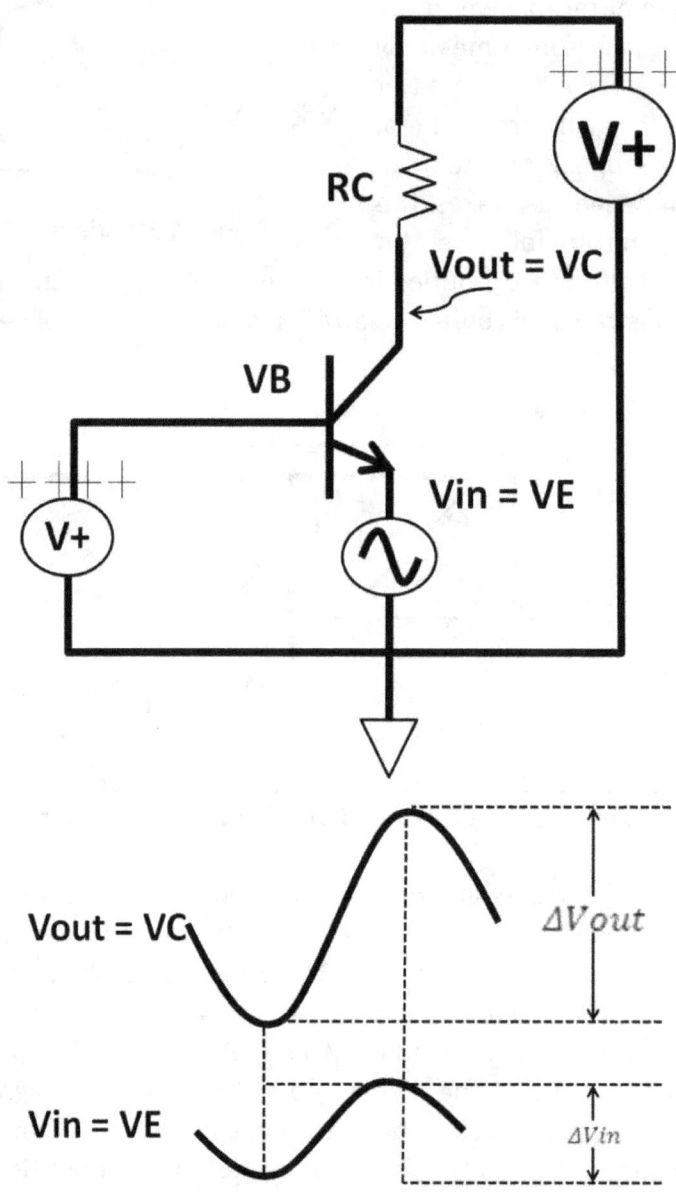

Figure 4.21: Common base amplifier

Single-Ended Amplifier Topologies Summary
Table 4-2 below summaries 3 singled-ended amplifier topologies.

	Common emitter	Common collector (Emitter follower)	Common Base
Base	Input	Input	Fixed voltage potential
Emitter	Ground or emitter resistor	Output	Input
Collector	Output	Connects to positive supply	Output

Table 4-2: Amplifiers' input and output configurations

Tranconductance (Gm), Small-Signal Models

You may wonder how you can precisely figure out the specific gain of the amplifiers. Utilizing a small-signal model facilitates this process for us. Small-signal models are simplified transistor circuits using ideal voltage, current sources to determine input, output impedances, and voltage gain. In order to use a small-signal model, transconductance (Gm) is introduced:

$$Gm = \frac{\Delta Iout}{\Delta Vin}$$

Gm is equal to the output current change divided by input voltage change. Multiplying Gm by Vin gives rise to output current, Gm X V. Recall that a transistor is an active device that produces a large current if certain requirements are met. Gm X V is used to model a transistor as a current source. Let's now apply these concepts back to a common emitter amplifier to derive voltage gain. First, we need to transform the original circuit into a small-signal model circuit using the Gm and superposition theorem.

Figure 4.22 is a small-signal model of a common emitter amplifier (hybrid-π) model.

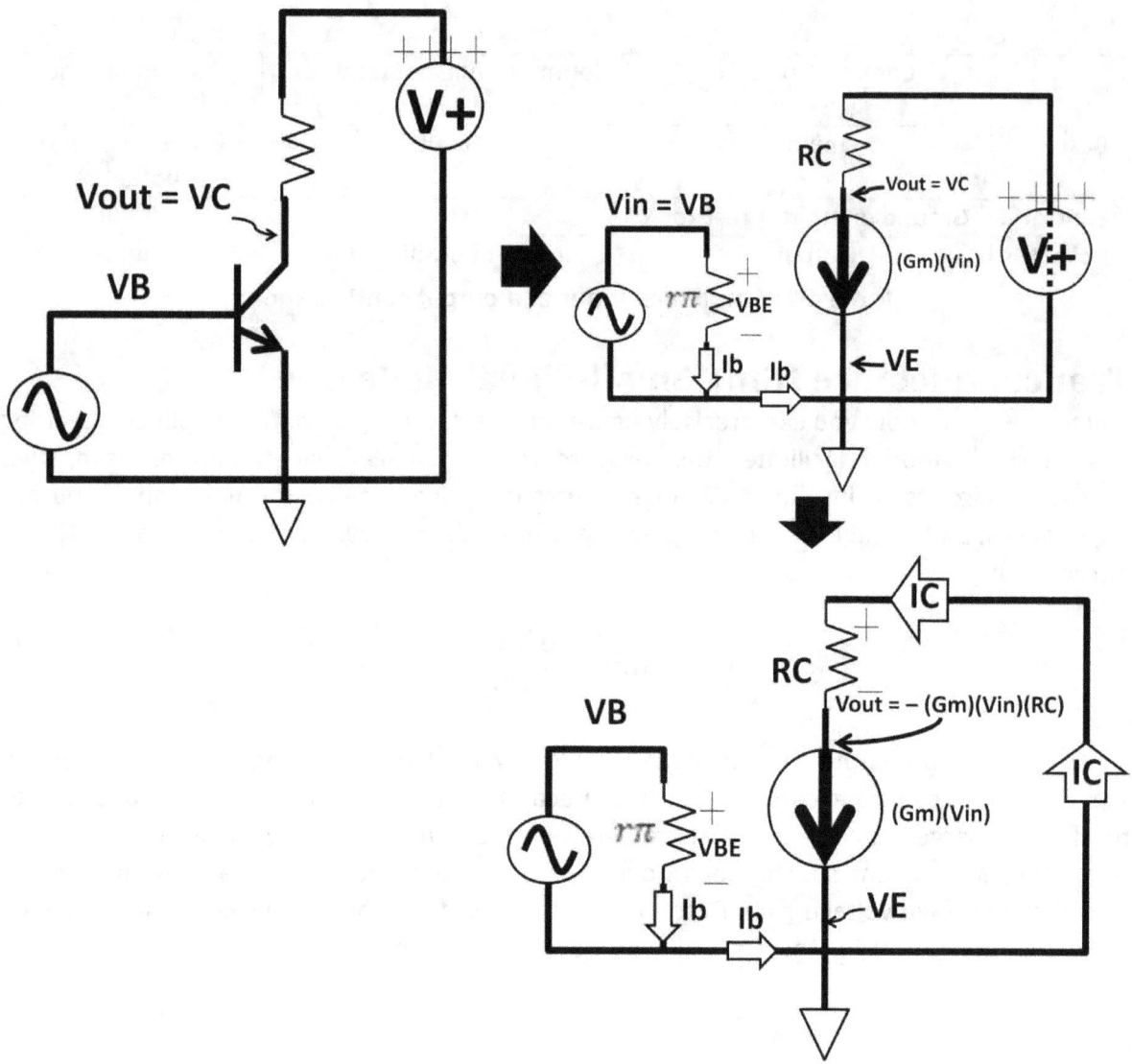

Figure 4.22: Common emitter amplifier small-signal model

Common Emitter Amplifier Input Impedance

On the previous page, the original common emitter amplifier is on the far left. After the transformation, r π connects to the Vin. r π is intrinsic (natural) base resistance. It is defined as the change of VBE over the change of base current (Ib):

$$r\pi = \frac{\Delta VBE}{\Delta Ib} = \frac{\Delta Vin}{\Delta Ib}$$

$$\beta = \frac{IC}{Ib}$$

$$Ib = \frac{IC}{\beta}$$

Substituting Ib to r π equation from above yields:

$$r\pi = \frac{\beta \times \Delta VBE}{\Delta IC}$$

ΔVBE = Vin,

$$Gm = \frac{\Delta IC}{\Delta VBE}$$

$$\frac{\Delta VBE}{\Delta IC} = \frac{1}{Gm}$$

r π in terms of β, Gm:

$$r\pi = \beta \times \frac{1}{Gm} = \frac{\beta}{Gm}$$

r π represents the input impedance of this circuit. For example, a bipolar transistor beta is 200 at room temperature (r.m.t.). For a 1 V VBE change, output current, IC changes by 1 mA:

$$Gm = \frac{1\,m}{1} = 1\,m$$

$$r\pi = \frac{200}{1\,m} = 200\,k\Omega$$

Once again, Gm in this circuit is:

$$Gm = \frac{\Delta IC}{\Delta VBE} = \frac{\Delta Iout}{\Delta Vin}$$

By multiplying Gm by VBE, it's left with output current IC represented by the current source.

$$\text{Gm} \times \Delta\text{VBE} = \text{Gm} \times \Delta\text{Vin} = \Delta\text{IC} = \text{Output current}$$

The positive voltage source ties to collector resistor (RC), and is converted to a short circuit shown on the far right. The voltage gain of the final small-signal model circuit is calculated as:

$$\text{Voltage Gain} = \text{hfe} = \frac{\Delta\text{Vout}}{\Delta\text{Vin}} = \frac{-(\text{Gm} \times \Delta\text{Vin})(\text{RC})}{\Delta\text{Vin}} = -\text{Gm}(\text{RC})$$

The negative voltage gain sign is because V+ was converted to ground while current continues to flow from ground towards RC. The voltage drop across RC would have to be below ground by **(− (Gm X ΔVin) X RC)**. This negative sign agrees with previous assessment that the common emitter amplifier's input and output are out of phase, i.e., when input is "+" output is "−". If **RC = 100 kΩ, ΔIC = 1 mA, ΔVBE = 1 V**:

$$\text{Gm} = \frac{\Delta\text{IC}}{\Delta\text{VBE}}$$

$$\text{Gm} = \frac{1\,\text{m}}{1} = 1\,\text{m}$$

hfe = − Gm X RC = − 1 m X 100 kΩ = − 100

Common Emitter Amplifier Output Impedance

As for output impedance, it is equally important from a circuit performance standpoint. The common emitter amplifier's output at the collector usually connects to a load. In AC analysis (sinusoidal input), this load presents finite resistive and capacitive reactance seen in parallel with the collector resistor (see figure 4.23). In AC small-signal analysis, the DC voltage source is replaced by a short to ground. Thus the effective output impedance is the parallel of RC, RLoad, and Cload (see figure 4.24a).

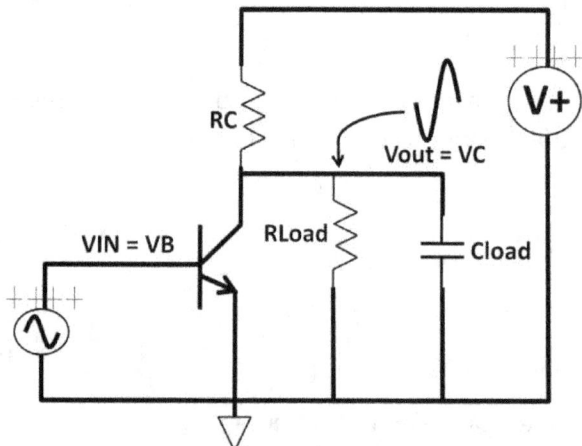

Figure 4.23: Common emitter with RLoad, CLoad at collector output

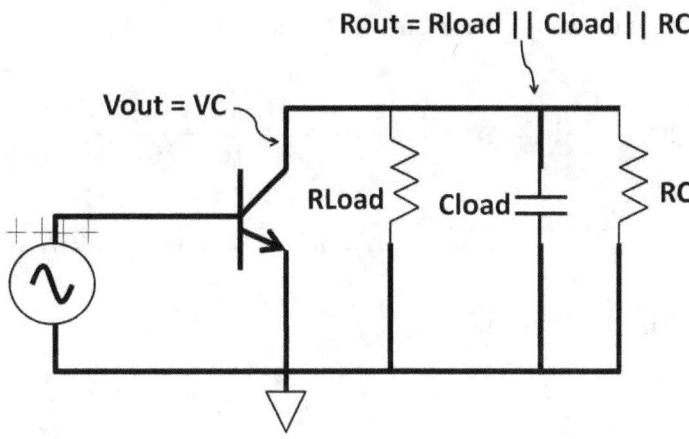

Figure 4.24a: Common emitter output impedance

RC is typically much larger than RLoad. Consequently, the output impedance is roughly equal to RLoad according to parallel resistor rules in chapter 1, DC. The result of this small-signal model concludes that the voltage gain is controlled largely by the Gm and RC sizes. The higher Gm and RC, the higher the voltage gain would be without any phase shift. Use of small-signal model applies to any transistor circuit types including the two previously discussed single-ended amplifiers. There are other transistor models such as Gummel-Poon and Ebers-Moll models describing transistor circuit behaviors. Regardless of the models you choose, always follow AC analysis rules and basic electronic principles. In the original common emitter small-signal model, there was a small re (internal emitter resistance) that was excluded in the model. There is the intrinsic resistance in the emitter. The re value is a function of emitter current and doping level. Some use 25 Ω (1 mA / IC) to model re resistance. The revised model is shown in figure 4.24b.

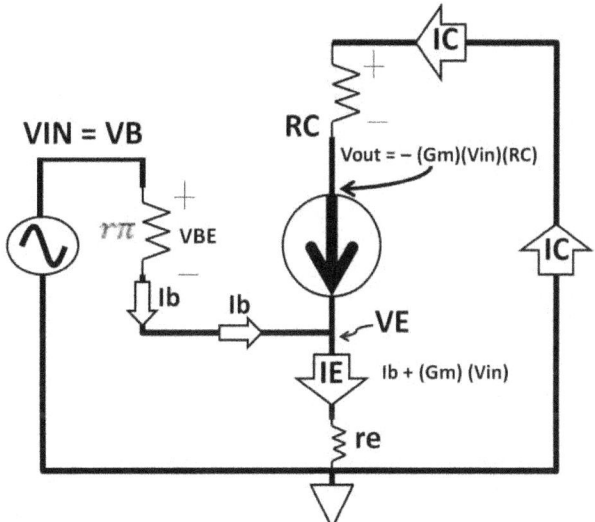

Figure 4.24b: Revised common emitter small-signal model

The voltage gain also needs to be revised from the small re as follows:

$$\text{Voltage Gain} = \text{hfe} = \frac{-(\text{Gm X Vin})(\text{RC})}{(\text{Ib})(\pi) + [\text{Ib} + (\text{Gm X Vin})]\text{re}}$$

$$\text{Voltage Gain} = \frac{-(\text{Gm X Vin})(\text{RC})}{(\text{Ib})[(\pi) + (1 + (\text{Gm X Vin}))\text{re}]}$$

Gm X Vin $\gg$ (Ib)(π):

$$\text{Voltage Gain} \approx \frac{-\text{Gm X Vin}(\text{RC})}{\text{re}}$$

From the above, you can see that the voltage gain is reduced by the re in the denominator. The voltage gain is further reduced if an external resistor is connected at the emitter terminal. You may wonder why anyone would want to design an amplifier with lower gain. The reason is to keep the amplifier stable. By adding an external resistor at the emitter, which is called emitter de-generation, it helps prevent the amplifier from going into oscillations. The details of the emitter de-generation relate to circuit design techniques and beyond the scope of this book. The readers should however, at least take note of their existence.

Common Collector Amplifier Small-Signal Model

The AC small-signal model of the common collector amplifier (emitter follower) is shown in figure 4.25. Just like a common emitter amplifier, $r\pi$ is the input impedance. The positive voltage source is converted to a short circuit shown in figure 4.25. Once again, Gm in this circuit is:

$$Gm = \frac{\Delta IC}{\Delta VBE} = \frac{\Delta Iout}{\Delta Vin}$$

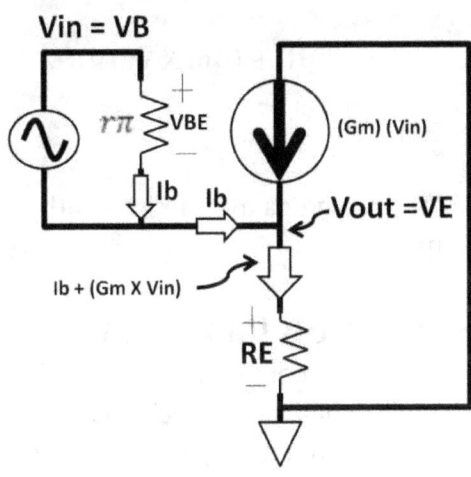

Figure 4.25: Emitter follower small signal model

By multiplying Gm by VBE, it's equal to current source:

Gm X ΔVBE = Gm X ΔVin = ΔIC = Output current

Vin is the voltage across $r\pi$ (VBE) plus voltage across RE (VE):

Vin = VBE + VE

VE = (Ib + IC) (RE), IC = (Gm X Vin):

Vin = VBE + (Ib + (Gm X Vin)) (RE)

VBE = (Ib) (r π):

Vin = (Ib) (r π) + (Ib + (Gm X Vin)) (RE)

IE = Ib + IC = Ib + (Gm X Vin):
VE = Vout = IE (RE):

Vout = VE = Ib + (Gm X Vin) (RE)

The voltage gain of the final small-signal model circuit is calculated as:

$$hfe = \frac{\Delta Vout}{\Delta Vin} = \frac{\cancel{Ib + (Gm\ X\ Vin)(RE)}}{(Ib)(r\ \pi) + \cancel{(Ib + Gm\ X\ Vin)(RE)}}$$

$Ib + (Gm\ X\ Vin)(RE) \gg (Ib)(r\ \pi)$:

$$hfe = \frac{1}{Ib\ X\ r\ \pi} < 1$$

$$\text{hfe} = \frac{\Delta \text{Vout}}{\Delta \text{Vin}} = \frac{\text{Ib} + (\text{Gm X Vin})(\text{RE})}{(\text{Ib})(r\,\pi) + \cancel{(\text{Ib} + \text{Gm X Vin})(\text{RE})}}$$

$\text{Ib} + (\text{Gm X Vin})(\text{RE}) \gg (\text{Ib})(r\,\pi)$:

$$\text{hfe} = \frac{1}{\text{Ib X } r\,\pi} < 1$$

The voltage gain came in slightly less than 1 without any phase shift (the output follows the input).

Common Base Amplifier Small-Signal Model

As for the common base amplifier, the same small-signal model technique can be applied to figure out voltage gain, input, and output impedances. Figure 4.26 shows the small-signal model of a common base amplifier. The DC source at the base has been replaced by a short circuit to ground. The small re is the intrinsic impedance of the emitter. The transistor is now represented by the output current source, **Gm X VBE**. Vout remains at the collector with effective output impedance equal to RLoad in parallel with RC and CLoad. To calculate gain, we need to derive Vout and Vin.

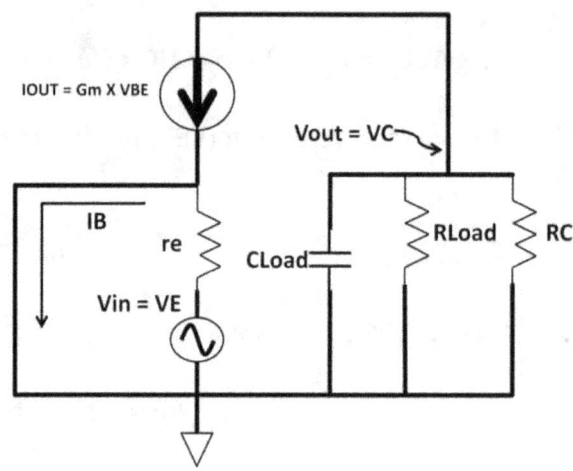

Figure 4.26: Common base amplifier small-signal model

$$\text{Voltage Gain} = \text{hfe} = \frac{\Delta \text{Vout}}{\Delta \text{Vin}}$$

IOUT = Gm X VBE:

$$\text{Vout} = (\text{Gm X VBE}) \text{ X } (\text{RLoad} \parallel \text{CLoad} \parallel \text{RC})$$

The current going through re came from the base; re is the effective input impedance. **Vin = IB X re = VBE,**

$$\text{Voltage gain} = \text{hfe} = \frac{\text{Vout}}{\text{Vin}} = \frac{\text{Gm X } \cancel{\text{VBE}} \text{ X } (\text{RLoad} \parallel \text{CLoad} \parallel \text{RC})}{\cancel{\text{VBE}}}$$

$$\frac{\text{Vout}}{\text{Vin}} = \text{Gm X } (\text{RLoad} \parallel \text{CLoad} \parallel \text{RC})$$

Notice the gain is positive, i.e., there isn't any phase shift from the input to the output. This agrees with previous assessment. The drawback of the common base amplifier is that the input impedance re is quite low. Be sure that the voltage source driving the emitter input is high impedance or else the input level at the emitter will be degraded. The output impedance is largely dependent on the RLoad just like the common emitter amplifier. From a design perspective, a small-signal model is a nice tool to check if the circuit makes sense first before the actual design.

Single-Ended Amplifier Summary

Table 4-3 below sums up the single-ended amplifiers' characteristics. Some textbooks assign these commonly used amplifiers in classes. The common emitter amplifier is considered a class A amplifier. It defines the amplifier is on during the entire input period, supplying output with an active signal 100% of the time, though 180 degrees out of phase. Class A amplifiers are inherently power inefficient due to the fact that the transistors are never turned off. Increasing power loss results in low power efficiency.

$$\text{Power Efficiency} = \frac{\text{Power Out}}{\text{Power In}} \times 100\,\%$$

An emitter follower (common collector) is considered a class B amplifier, which means the output is active only 50% of the time. If a common collector amplifier is emitter-ground, when the input goes below ground during half of the period, the transistor is said to be off resulting in a rectified half-wave output. Class B amplifier is more efficient (only 50% of time on). It however lacks the ability to drive the load at 100% duty cycle. Class AB is another amplifier type that combines the best of both class A and B. Class AB conducts between 50 to 100% of the time. A common class AB example is a push-pull topology using combinations of NPN and PNP transistors. This topology, though it increases efficiency and load driving capability, comes at the expense of circuit complexity.

Characteristics	Common base	Common emitter	Common collector
Input impedance	Low	High	High
Output impedance	High	High	Low
Phase shift	0 degree	180 degrees	0 degree
Voltage gain	High	High	< 1 (negative dB)

Table 4-3: Single-ended amplifiers characteristics

NMOS and PMOS

Similar to bipolar transistors, CMOS transistors (MOSFETs) are 3-terminal devices with complementary N- and P-types. Some refer to the N-type device as NMOS (NFET) and the P-type device as PMOS (PFET). The MOSFETs' structures are fundamentally different than the bipolar ones despite sharing similar circuit behavior. Both NMOS and PMOS are made of N- and P-junctions and poly-silicon gate combination through chip manufacturing process. By shrinking transistor sizes and with the concept of mass production, manufacturing throughput could increase substantially. Figure 4.27 shows the top-level view of a silicon wafer containing chips "printed" on them. The thickness of the wafer is in the order of 100 um to 200 um. A single chip is called a "die." The more advanced process on the right houses more chips than the older one on the left. This increases the total production number for the same process time amount and silicon space.

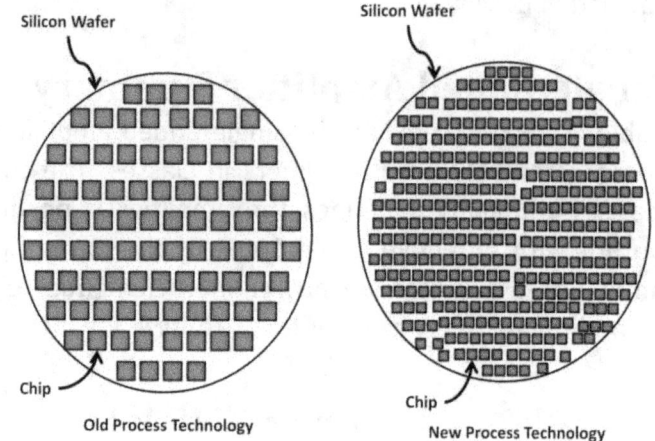

Figure 4.27: Top view silicon wafer

In addition to scaling down device sizes, wafer diameter had increased from 6-inches (150 mm) to 12-inches (300 mm) to 18-inches (400 mm) in just two decades. With rising computing power demand, many electronic circuits are integrated onto a single chip, noticeably in portable devices, like smartphones where longer battery life is required. Running these devices at lower voltages help increase battery life. Since 1997, CMOS voltage supply had since scaled up from 5 V to 3.3 V to 2.5 V to 1.8 V to 0.9 V in recent years. It's only feasible to run electronic circuits at lower voltages if the transistors are smaller, or else higher voltages will break down smaller-sized transistors. The trend of making transistors smaller closely echoes Moore's Law. It states that transistor size will shrink and numbers will increase twofold every year. Chip companies like Intel, Advanced Micro Device (AMD), IBM Microelectronics, and others with manufacturing capabilities tend to develop their own proprietary processes to gain and maintain a competitive edge. The high cost of building chip manufacturing plants and of fabrication (fab), often in billions of dollars, pose huge capital expenses to these companies. For others that don't have such financial strength, contract manufacturing facilities are available. There are over twenty contract manufacturing companies (foundries) worldwide. The top 4 IC foundries are Taiwan Semiconductor Manufacturing Company (TSMC), United Microelectronics Corporations (UMC), Global Foundries, and Samsung Semiconductor. In 2011 they brought in, combined, over US $20 billion in revenue. The foundry business model has made "fabless" design companies a means to manufacture chips without building a fab. These chip manufacturing capabilities not only apply to CMOS, but also to bipolar and any other discrete devices.

3D NFET

A NMOS 3-dimentional cross section model is shown in figure 4.28.

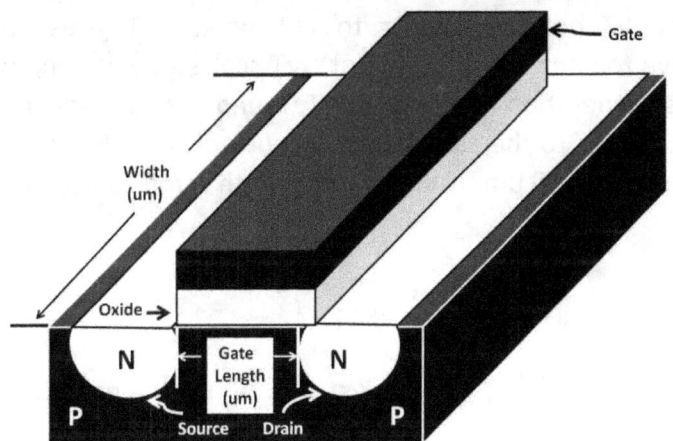

Figure 4.28: A NMOS 3-dimentional cross section model

The height of this 3D model is less than 20 um thick. The CMOS gate is made of polysilicon doped positively. Beneath the gate is an oxide layer made of silicon dioxide (SiO2), which is used as an insulator. The oxide layer thickness and quality determine the performance of the MOSFET. Advanced CMOS process strives for making the oxide layer as thin as possible while maintaining high quality for size-shrinking, lower voltage domain, and increasing speed reasons. The oxide in the diagram was drawn out of proportion just to show that in reality, the oxide layer is much thinner. Oxide thickness found in state-of-the-art CMOS processes can be as thin as 50 angstroms (1 angstrom = 1 X 10^{-10} meter). The advanced process's gate length can be as low as 30 nm and below. The latest finFET technology (still being developed at the time of publication) pushes device geometry down to 10 nm gate length. Circuit designers do not have control over this parameter due to the fact that it's fixed by the manufacturing process. The two N-junctions, called source and drain, are located above the P-substrate region. The doping levels of these junctions are entirely dictated by the process. The only parameters that circuit designers could vary for adjusting circuit performances are the transistor's width and length. The drain current (Id) transfer function is modeled as below:

$$Id = \left(\frac{uCox}{2}\right) \times \left(\frac{W}{L}\right) (VGS - VT)^2$$

u: effective mobility, Cox: gate-oxide capacitance per unit area. W, L: CMOS transistor's width and length:

VGS = (Voltage across gate and source) = VG – VS

Drain Current and Threshold Voltage

Similar to the base-emitter diode forward-biased voltage in bipolar, VT is the threshold voltage. This parameter is a constant for a specific process although it varies strongly with temperature. In fact, VT behaves similar to VBE where VT goes down with increasing temperature (negative temperature coefficient). VT scales down along with shrinking device size from one process generation to the next. VT, found in many advanced CMOS processes, is about 0.9 V. For example, to the calculate drain current of an NFET in a 2 um (gate length) process, its uCox is roughly 100 um. If **W = 5 um**, length remains at minimum 2 um, **VGS = 2**, **VT = 0.9 V**:

$$Id = \left(\frac{uCox}{2}\right) \times \left(\frac{W}{L}\right)(VGS - VT)^2$$

$$Id = \left(\frac{100\ u}{2}\right) \times \left(\frac{5\ um}{2\ um}\right)(2\ V - 0.9\ V)^2 = 275\ uA$$

NFET and PFET Symbols

NFET and PFET schematic symbols (see figure 4.29a) have arrows to represent the current flow directions. Similar to bipolar transistors, MOSFETs are three-terminal devices. Gate, drain, and source correspond to a bipolar transistor's base, collector, and emitter.

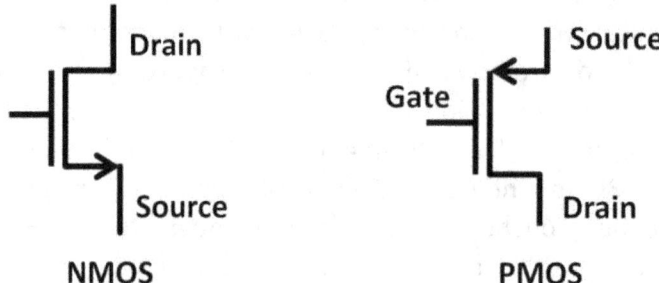

Figure 4.29a: NFET, PFET schematic models

There are other alternatives to MOSFET's symbols. Figure 4.29b shows an example. In this book, we will use the symbols in figure 4.29a.

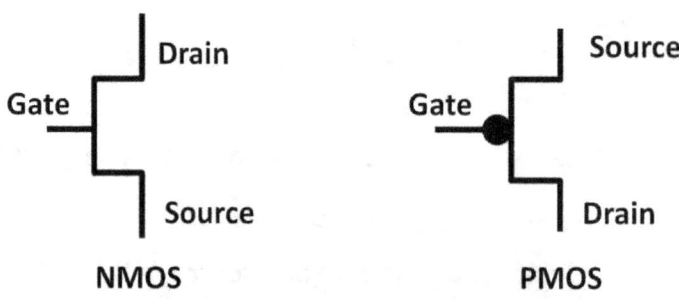

Figure 4.29b: Alternative MOSFET symbols

There is actually a fourth terminal in MOSFET, called a substrate terminal as indicated in figure 4.30. This terminal contacts the transistor substrate. If it connects to the source of NMOS, it forms a diode called the "body diode." There are design trade-offs when deciding whether or not to connect the substrate terminal to the source. One diode application is a catch diode mentioned in the buck regulator circuit in chapter 3, AC.

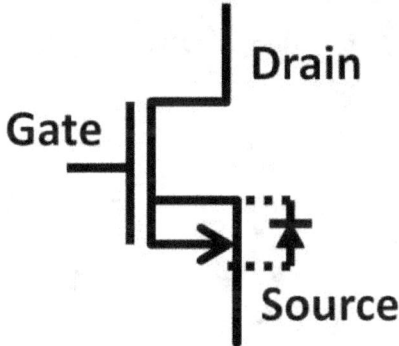

Figure 4.30: Body diode

From a top view of the actual device printed on silicon, the CMOS transistor would look similar to figure 4.31. These shapes were printed through hundreds of IC manufacturing steps involving the use of numerous chemicals and expensive equipment. Although the details of these steps are beyond the scope of this book, you should at least recognize that transistors are produced by highly efficient, large-scale, complex manufacturing processes.

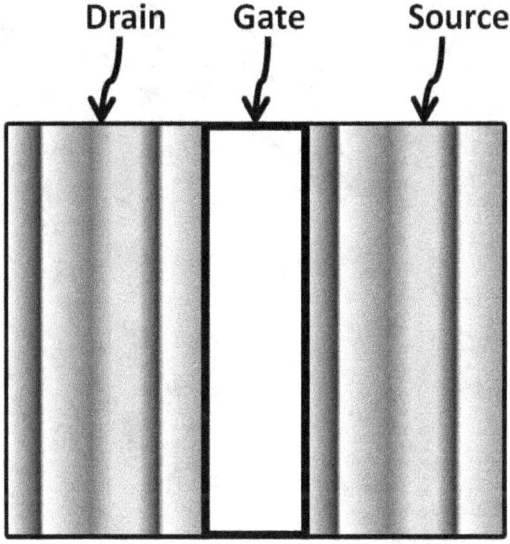

Figure 4.31: CMOS transistor top view

IC Layout

The actual device's shapes "printed" on the semiconductor chip are the device layout. Figure 4.32 shows the top view of a silicon chip layout comprising transistors and resistors created by IC schematic capture software (top). An Intel i7 core silicon chip (About 215 mm^2) seen under a microscope (bottom) contains over one billion transistors.

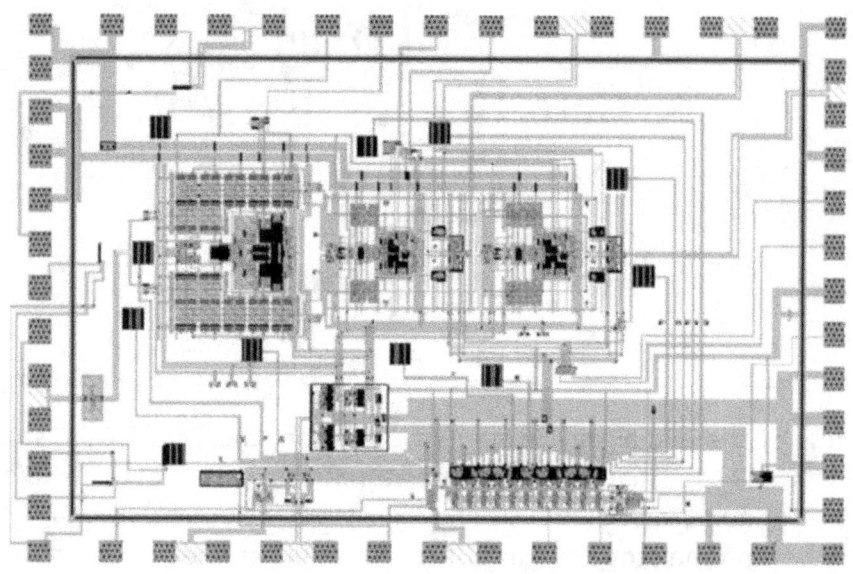

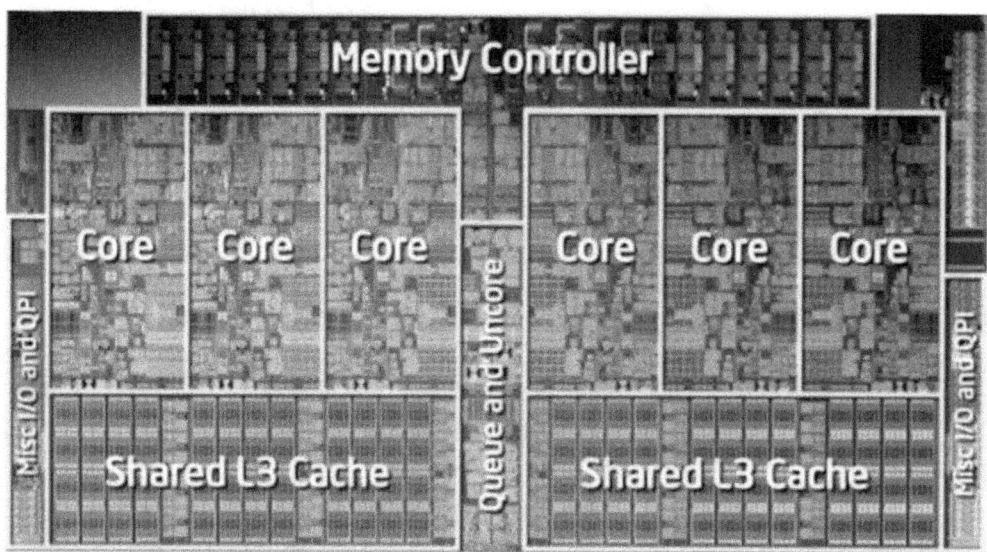

Figure 4.32: IC layout in software (Top), Intel I7 core silicon chip (Bottom)
Courtesy of Dr. Bruce Wooley and Tallis Blalack (Stanford University)

VHDL and Verilog

For high-density design like Application Specific Integrated Circuit (ASIC), Field Programmable Gate Array (FPGA), and CPUs, it is impossible to place millions of transistors manually one by one during the design process. Instead, digital designers use programming techniques to write programs (scripts) to represent digital functions as behavioral models. The scripts are called Very High Level Descriptive Language (VHDL). One popular scripting language is Verilog. Below is a script example for a 2-input AND gate. This logic AND gate (further discussed in chapter 5, Digital Electronics) consists of A and B inputs and F output.

```
module AND2gate (A, B, F);

input A;
input B;
output F;
reg F;
always @ (A or B)
begin
F <= A & B;
end
endmodule
```

The Verilog scripts would be processed by sophisticated software algorithm in the form of simulations and verifications. Digital designers would use timing waveform tools that are built in the software to verify functionalities performing timing analysis of the design. Upon digital design completion, the final steps are synthesis using a computer-aided design (CAD) tool, which generates final schematics and physical layout automatically via automatic-synthesis function. These chip-design methodologies are very complex. There are only a handful of companies supplying software in this area. Cadence Design Systems, Mentor Graphics and Synopsis are market leaders in this field. Once the circuits were constructed in the schematic capture software using schematic symbols, such design would need to be converted to layout using layout software. High-density design incorporates automatic layout generations. The verification process, Layout versus Schematic (LVS), checks if both schematic and layout match or not. Layout designers will correct any discrepancies if found, between schematic and layout. Design Rule Checking (DRC) is another verification tool. DRC checks if the physical layout meets the design rules set by the manufacturing processes. A design rule example is the minimum gate length. If the gate in the layout is drawn shorter than the minimum gate length specified by the manufacturing process, the DRC will flag indicating a DRC fail. The layout designers can then correct the errors accordingly. Once LVS and DRC are complete, the layout will be sent electronically to manufacturing. This process historically is called tapeout. The time it takes to manufacture ICs differs by companies. Generally speaking, it takes several weeks to complete the entire process. ICs printed on the silicon wafers are processed in batches. They often are counted in lots (boxes), in which each lot contains number of wafers (10 to 12 typically). The number of lots depends on the order sizes. Electrical tests are performed throughout the manufacturing process to make sure devices are within test spec.

MOSFET Cross Section and Operations

The mechanics of MOSFET's operations are best described using the conceptual NMOS cross sectional diagram (see figure 4.33).

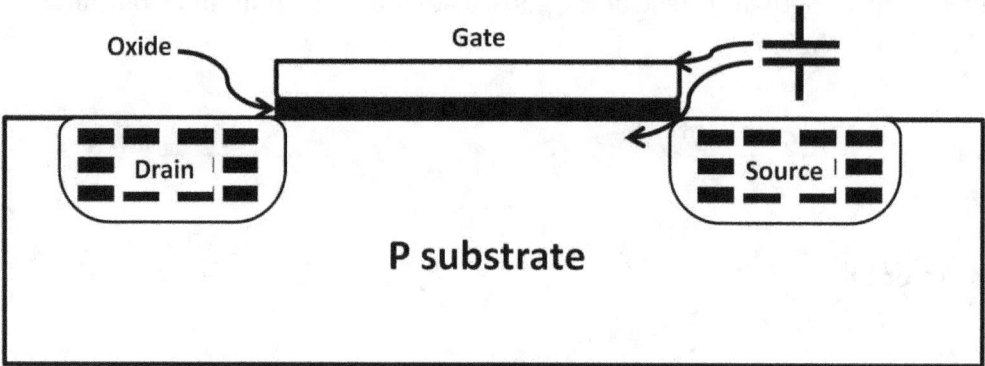

Figure 4.33: MOSFET cross section

The gate (top plate), oxide (insulator), and p-substrate (bottom plate) underneath are modeled as a capacitor. This capacitor essentially gave the meaning to MOSFET. FET stands for field-effect transistor. This field refers to the electric field of the capacitor. Applying two voltage sources at the gate and the drain strike this point clearly in figure 4.34.

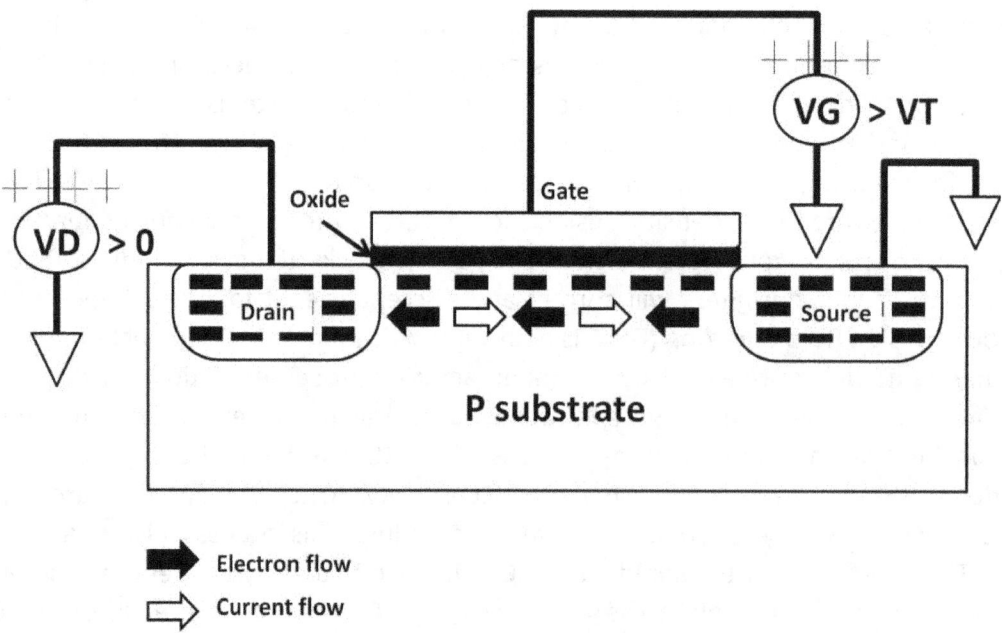

Figure 4.34: MOSFET operations

MOSFET On-Off Requirements

If voltage at the gate (VG) is slowly rising, electrons are attracted to the surface right underneath the oxide forming an electron passage called a channel. Electrons are minority carriers doped in the substrate. The channel is effectively inverted as electron concentration increases with increasing positive gate voltage (VG). This is a strong inversion phenomenon. When VG increases to at least equal to or larger than VT (VG ≥ VT) and the drain voltage (VD) is higher than ground (VD > 0 V), the NMOS is said to be "enhanced"; the channel is now "pinched off" giving rise to current flow (see figure 4.34). At this point, the transistor is active (on). If the NMOS is modeled as a switch, it's now closed with finite impedance. To turn on PFET, the polarities would be reversed. Table 4-4 below summarizes the on and off conditions and the requirements of turning on and off N and PFETs.

	NFET	PFET
VGS > VT or VGS – VT > 0 VDS > 0	ON	OFF
VSG > VT or VSG - VT > 0 VSD > 0	OFF	ON

Table 4-4: N/PFET On and off requirements

Convert figure 4.34 to a schematic. Figure 4.35 shows us the difference between N and PFETs on-conditions (ID > 0).

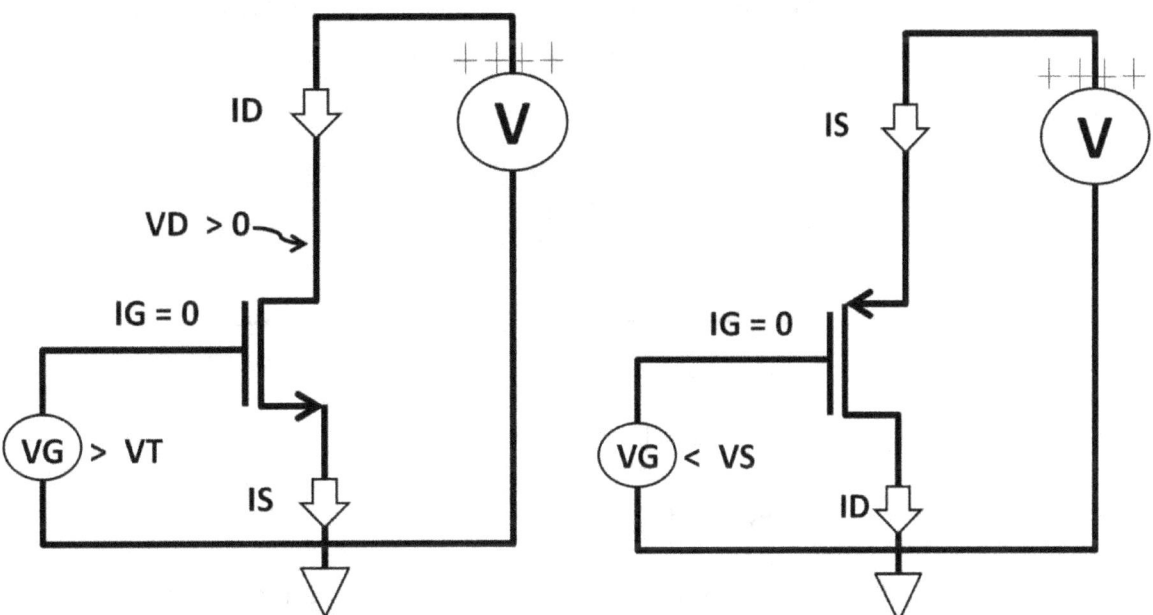

Figure 4.35: NFET and PFET different in operations

Unlike a bipolar transistor's base, CMOS gate, at DC, has no gate current because of the capacitor (gate, oxide, and substrate). As a result, ID and IS are almost equal to each other except for small leakage current. The bad news is that the leakage current increases exponentially over temperature change. These leakage currents come in the form of dynamic gate current. Despite zero DC gate current, during AC (switching signal), there would be current flowing towards and out of the gate. These currents flow towards and out of the gate during AC signal transition. This is called dynamic gate current. Figure 4.36 demonstrates this event. During voltage square wave transition, currents are shooting up, down, to, and from the CMOS gate. This current contributes noises (glitches) propagating throughout the systems. Extra care is required to minimize these glitches. They adversely impact the overall system noise performance. One of the techniques to reduce dynamic gate current is to add a resistor at the gate (RG). The resistor suppresses the gate current and helps protect the gate from transient voltage-spike damage at the expense of slower transition time. The size of RG depends on timing transition requirement. Typical size varies from 10's to 100's of ohms.

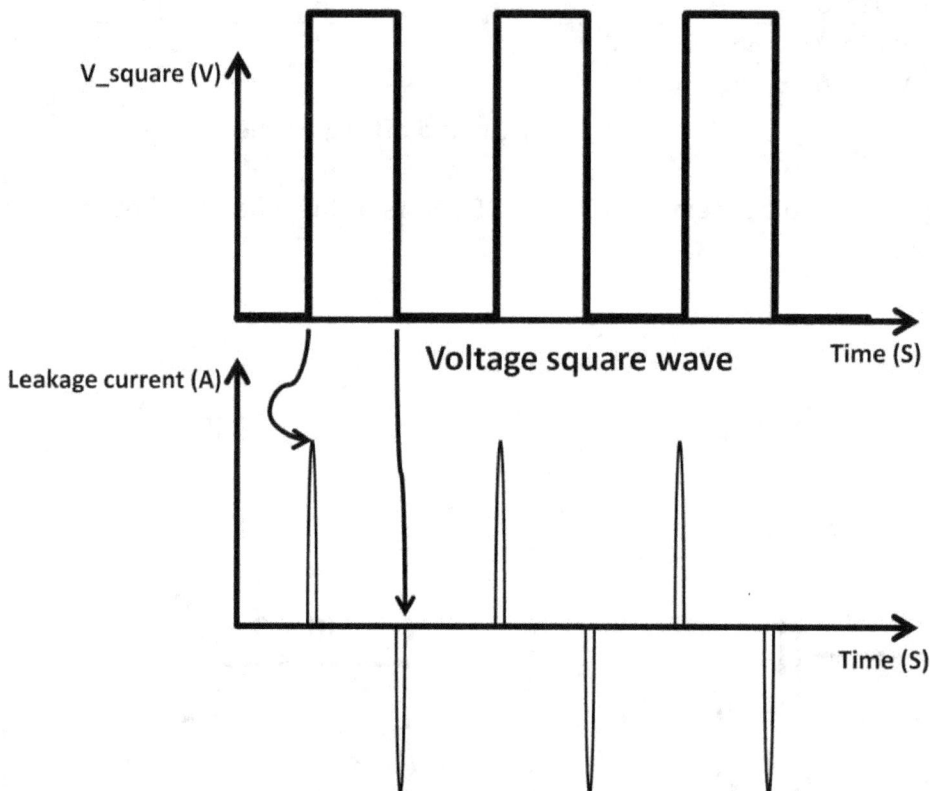

Figure 4.36: Dynamic gate current

The other fact about CMOS transistors is that there are no active diodes present in them except the body diode. The arrow of the schematic symbol simply indicates the current directions. The ID versus VDS curves on the next page shows the CMOS transistors' operating regions (see figure 4.37).

ID versus VDS Curve

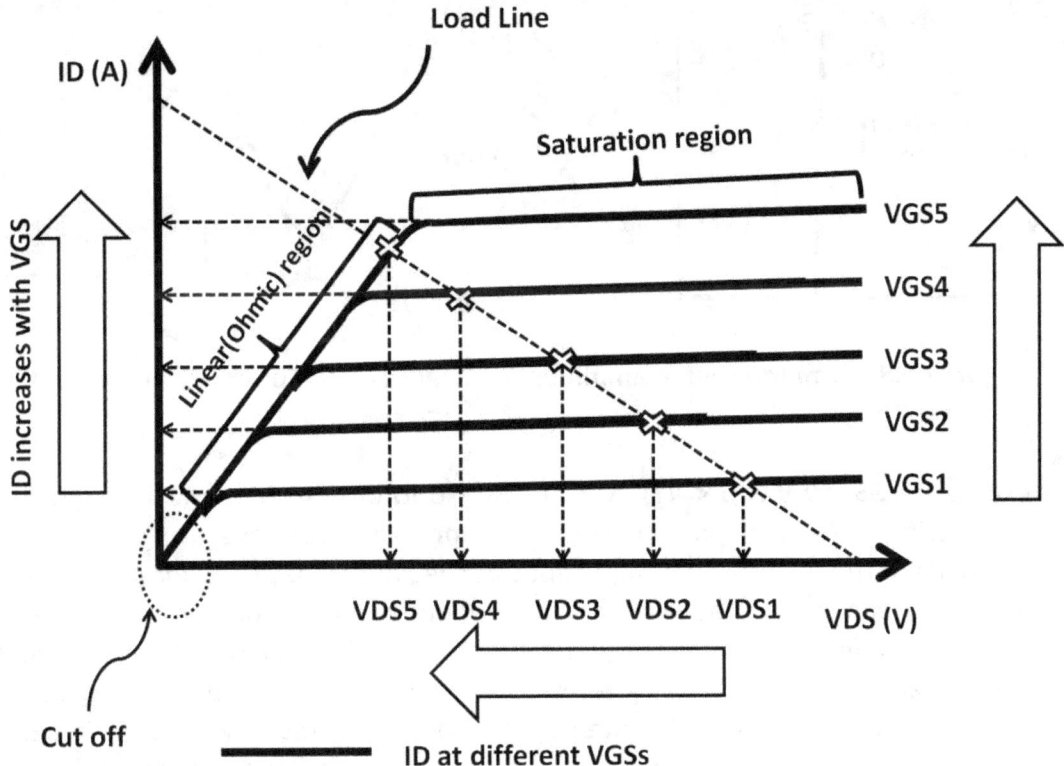

Figure 4.37: ID vs. VDS

They look almost identical to bipolar transistors' IC versus VCE curves. The exceptions are the region definitions where the constant current region is now called saturation. This is the exact opposite of bipolar transistor. The region where VDS is low is the Linear (Ohmic) region. Cut off occurs when the transistor is turned off without any drain current.

CMOS Source Amplifier

If we modify the NMOS circuit in figure 4.35 to figure 4.38, we obtain a common source amplifier that performs an inverter function just like the bipolar inverter (common emitter) without the base. CMOS transistors can be constructed in common source, common drain (source follower) and common gate amplifiers similar to those discussed in the bipolar sections. Similar to the common emitter amplifier, the common source amplifier's input is at the gate while the output is located at the drain. The input and output waveforms are shown in figure 4.39. They are 180 degrees out of phase with positive voltage gain.

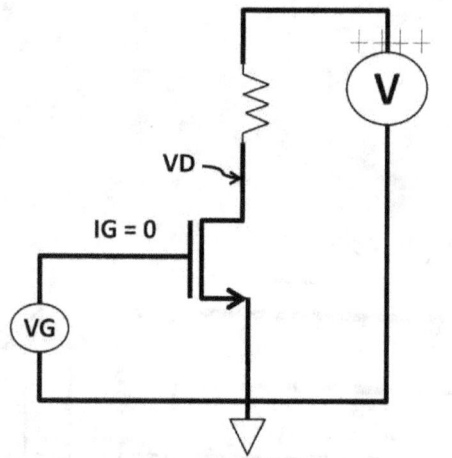

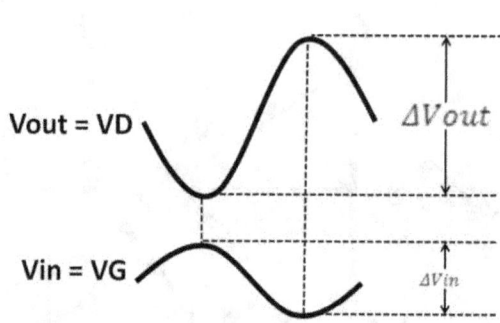

Figure 4.38: Common source amplifier

Figure 4.39: Common source amplifier input, output

If, for example, **VGS = 0 V (VG < VT)**, NFET is off. No ID or IS would flow. No voltage drops across the resistor. VD is at positive voltage supply. One incentive using MOSFET as an amplifier choice is the infinite input impedance at the gate (capacitor). Recall chapter 3, AC, capacitors are open circuits at DC, i.e., infinite impedance. From an input impedance standpoint, it is preferable that the amplifier's input stage have extremely high impedances (maximum voltage at the input). This gives MOSFETs better edge over bipolar transistors, not to mention the benefit of lacking DC gate current. These features make CMOS circuits easier to build, test, and measure. The same small-signal model technique used to analyze bipolar amplifiers can be used on CMOS circuits as well. The original common source amplifier in figure 4.35 was transformed to the small-signal model in figure 4.39a on page 141. The voltage at the gate, VG is the same as Vin. It is now facing an open circuit by the gate-oxide capacitor. Gm in this circuit is:

$$Gm = \frac{\Delta IOUT}{\Delta VIN} = \frac{\Delta ID}{\Delta VGS}$$

By multiplying Gm by VGS, which is the input, it's left with output current, ID, represented by the current source.

$$Gm \times \Delta VGS = Gm \times \Delta Vin = \Delta ID = \text{Output Current}$$

The positive voltage source ties to the drain resistor (RD), is converted to a short circuit shown in the bottom of figure 4.39a. The voltage gain of the final small-signal model circuit is calculated as:

$$\text{Voltage gain} = hfe = \frac{\Delta Vout}{\Delta Vin} = \frac{-(Gm \times \Delta Vin) \times RD}{\Delta Vin} = -Gm(RD)$$

The "−" voltage gain sign came from the fact that V+ was converted to ground while current continued to flow from ground towards RD. The exact same analysis technique in the bipolar common emitter amplifier applies to common source. The voltage drop across RD is below ground by **− (Gm X ΔVin) X RD**. The result of this small-signal model indicates that the voltage gain is controlled largely by Gm and the RD size. The higher the Gm and RD, the more voltage gain you could achieve. For example, a common source amplifier's **ΔID = 1 mA, ΔVGS = 1 V, RD = 10 kΩ**. Voltage gain:

$$\text{Voltage gain} = \text{hfe} = -\text{Gm}(\text{RD}) = \frac{-(1\text{ mA})}{1\text{ V}} \times 10\text{ k}\Omega = -10$$

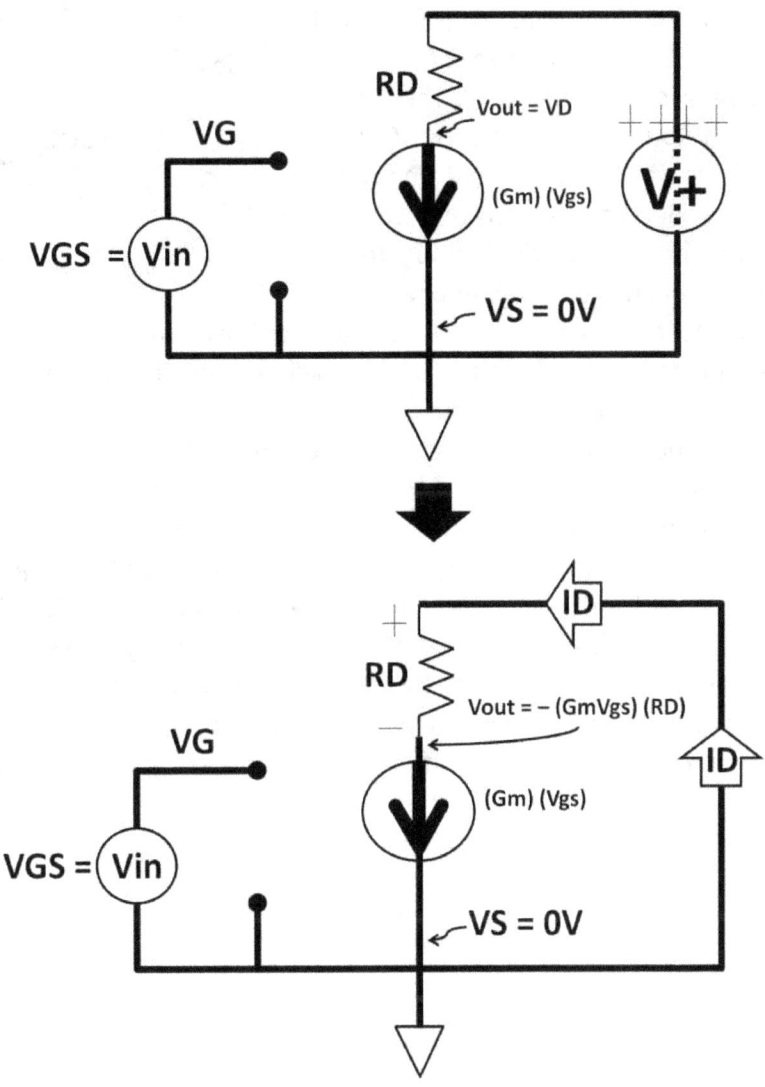

Figure 4.39a: Common source amplifier small-signal model

MOSFET Parasitic

Small-signal analysis applies to any amplifier design. It simplifies design tasks and gives first-order confidence the design works to your expectations. However, many real world circuits process AC signals. This AC requirement complicates transistor behavior. The NFET model examines what it means in figure 4.40. Parasitic capacitances are distributed around both bipolar and CMOS transistors. The NFET example includes capacitors from gate-to-drain capacitor (CGD), gate-to-source (CGS), drain-to-substrate (CDSub), drain-to-source (CDS), and source-to-substrate (CSSub). The capacitances of these parasitic caps are relatively low. They have little effects in DC. If you recall chapter 3, AC, **Xc = 1/ 2 π f C**, Xc is infinite at DC. If the signal frequency is high, Xc starts to decrease generating current paths via the capacitors. This Xc has profound effects on transistor functions including changes in

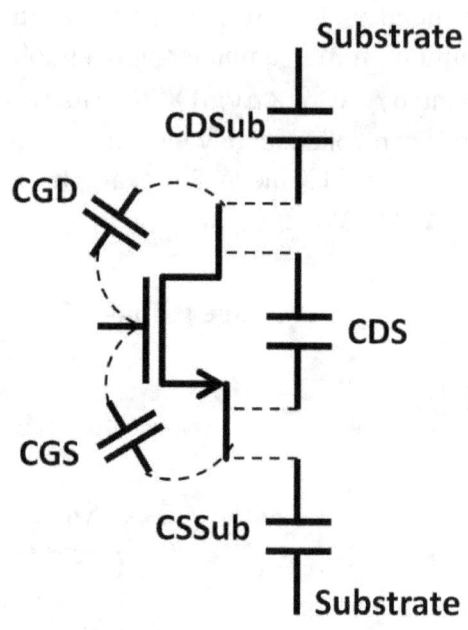

Figure 4.40: NFET model

gain, leakage current, input, and output impedances. This somewhat explains why analog design is a challenging task in addition to numerous changing parameters, from power supply values, temperature fluctuation, frequency ranges, or bandwidth. In most cases, a transistor datasheet only list specifications as a range of numbers at some pre-defined conditions. These capacitances are strong function of VGS, VDS, temperature, and switching frequency. MOSFETs datasheets sometimes list them as input capacitance (CISS). They can be in the order of 100s of pico farads (pF). For high speed applications, gate charge is an important parameter that describes how much charge (Q unit in coulomb) the MOSFET needs to switch at certain conditions. These parameters become significant at high speed, which could slow down the overall speed. Gate charges are heavily depending on MOSFET's threshold voltage (VT) as well as the type of load connected to it. To select the right MOSFET for an application, engineers need to understand the trade-off among parameters and performance.

Common Drain Amplifier (Source Follower)

Let's now analyze the second CMOS amplifier type: the common drain amplifier. Its input is at the gate. Output is at the source. The connections are similar to the common collector amplifier (emitter follower). This is why the common drain amplifier is called the source follower. Figure 4.41 shows two source followers: NFET and PFET types. Figure 4.42 shows the input and output waveform. Both are in phase with Vout slightly less than the input.

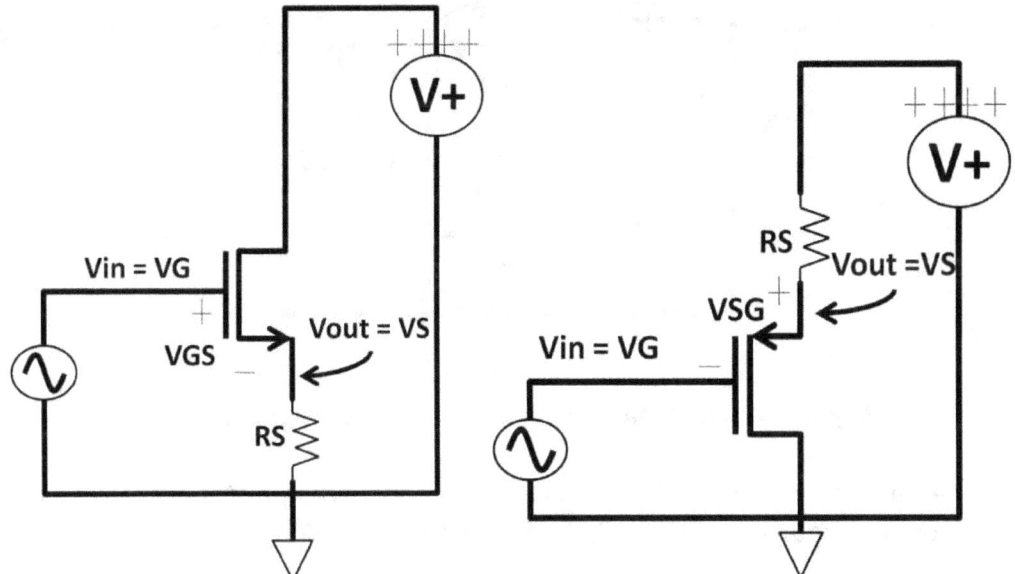

Figure 4.41: Common drain (source follower) amplifier

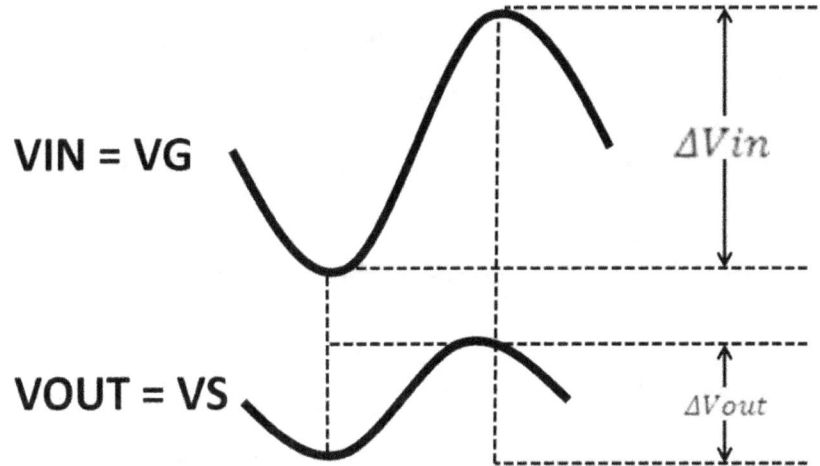

Figure 4.42: Source follower input, output

Using a NFET-based source follower, when VG is at ground **(VGS < VT)** and NFET is off, **ID = IS = 0 A**. No voltage drops across the RS. VD is measured at positive voltage supply.

Same small-signal model technique used to analyze common source amplifier can be used on common drain amplifier. Figure 4.41 was transformed to small-signal model in figure 4.43.

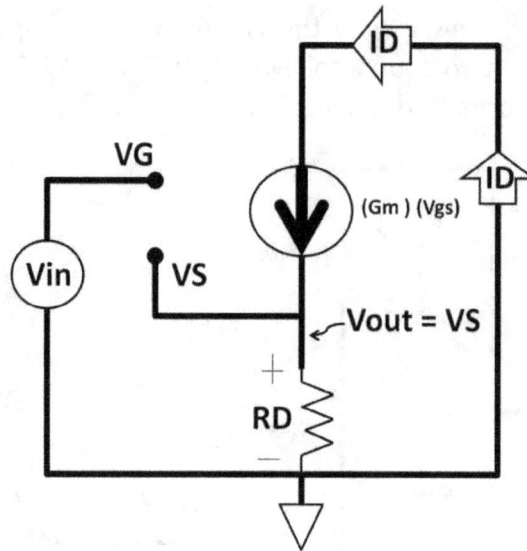

Figure 4.43: Common drain amplifier small-signal model

The voltage at the gate, VG, is Vin. It is now facing an open circuit by the gate-oxide capacitor. Similar to a common source amplifier, Gm in this circuit is:

$$Gm = \frac{\Delta IOUT}{\Delta VIN} = \frac{\Delta ID}{\Delta VGS}$$

By multiplying Gm by VGS, which is the input, we're left with output current, ID, represented by the current source.

$$Gm \times \Delta VGS = Gm \times \Delta Vin = \Delta ID = \text{Output Current}$$

The drain ties to V+, which is converted to a short circuit shown on the far right.
Vin = VGS + Vout. Vout = VS = (Output current X RD) = (Gm X VGS) X RD

$$\text{Voltage gain} = hfe = \frac{\Delta Vout}{\Delta Vin} = \frac{(Gm \times VGS \times RD)}{VGS + (Gm \times VGS \times RD)}$$

$$\text{Voltage gain} = hfe = \frac{\Delta Vout}{\Delta Vin} = \frac{(Gm \times \cancel{VGS} \times RD)}{\cancel{VGS}(1 + Gm \times RD)}$$

$$\text{Voltage gain} = hfe = \frac{\Delta Vout}{\Delta Vin} = \frac{(Gm \times RD)}{(1 + Gm \times RD)} \approx 1$$

The voltage gain of the source follower is slightly less than 1 (Negative dB). For example, a common drain amplifier's ΔID = 1 mA, ΔVGS = 1 V, RD = 10 kΩ. Gm = 1 m / 1 = 1 m. The voltage gain is thereby:

$$\frac{(Gm \times RD)}{(1 + Gm \times RD)} = \frac{(1\,m \times 10\,k\Omega)}{(1 + 1\,m \times 10\,k\Omega)} = 0.909$$

To calculate gain in dB:

$$\frac{Vout}{Vin} \text{ in dB} = 20 \log\left(\frac{Vout}{Vin}\right)$$

= 20 log (0.909) = – 0.83 dB

Common Gate Amplifier

The third and last single-ended CMOS amplifier is the common gate amplifier. Figure 4.44 shows NFET- and PFET-based common gate amplifiers. The NPN-based common gate amplifier input is at the source, the output at the drain, and the gate ties to a DC voltage source. The common gate amplifier provides high gain without phase shift, as shown in figure 4.45.

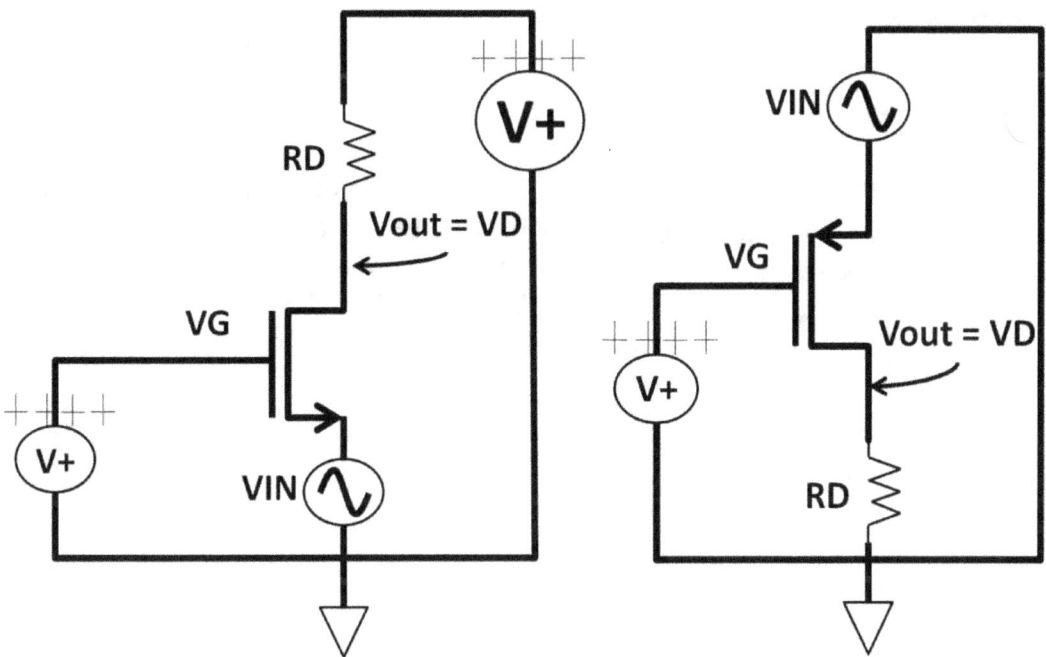

Figure 4.44: Common gate amplifiers (N/PFETs)

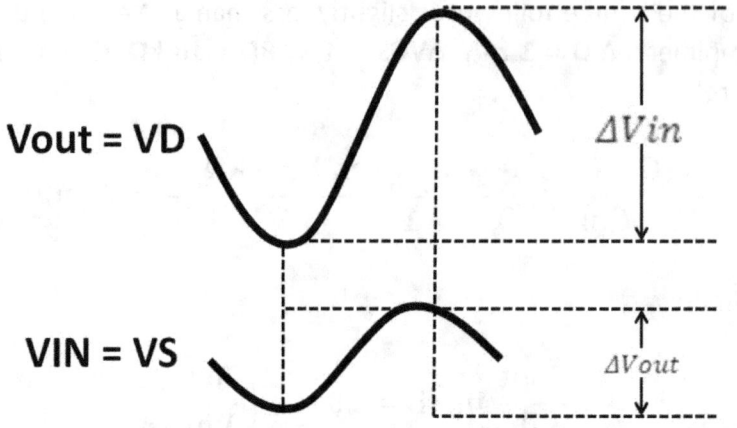

Figure 4.45: Common gate amplifier, NFET, PFET, waveform

The common gate amplifier gain can be realized by small-signal model in figure 4.46.

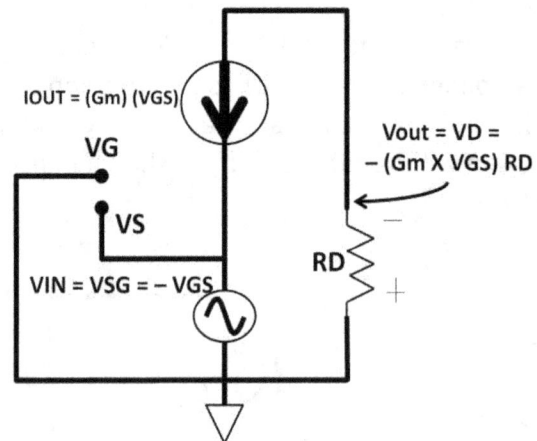

Figure 4.46: Common gate amplifier small-signal model

$$VIN = VS - VG = VSG = -VGS$$

$$Vout = VD = -(Gm \times VGS)\, RD$$

$$\text{Voltage gain} = hfe = \frac{\Delta Vout}{\Delta Vin}$$

$$hfe = \frac{-(Gm \times VGS \times RD)}{-VGS} = Gm \times RD$$

From the gain equation, you can see that gain is positive indicating there isn't any phase shift. The amount of gain is mainly controlled by Gm and RD.

Bipolar versus CMOS

Among the six different types of amplifiers, there are trade-offs among them when it comes to choosing one that meets your design target. Table 4-5 details the differences, pros, and cons between bipolar and CMOS transistors. The differences between MOSFETs and bipolar transistors create interesting dynamics when it comes to designing and analyzing electronic circuits. Understanding the trade-offs saves you time, gets your final system within specs.

	Bipolar	CMOS
Transconductance (Gm)	Higher	Lower
Base or gate currents	Higher	Zero gate current at DC
Design complexity	More complex	Simpler
Size (Same Gm)	Smaller	Larger
VBE or VT mismatch	Better	Worse
Input impedance	High	Infinite

Table 4-5: Bipolar and CMOS electrical performance

Differential Amplifiers

So far, we covered single-ended amplifiers. Differential amplifiers (diff amp) are well-suited to many applications. They are in many cases superior to singled-ended design. This section describes what diff amps are and the motivation behind using them. Look back at single-ended amplifiers. The inputs and outputs are referenced from positive to ground or the most negative rail voltages. We know by now there is no such thing as a perfect voltage source or ground. A DC power supply, even ground, has noise riding over it (see figure 4.47). These imperfect qualities lead to inaccuracies. These noises become more

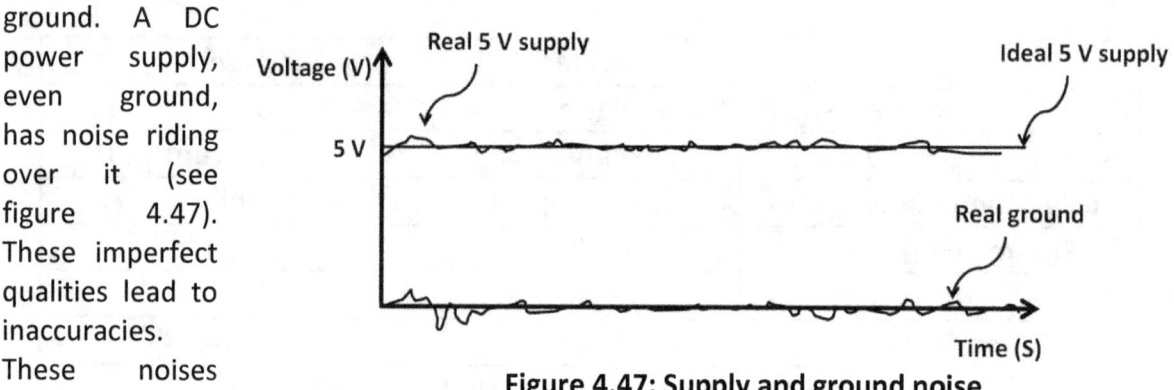

Figure 4.47: Supply and ground noise

apparent in microelectronics when transistors are measured in the sub-micron level running in extremely low voltages. Any substantial noise could falsely trigger a transistor to give wrong results. One solution to this problem is to not use ground or the negative rail as a reference to the input but rather to use two (differential) inputs instead. Figure 4.48 consists of two NFETs (Q1, Q2) and two drain resistors. The input voltage is no longer a single-ended input reference to ground. It consists of two inputs, V1 and V2. More precisely, the inputs are the difference between V1 and V2 **(V1 − V2 = Vdiff)**. This topology eliminates ground as voltage reference.

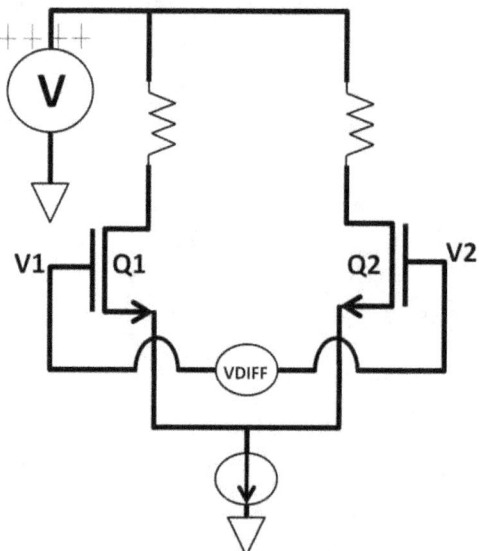

Figure 4.48: Differential amplifier

Common Mode

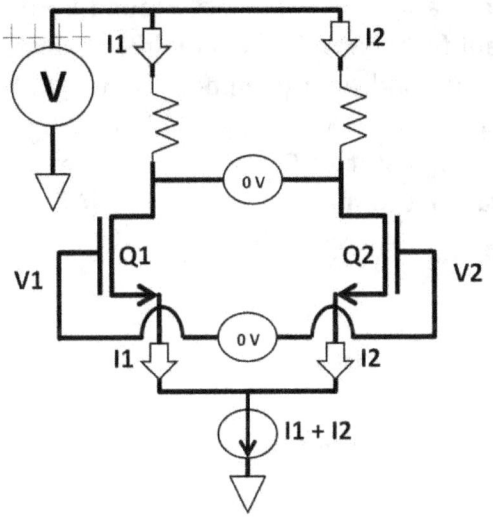

Figure 4.49a: Current split in differential amplifier

In terms of diff amp voltage gain, we first need to introduce common mode. **Common mode voltage = (V1 + V2) / 2**. When both V1 and V2 are the same (see figure 4.49a), there is no voltage difference between the two inputs. Currents (I1, I2) are split equally between two drain resistors (KCL in chapter 1, DC). The outputs are at the drain between Q1 and Q2. They are exactly the same, i.e., the voltage output difference is zero. The amplifier now has zero differential voltage gain. When both inputs are the same, diff amp exhibits zero common mode gain. The zero common mode gain is yet another advantage over single-ended topology. Any noise appearing at both inputs, if they are equal in value, would not get amplified showing up at the output. However, even with diff amp, there would be some small common mode gain. This gain is caused by the mismatch between resistor and transistor sizes even if their designed values are the same. What it means is that during the manufacturing of electronic components, no two physical electronic components can be manufactured identically to each other due to process technology limitations (gradients). For example, for a resistor with 10 kΩ in value, there could be plus or minus percentage (e.g., 5%) difference between the actual resistors and the resistors' design value. In other words, there could be as much as 500 Ω difference between the two physical 10 kΩ resistors. This percentage difference varies from process to process. A process spec sheet would tell you such information. Besides the resistors, transistors have same inaccuracies where no two transistors are the same even if they are identical in the schematic. Their width, length, and VT may be slightly different. These non-ideal device characteristics lead to uneven current between I1 and I2, even when V1 and V2 are the same. These differences in current results in voltage difference at the output, hence common mode gain. The current (I1 + I2) combined at sources is called tail current (tail-like shape of NMOS). These currents are sunk by the current source. Numerous academic textbooks use a resistor at the source to produce the current. This is impractical because the current generated by the voltage drop across the resistor is constantly changing when V1 and V2 are AC signals. This leads to differences in I1, I2, and drain voltages. The voltage gain will be changing constantly, unacceptable for stable gain operations. Most real world diff amps use current source to supply constant current to the circuit (see figure 4.49b).

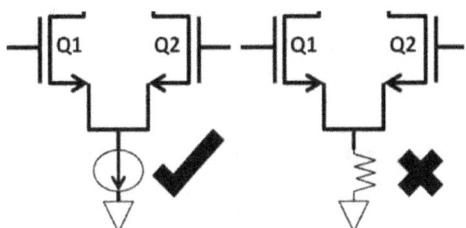

Figure 4.49b: Tail current source vs. resistor

CMRR and Differential Gain

To quantify this small gain, the common mode rejection ratio (CMRR) is easily found in the differential amplifier datasheet. It's a means to quantify how well the amplifier rejects common mode signal. Using Adm (differential mode gain) and Acm (common mode gain), **CMRR = Adm / Acm**. In an ideal case, **Acm = 0** and **CMRR = infinite**. In reality, CMRR would be large although not infinite due to small Acm in the denominator of CMRR. The differential amplifier gain thereby is equal to **(differential mode gain + common mode gain)** or **(Adm + Acm)**. The differential mode gain is simply a measure of voltage change (difference) at the output versus input:

$$\text{Adm} = \frac{\text{Vout\_diff}}{V1 - V2} = \frac{\text{Vout\_diff}}{\text{Vin\_diff}} = \frac{\Delta \text{Vout}}{\Delta \text{Vin}}$$

We could use figure 4.50 to show the meaning of voltage difference at the output. If **V1 > V2**, current is steered towards Q1 so that **I1 > I2** (larger I1 arrow). This difference in current generates a voltage difference (Vout_diff) between Q1, Q2's drain voltages **(VD_Q1 < VD_Q2)**. To find out the exact differential amplifier gain, we could use the same small-signal model technique in the previous section and apply a half-circuit analysis. A half-circuit takes the half of the differential circuit where the common mode voltage input is zero:

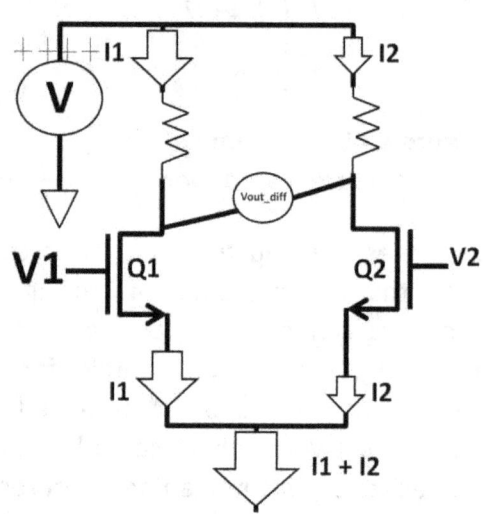

Figure 4.50: Voltage output difference

$$\frac{V1 + V2}{2} = 0$$

$$V1 = -V2$$

If both inputs are the same, **V1 = − V1**, for differential voltage input:

$$\text{Vin\_diff} = V1 - (-V1) = 2 \times (V1)$$

We split the differential circuit in half. V1 would be equal to Vin_diff divided by:

$$V1 = \frac{\text{Vin\_diff}}{2}$$

Using the small-signal model, output voltage at Q1:

$$\text{Vout\_1} = -\text{Gm (V1)} \times \text{RD1}$$

$$\text{Vout\_1} = -\frac{(Gm)(Vin\_diff)(RD1)}{2}$$

Go through the same steps on the right-hand side for Q2:

$$\text{Vout\_2} = -\frac{(Gm)(Vin\_diff)(RD2)}{2}$$

The final differential amplifier voltage gain, Adm or Av(diff):

$$\text{Av(diff)} = \frac{\text{Vout\_1} - \text{Vout\_2}}{\text{Vin\_diff}}$$

$$\text{Av(diff)} = -\frac{\frac{Gm\,(Vin\_diff)\,(RD1)}{2} - \left(-\frac{Gm\,(Vin\_diff)\,(RD2)}{2}\right)}{Vdiff\_in}$$

$$= \frac{-Gm(RD1 + RD2)}{2}$$

The differential gain is a function of **Gm X (RD1 + RD2)** where **(RD1 + RD2)** is the output impedance. To visualize how diff amp works, we use Vout_diff vs. Vin_diff DC graph in figure 4.51a. When **V1 = V2** (Vin_diff = 1), Vout_diff = 0 (no voltage gain). When V1 > V2, Vin_diff is positive (right-hand side of the graph). V1 >> V2 indicates Q1 VDS approaches 0 V (cut off). The most negative the curve can go is **– V++**. When V1 << V2, Vin_diff is negative (left-hand side of the graph). If V1 << V2, Q2 VDS approaches 0 V (cut-off), and the highest the curve can go is V++. The resistors in the diff app are called passive load resistors. In practical diff amp circuits, load resistors are seldom used. With a transistor's high drain impedance, transistors are used as active loads to achieve high gain in diff amps. An example is shown in figure 4.51b. This differential amplifier topology is the basic building block of the operational amplifier. The two PFETS are constructed as a current mirror with high drain impedance. Current mirror is discussed in the next section.

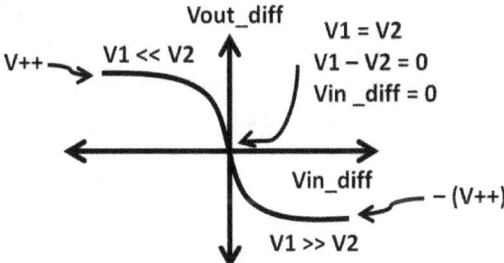

Figure 4.51a: Vout_diff vs. Vin_diff

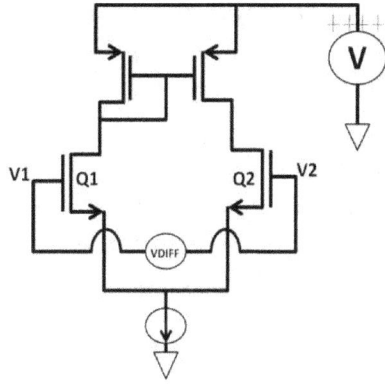

Figure 4.51b: Active load

Current Mirror

Let's examine the current mirror (see figure 4.52a). They are widely used in microelectronic design to generate current references. The purpose of the current mirror is to first generate a reference current (IREF), then replicate it to create multiple copies. The copies of the current can be manipulated by varying transistor sizes. The current copies can then be used in other places throughout the design. Its simplicity makes a powerful circuit to easily generate any number of currents from only few components. Figure 4.52a is a current mirror example made up of a 10 kΩ resistor and two NPNs (Q1, Q2). This circuit is only practical when implemented in an IC design; discrete components exhibit too many parameter mismatches making it difficult to achieve descent accuracy. The Q1 collector is shorted to Q1 and Q2's bases, i.e., **VC = VB**. This forces both Q1 and Q2's VBEs to be equal. If you recall the collector current equation:

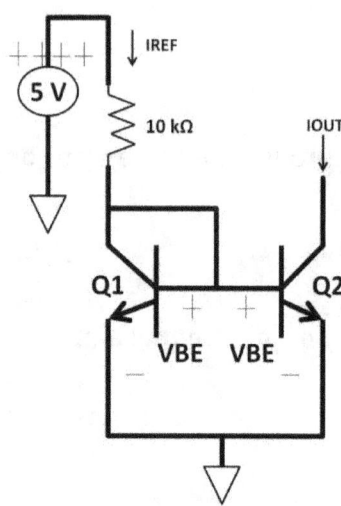

Figure 4.52a: Current mirror

$$IC = A \times IS \times e^{(VBE / VT)}$$

Both collector currents are identical if Q1, Q2 area, VBE, and temperature are the same. The current reference (IREF) is found by KVL on the 10 kΩ resistor:

$$IREF = \frac{(5-1)\,V}{10\,k\Omega} = 0.4\,mA = IOUT$$

The PMOS current mirror used in the differential amplifier as active load works similarly. Figure 4.52b shows an example. For 1 V VGS, ID assumes to be 0.4 mA (process dependant). The reference current:

$$IREF = \frac{(5-1)\,V}{10\,k\Omega} = 0.4\,mA$$

VGSs for two PFETs on the right have are the same (gates and sources are tied together). Because their sizes are 2X and 3X larger than the reference PFET:

        IOUT1 = 2 X 0.4 mA = 0.8 mA
        IOUT2 = 3 X 0.4 mA = 1.2 mA

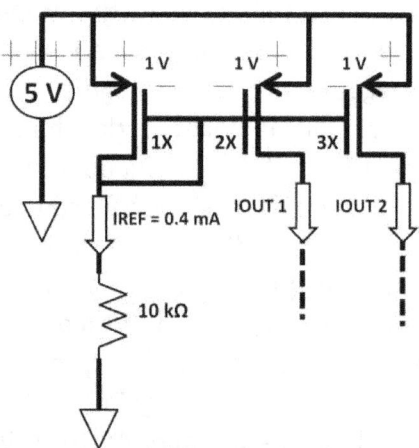

Figure 4.52b: PMOS current mirror

PFET's drain is high impedance. As discussed in the prior section, this is advantageous in a differential amplifier where current mirror is frequently used as an active load to provide higher voltage gain.

Op-Amp

Operational amplifier (op-amp) inputs are differential; output is typically single-ended although differential outputs are fairly common. Differential amplifiers in the previous section are great choices for op-amp input. All amplifiers discussed so far are open-loop where the amplifier output is not connected back to the input (feedback). An op-amp connected in open-loop offers extremely high gain. To achieve such impressive gain, multi-staged amplifiers are needed. Total system multi-stage amplifiers' gain is determined by the multiplication of individual gains. Assume individual stage gains are A1, A2, and A3, **total op-amp gain = A1 X A2 X A3.** An op-amp with reasonable bandwidth could have open-loop gain of 90 dB to 100 dB.

$$100 \text{ dB} = 20 \log \left(\frac{\text{Vout}}{\text{Vin}}\right)$$

$$\frac{\text{Vout}}{\text{Vin}} = 100,000$$

This means if Vin changes by 1 V, theoretically, Vout changes by 100,000 V. 100,000 V is not a practical number to use or list in the datasheet. For this reason, open-loop gain is written as V / mV (1 V / 0.001V). From the above example, the open-loop gain is:

$$\frac{100 \text{ V}}{1 \text{ mV}} = \frac{100 \text{ V}}{0.001 \text{ V}} = \frac{100,000 \text{ V}}{1 \text{ V}}$$

To achieve reasonable, manageable gain, feedback or closed-loop configuration is necessary. As for unity-gain bandwidth, it's the frequency where the op-amp drops to a gain of 1 (0 dB):

$$\text{Vout} = \text{Vin}$$

$$\frac{\text{Vout}}{\text{Vin}} = 1$$

$$20 \log 1 = 0 \text{ dB}$$

Recall the low-pass filter in chapter 3, AC. We could use a bode plot to explain op-amp gain vs. frequency (see figure 4.53). Ideally, the higher the unity-gain bandwidth, the better the op-amp will be. Transistors are active devices. An op-amp is made primarily of transistors. At low frequency (DC), gains in dB would be much more than 0 dB as opposed to a low-pass filter, where the highest

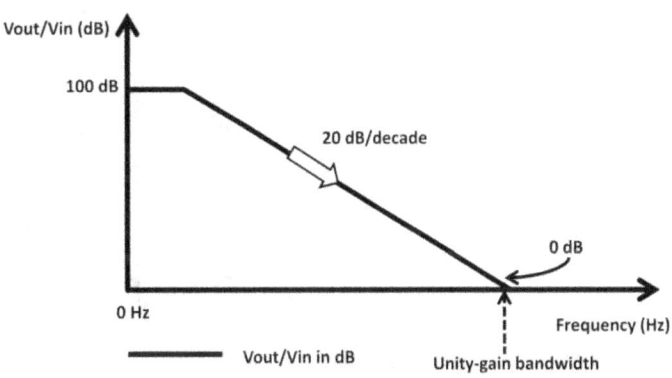

Figure 4.53: Op-amp gain vs. frequency

filter "gain" is 0 dB because filters do not have gain; they have attenuation instead. The low-pass filter attenuation and op-amp are both – 20 dB/decade (– 20 dB per decade) with increasing frequency. A closed-loop op-amp configuration means that the output is feeding back to the input providing lower but controllable gain. You may wonder why anyone would want to design an amplifier with lower gain. The reason lies in analog signal processing. Many analog applications need precise and controllable gains. This is where closed-loop and feedback network come in. We will discuss them shortly. The op-amp has its own schematic symbol (see figure 4.54).

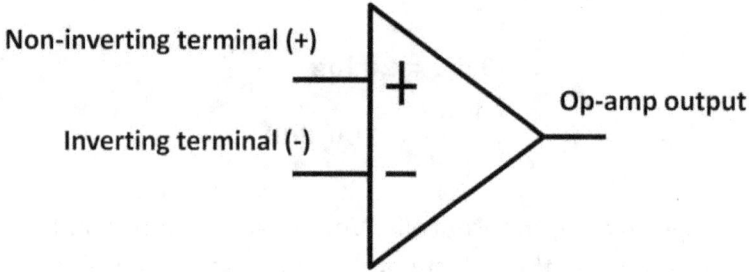

Figure 4.54: Op-amp schematic symbol

The triangular symbol includes two input terminals, denoted by "+" (non-inverting terminal) or "–" (inverting terminal). Singled-ended output is located on the right. The differential amplifier discussed in the previous section is a good example being used as the op-amp's input stage. Figure 4.55 shows an op-amp input stage example using PFETs Q1 and Q2 as the input stage, Q3, and Q4 (active loads). The NMOS in this circuit are the active loads.

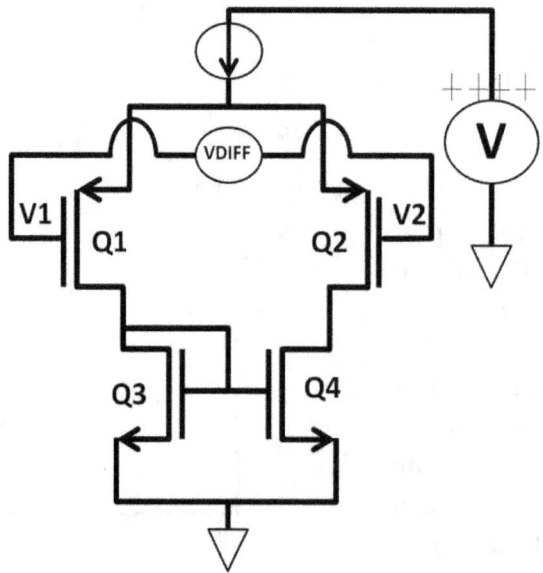

Figure 4.55: PFET input stage different amplifier

Op-Amp Rules

These are the rules associate with op-amp worth noting:
1) Input impedance: infinite
2) Output impedance: zero
3) Input offset voltage: zero

Let's take a complete op-amp circuit as an example to make these rules clear in figure 4.56.

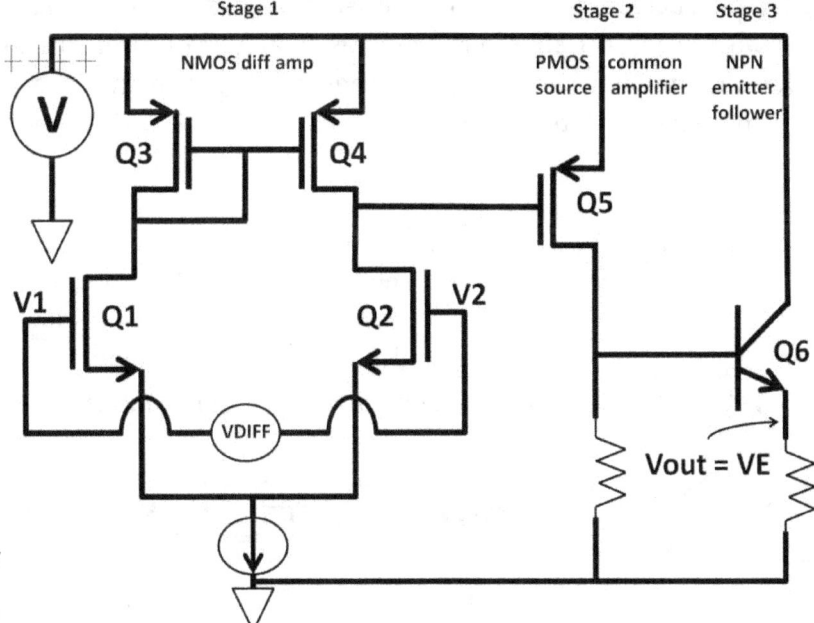

Figure 4.56: Multi-stage amplifiers

Q1, Q2, Q3, Q4, and the current source make up stage 1 of the op-amp. Stage 1 is a CMOS-based amplifier with V1 and V2 as inputs at Q1 and Q2's gates. This makes the input impedance infinite at DC from rule 1. Q3 and Q4 form a PMOS current mirror. The output is located at the Q4 drain (PFET) and Q2 (NFET) causing a 180-degree phase shift. This output is taken to Q5's gate. Its drain resistor forms stage 2: A PMOS common source amplifier with the output located at Q5 drain. Stage 2 provides a 180-degree phase shift netting a zero-degree shift from stage 1. Finally, Q6 is the final stage, 3: The emitter follower offers low output impedance (ideally zero Ω from rule 2) and with a zero-degree phase shift. In reality, we know that the emitter has low, finite impedance. The majority of the op-amp gain comes from stages 1 and 2 (both common source amplifiers) with slight trade-off from stage 3's small emitter follower's voltage loss. The zero offset voltage means when V1 and V2 are the same, both Q1 and Q2 gate voltages are identical. In reality, this is not true because no two transistors could be manufactured exactly the same in sizes resulting in different threshold voltages (VT), i.e., different Q1 and Q2 gate voltages. When **V1 >> V2**, Q1 is enhanced and current steers to the left-hand side of stage 1. Q3 VGS is established equaling Q4's VGS. With this enhanced Q4, however, no current is able to flow downward from Q4 because Q2 is off (open circuit). This lifts Q4 drain up towards the positive supply. This high output collapses Q5 VGS shutting it off. No current flows to the drain resistor yanking Q6 (NPN) base to ground. VBE is zero voltage so the output is low. In summary, when V1 is high, V2 is low, and Vout is low. Now let's consider when **V2 >> V1**. Q1 is now off, and Q3 then cuts off. Without adequate VGS of Q3, Q4 is off. Even V2 is high enhancing Q2; there isn't any current flowing through it. This pulls Q2's drain down which in turn causes Q5 VGS larger than VT turning it on. If VGS is known with a drain resistor size designed properly to be higher than VBE voltage plus the I R drop (voltage across) of the emitter resistor, Vout is lifted up. Table 4-6 shows the op-amp operation with terminal definitions. One easy way to interpret this summary is that an open-loop inverting amplifier simply means when the input of the inverting terminal is higher than of the non-inverting terminal, the output goes low and vice versa (see figure 4.57a).

	Inverting amplifier (-)	Non-inverting amplifier (+)	Comment
V1	Hi	Low	Inverting terminal (-)
V2	Low	Hi	Non-inverting terminal (+)
Vout	Low	Hi	Output

Table 4-6: Op-amp operation with terminal definitions

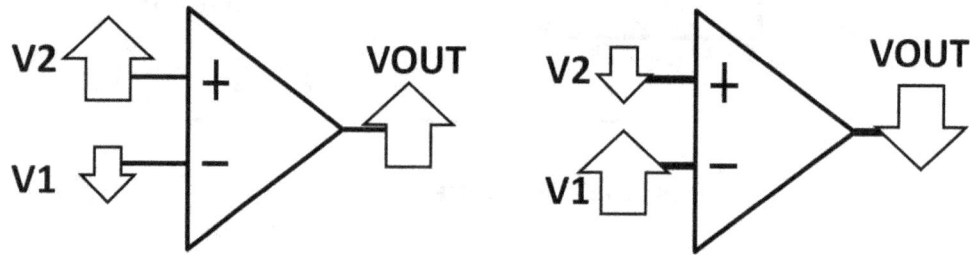

Figure 4.57a: Inverting, Non-inverting, output relationship

From the buck regulator in chapter 3, AC, the op-amp was used as a comparator connected in open-loop. In that example, the non-inverting terminal is fixed with a voltage source. If the non-inverting terminal is higher than the voltage source at the negative terminal, the output will lift up to the rail. This can be realized by the high open-loop gain. Suppose the op-amp has 100 dB open-loop gain with positive rail voltage at 5 V to ground.

$$100 \text{ dB} = 20 \log \left(\frac{\text{Vout}}{\text{Vin}}\right)$$

$$5 = \log \left(\frac{\text{Vout}}{\text{Vin}}\right)$$

$$\frac{\text{Vout}}{\text{Vin}} = 100,000$$

This means that it only takes 50 uV difference between the positive and negative terminals to flip the output to either 5 V or ground. This is a comparator circuit comparing the voltage difference between two terminals. The outputs are either going up to the positive rail or ground. Open-loop is perfect in comparator topology because of its high gain. This high gain directly relates to high slew rate, which specifies how fast voltage changes over time, i.e., **ΔV / ΔT**. For example, a step response at the input produces an output with slew rate at **10 V / 1 us** (see figure 4.57b). This means the device would be able to change 10 V at the output in one microsecond (us). A high slew rate is desirable because it reduces the time for the input and output to reach to its intended levels. Settling time is another important op-amp spec that is commonly found in a datasheet. In figure 4.57c, settling time specifies how quickly the op-amp responds to an input and output settles in 2 us when the output value stays within the predefined error band.

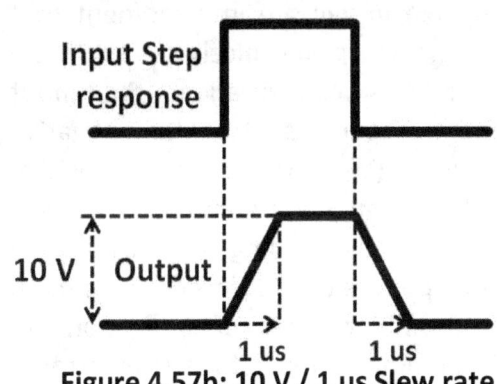

Figure 4.57b: 10 V / 1 us Slew rate

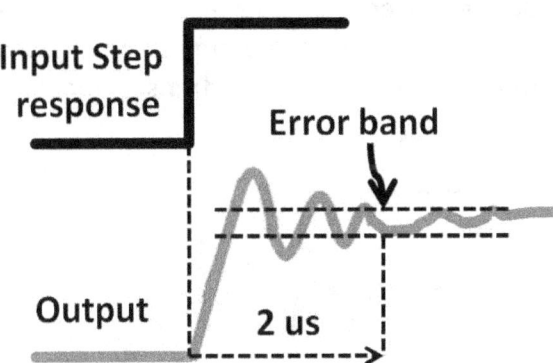

Figure 4.57c: 2 us settling time

Apart from voltage gain, many analog circuits require current gain. An op-amp used to drive a high-current motor is one example. It requires the op-amp output stage to provide sufficient current for the motor to turn. We will summarize all major op-amp parameters at the end of this section. As nice as high open-loop gain sounds, in many cases, we would like the gain to be lower and in a more controlled manner. In many cases, an audio amplifier only needs to amplify the input signals 3 to 5 times. If lower and controlled gain is required, closed-loop op-amp configuration can be implemented. There are two popular ways (inverting and non-inverting amplifiers) to connect an op-amp in closed-loop. Both are covered in the next few sections and can be easily explained by Ohm's law, KVL, and KCL.

Inverting Amplifier

Inverting amplifier is first shown in figure 4.58. Our goal is to develop the Vout vs. Vin transfer function, i.e., closed-loop gain, with respect to the Rf and Ri. You will see in a moment that the amplifier gain is nicely set by Rf and Ri. In this configuration, the input voltage connects to the negative terminal (V–). There is a resistor feedback network, Rf and Ri. Part of the output is now feeding back to the op-amp input creating a closed-loop circuit. For closed-loop op-amp connections, if the op-amp output stage (transistors)

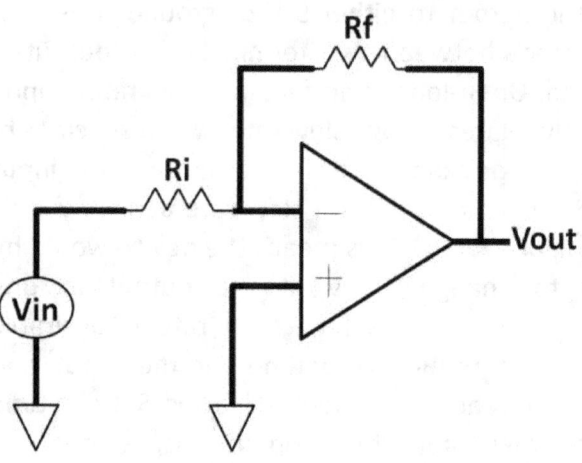

Figure 4.58: Inverting amplifier

isn't driven in saturation (out of range), Vout would do whatever it takes to force the difference between positive and negative terminals to be zero, **V+ – (V–) = 0**. This rule means that V– has the same voltage as the V+ at 0 V. This ground potential is called "virtual ground." Once we have obtained the virtual ground connection, we can apply the infinite input impedance op-amp rule. A transformed circuit inside the dotted line is developed (see figure 4.59). The infinite input impedance prevents any current going into the op-amp turning it into an open circuit. This op-amp literally can be taken out of the picture for the transfer function gain analysis.

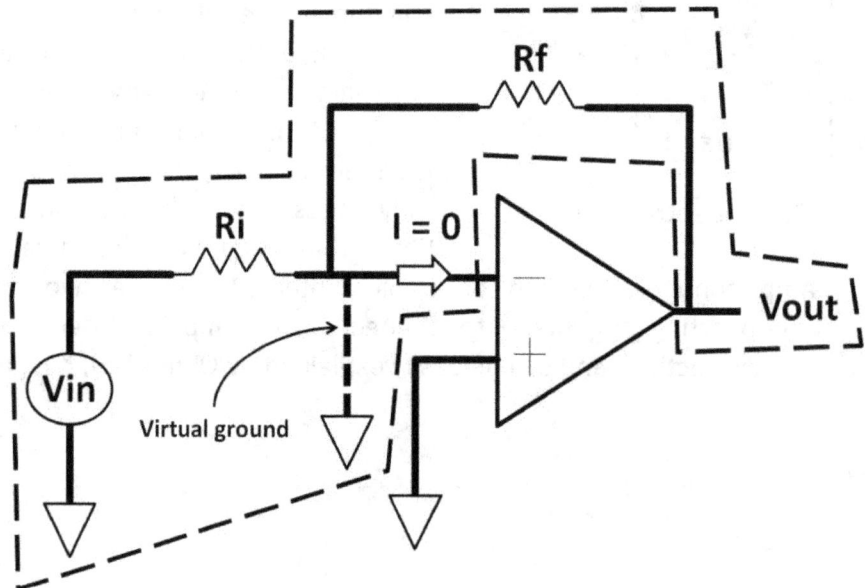

Figure 4.59: Transformed op-amp

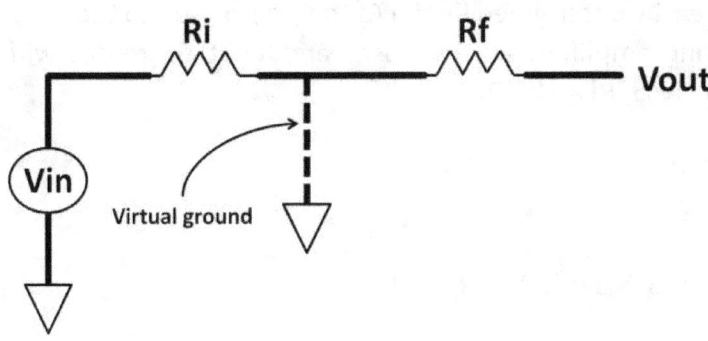

Figure 4.60: Simplified op-amp circuit

The simplified version of the circuit is shown in figure 4.60. This circuit can further be realized as two individual circuits. One from Vin and Ri to ground, the other from Vout and Rf to ground (see figure 4.61). The current of both circuits go to virtual ground. These currents can't go to op-amp's infinite impedance negative terminal. The amount of current goes from Vin to ground is identical to current from Vout. The only difference is current direction as both currents are flowing towards each other resulting in a negative sign. The individual circuits are shown in figure 4.61.

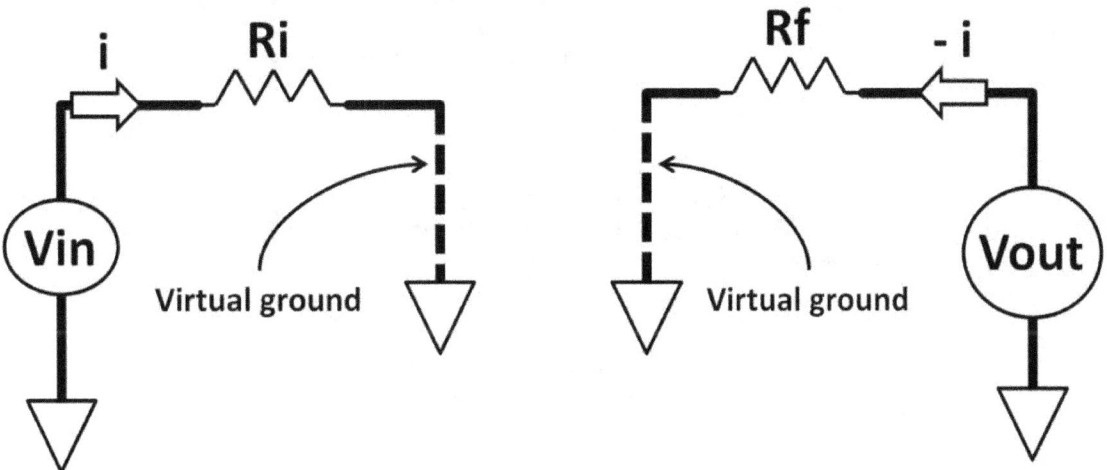

Figure 4.61: Individual op-amp circuits

Using Ohm's law and current, Vout / Vin equations are derived:

$$I = \frac{Vin}{Ri} = -\frac{Vout}{Rf}$$

$$\frac{Vout}{Vin} = -\frac{Rf}{Ri}$$

The voltage gain, Vout / Vin, is now easily determined by the Rf to Ri ratio. Due to the "−" (Vout / Vin) sign, this is an inverting amplifier. Vin increases and Vout decreases with a controllable gain. If the desired gain is − 5, **Ri = 10 kΩ**:

$$\frac{Vout}{Vin} = -5 = -\frac{Rf}{10\ k\Omega}$$

$$Rf = 5 \times 10\ k\Omega = 50\ k\Omega$$

Non-Inverting Amplifier

What if you want a positive gain at the output? One could add a common source or emitter amplifiers at the op-amp output to revert the phase, or could use a non-inverting amplifier (see figure 4.62).

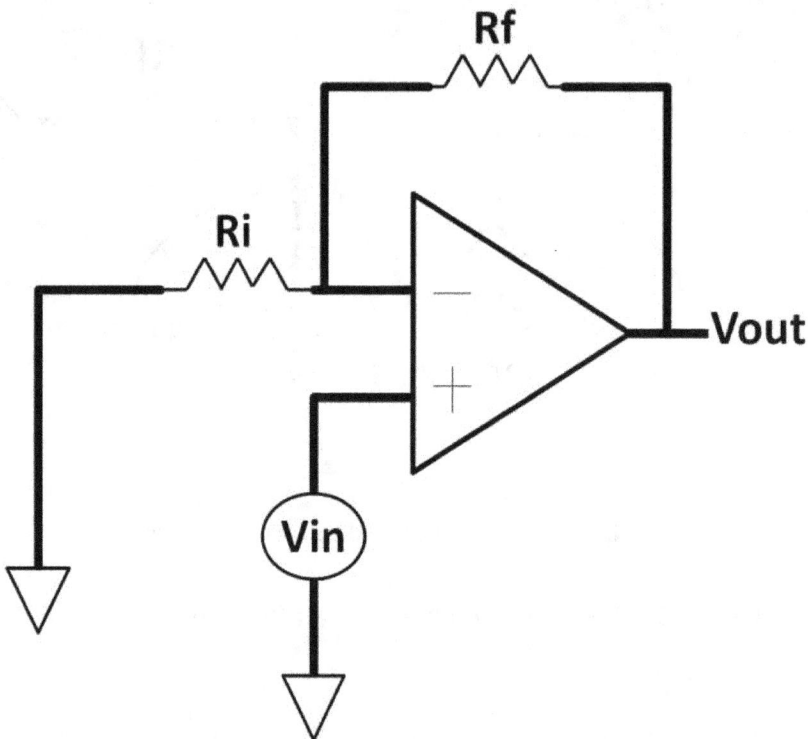

Figure 4.62: Non-inverting amplifier

The only difference between this amplifier and the inverting one is that Vin is connected to the positive terminal. The left side of Ri is connected to ground. The resistive feedback network, however, remains on the negative terminal side. The Vout / Vin gain transfer function is examined through the modified circuit (see figure 4.63).

The same rule is applied to the non-inverting amplifier. Both positive and negative terminals are at the same voltage potentials. This makes **V− = Vin**. This op-amp rule converts figure 4.63 to the final modified version (see figure 4.64). This circuit strikingly resembles a voltage divider. The voltage transfer function is thereby:

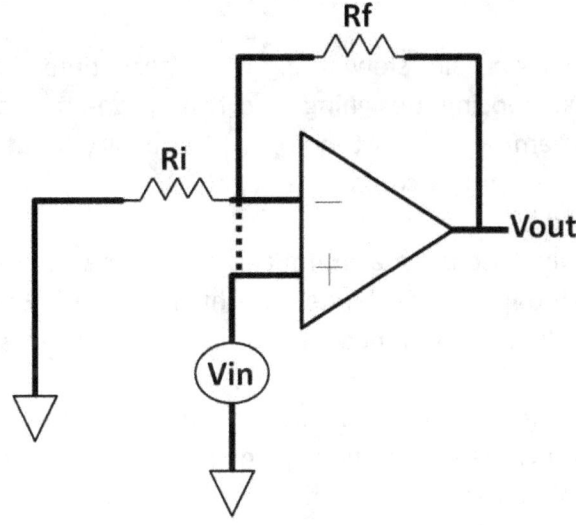

Figure 4.63: Modified non-inverting amplifier

$$\text{Vin} = (\text{Vout}) \times \frac{\text{Ri}}{\text{Rf} + \text{Ri}}$$

$$\frac{\text{Vout}}{\text{Vin}} = \frac{\text{Rf} + \text{Ri}}{\text{Ri}} = 1 + \frac{\text{Rf}}{\text{Ri}}$$

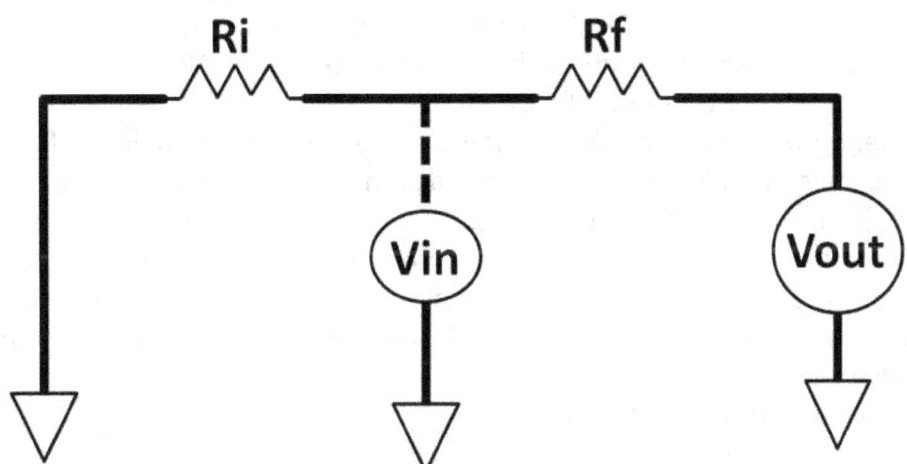

Figure 4.64: Final modified non-inverting amplifier

The closed-loop voltage gain of a non-inverting amplifier is conveniently set by the Rf to Ri ratio. The closed-loop gain is positive, hence the non-inverting amplifier name convention. If a gain of 10 is your design target, **Ri = 10 kΩ**:

$$\frac{\text{Vout}}{\text{Vin}} = 10 = 1 + \frac{\text{Rf}}{10\ \text{k}\Omega}$$

$$\text{Rf} = 9\ (10\ \text{k}\Omega) = 90\ \text{k}\Omega$$

Op-Amp Parameters

There are many op-amp parameters in addition to gain, slew rate, and settling time. Get familiar with these op-amp parameters makes choosing, designing, and testing op-amps an easier task. The other major significance of op-amp feedback topology is the ability to alter the op-amp input or output impedances. For the input, it could be a negative effect. Ideally, op-amp inputs are infinite, which is now lowered by the Ri and Rf.

-Supply and input voltage: Supply voltage defines absolute maximum and minimum values of power supply you can apply to the op-amp. Input voltage defines the highest and lowest voltage you can apply to the input terminals. Unless the op-amps are rail-to-rail, input voltage is less than supply voltage.

-Supply current: It tells you how much current the op-amp will be sourced from the op-amp power supply. When the op-amp is not driving any load or amplifying any signal, the op-amp still draws current to keep its operations. This current is specified as quiescent current. Quiescent current applies to any electronic device such as voltage regulators or controllers.

-Common mode rejection ratio (CMRR): It has the same definition as described in previous section. It tells how well the op-amp rejects the common mode signal from noise. The CMRR unit is in dB. The larger the dB, the better CMRR performance would be.

-Power supply rejection ratio (PSRR): It specifies how much output changes from power supply changes. It's measured in dB with transfer function as: **(ΔPower supply) / (ΔVout)**. PSRR is infinite in an ideal case **(ΔVout = 0)**.

-Output voltage swing: It defines the voltage range the output could go from the most positive to the most negative levels. The range depends on the load. With a smaller load (big load resistor, lower current), the output can go higher than a larger load.

-Output source and sink current: This is the maximum current op-amp could supply and receive. Figure 4.65 depicts what source and sink current mean. There are two electronic loads (circle symbols) in this circuit. The top load turns on when op-amp output goes low sinking current towards the op-amp. When the op-amp output goes high, the top load turns off, and the bottom load turns on. The op-amp is now sourcing current to the bottom load. The current amount capable of sourcing and sinking to and from the load is the op-amp source/sink current parameter.

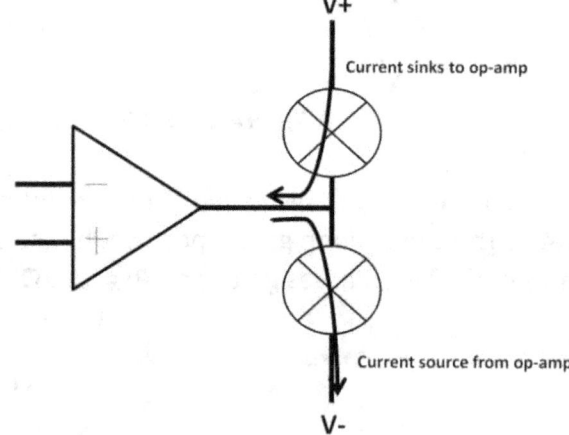

Figure 4.65: Op-amp source, sink current

-Input offset voltage: This is the voltage difference between the positive and negative terminals that is needed to bring ΔVout to zero. Ideally, input offset is zero, meaning when the difference of the two input terminals is the same, there is zero voltage output change.

-**Input offset current**: This current goes in or out of the op-amp's input terminals. An op-amp with a CMOS input stage doesn't have such spec. Only bipolar carries this spec due to base currents. For NPN, input bias current goes into the op-amp. For PNP, base current comes out of the input terminals. This current adds to the offset voltage. For this reason, minimal input bias current is desirable.

-**Power consumption:** The maximum power in watts that op-amp consumes. This relates to power supply voltage and supply current.

-**Input impedance:** This is the input impedance looking into the op-amp. For a CMOS input op-amp, input impedance is infinite. For a bipolar-based op-amp, its base's impedance is high but not infinite. 5 MΩ input impedance is typical.

-**Open-loop gain, bandwidth**: Some use large signal voltage gain to represent open-loop gain. Instead of dB, some datasheets translate dB to **V / mV** to describe open-loop gain. For example, **100 V / mV** is equivalent to 100 dB. **20 log (100 / 1 m) = 100 dB**. Open-loop gain can be realized in a bode plot in frequency response (see figure 4.66). The open-loop gain is much larger than controlled closed-loop gain.

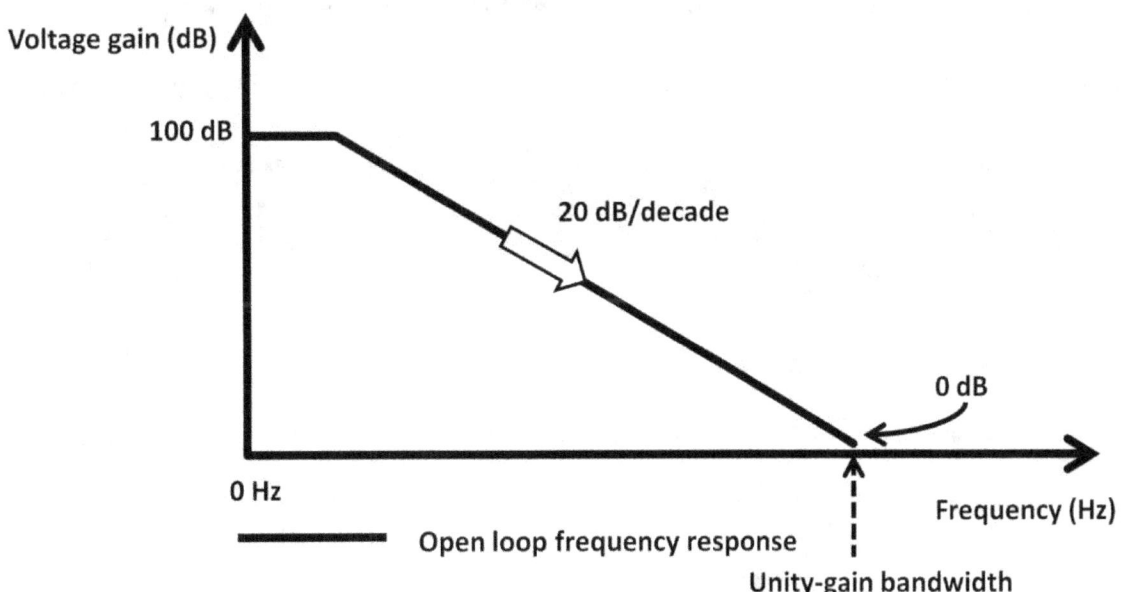

Figure 4.66: Op-amp frequency response

Be mindful that datasheets only list value ranges on a particular parameter from maximum to minimum. Most parameters are guaranteed only for a specific set of conditions, e.g., a specific temperature range (**− 55 °C ≤ T ≤ + 125 °C**) or supply voltage level.

LM741

Perhaps the most talked about op-amp in academics is the general purpose LM741 op-amp. It's an 8-pin bipolar transistor-based, differential input, single-ended output op-amp. Figure 4.67 shows LM741 in a metal can package; it also shows the pin names, and numbers. The diameter of the can is about 0.37 inch. Pins 1 and 5 are usually connected together with a 10 kΩ resistor to reduce the offset voltage.

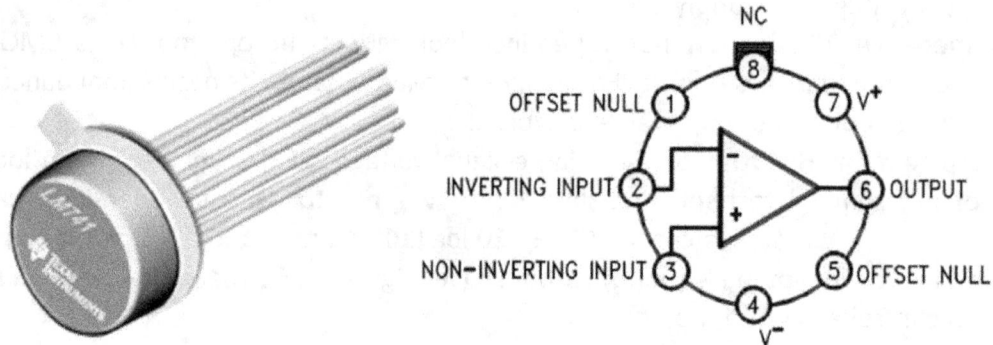

Figure 4.67: LM741 in a can package (left) and pin names, numbers (right)

Table 4-7 Texas Instruments part of LM741 datasheet. http://www.ti.com/product/lm741

Parameter	Conditions	Minimum	Typical	Maximum	Unit
Supply Voltage	TA = 25 °C	-22		22	V
Input Voltage	TA = 25 °C	-15		15	V
Input Offset Voltage	TA = 25°C, RS ≤ 50 Ω		0.8	3.0	mV
Input Bias Current	TA = 25 °C		30	80	nA
Input Resistance	TA = 25 °C, VS = ±20 V	1.0	6.0		V
Large Signal Voltage Gain	TA = 25 °C, RL ≥ 2 kΩ	50			mV/V
Output Voltage Swing	RL ≥ 2 kΩ	±15			V
Current	TAMIN ≤ TA ≤ TAMAX	10		40	mA
Common Mode Rejection Ratio	TAMIN ≤ TA ≤ TAMAX	80	95		dB
Slew Rate	TA = 25 °C	0.3	0.7		V/us
Bandwidth	TA = 25 °C	0.437	1.5		MHz
Power Consumption	TA = 25 °C, VS = ±20 V		80	150	mW

Table 4-7: LM741 datasheet (Partial)

Note: TA (Operating temperature), RS (Source impedance), VS (Source voltage), and RL (Load impedance)

Current Mirror Inaccuracies

Use the current mirror concept developed earlier (see figure 4.52a). The design goal is to create IOUT, an exact copy of IREF if both Q1 and Q2 sizes are the same. By just three components, a reference current and copies of currents are created. Changing Q2's size easily creates multiple current amplitudes. For example, doubling the Q2 size from Q1 makes IOUT twice as much as IREF. This circuit does have flaws. The IOUT would not be exactly equal to IREF. The main errors come from the base current and the size mismatch between the two NPNs.

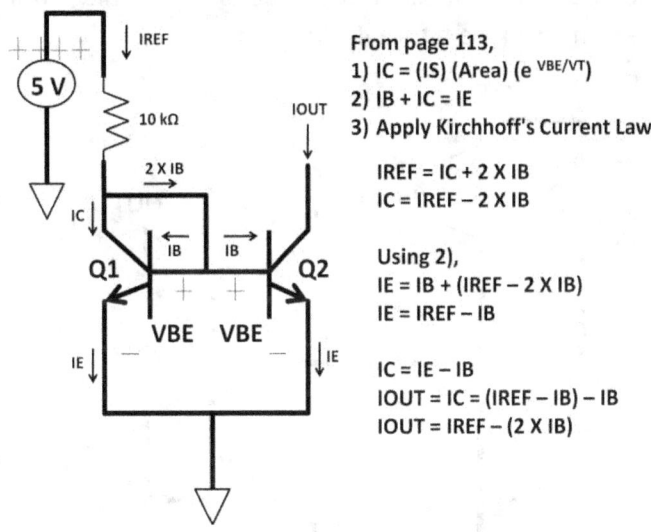

Figure 4.68: Current mirror error

Figure 4.68 examines this inaccuracy. Using the KCL, **IB + IC = IE** rule, it can be seen that IOUT is two IBs less than IREF. **IOUT = IREF − (2 X IB)**. For example, **VBE = 1 V, IC = 0.3 mA** (a specific transistor spec). **IREF = (5 V − 1 V) / 10 kΩ = 0.4 mA = 400 uA**. Suppose **beta (β) = 100, IB = 0.3 mA / 100 = 3 uA**. From the math derivation in figure 4.68: **IOUT = IREF − (2 X IB) = 400 uA − (2) X (3 uA) = 394 uA**

The current error in percentage:

$$\frac{IREF - IOUT}{IREF} \times 100\%$$

$$\frac{400\ uA - 394\ uA}{400\ uA} \times 100\%$$

$$\frac{6\ uA}{400\ uA} \times 100\% = 1.5\%$$

Despite 1.5% appearing to be a low number, recall that VBE and transistor beta are dependent on temperature. This error worsens with temperature and supply variations. The error percentage could go up quickly. For high accuracy design, it may be unacceptable. Because this error is mainly caused by the base current, you may be tempted to use a CMOS transistor to solve this problem, thinking that there is no gate current in MOSFET. However, VT matching of CMOS is worse than VBE in microelectronic design. Device matching quantifies how well two devices would be identical to each other. Comparatively, because CMOS VT matching is poorer than bipolar VBE due to the matching problem, the benefits of zero CMOS gate current are diminished.

Wilson Current Mirror

There are simple design techniques we could implement to improve the bipolar-based current mirror (see figure 4.69).

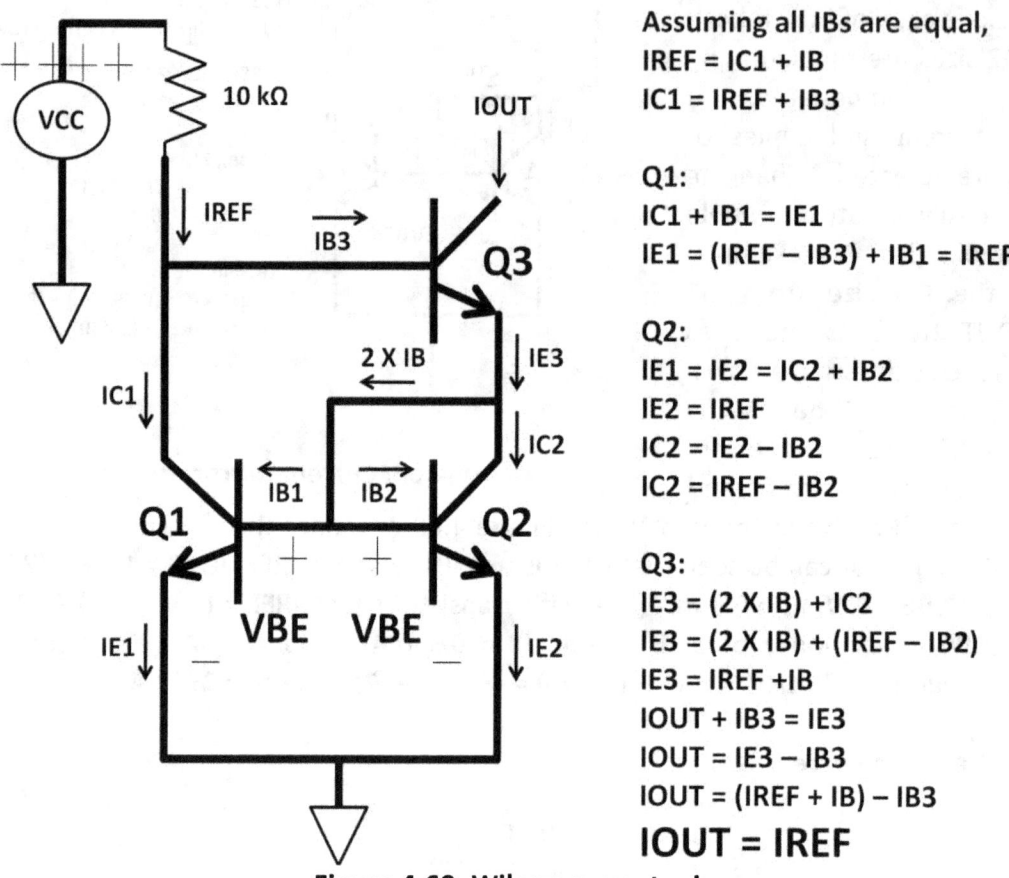

Assuming all IBs are equal,
IREF = IC1 + IB
IC1 = IREF + IB3

Q1:
IC1 + IB1 = IE1
IE1 = (IREF − IB3) + IB1 = IREF

Q2:
IE1 = IE2 = IC2 + IB2
IE2 = IREF
IC2 = IE2 − IB2
IC2 = IREF − IB2

Q3:
IE3 = (2 X IB) + IC2
IE3 = (2 X IB) + (IREF − IB2)
IE3 = IREF +IB
IOUT + IB3 = IE3
IOUT = IE3 − IB3
IOUT = (IREF + IB) − IB3
IOUT = IREF

Figure 4.69: Wilson current mirror

This is a Wilson mirror circuit, invented by Mr. George Wilson in 1960s. It's still popular today and used by many IC designers. By making two changes to the original circuit, Wilson's mirror IOUT is now equal to IREF. These simple changes are: **1)** Add Q3. **2)** Swap the Q1 and Q2 base to the Q2 collector instead of Q1. The mathematical derivation looks tedious. If you look closely, however, they are no more than KCL, IC, IE, and IB rules and simple arithmetic. This circuit serves another purpose. The Q1 collector voltage (Q1_VC) is now fixed at two VBEs **(VBE2 + VBE3)**. This fixed voltage at the Q1 collector ensures Q1 doesn't go into saturation (VCE being too low) and stay in the normal operating region (constant current). All these design "fixes" so far require good transistor understanding. Any electronic innovations are always backed by basic electronic principle no matter how complicated they turn out to be.

Bipolar Cascode

The technique of constant collector voltages is called cascode. Figure 4.70 is a current mirror using this technique.

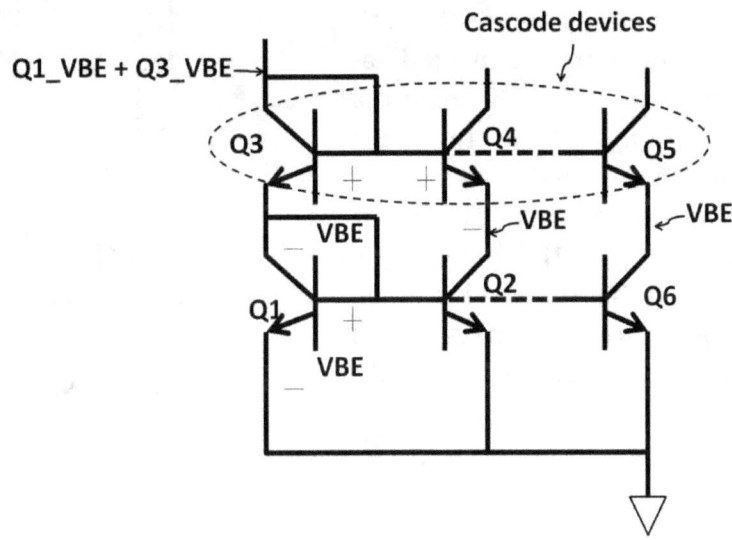

Figure 4.70: Cascode current mirror

Q3, Q4, and Q5 are cascode devices. All VBEs (Q_VBE) are identical. The cascode devices keep Q1, Q2, and Q6's collector voltages (Q1_VC, Q2_VC, Q3_VC) equal. Apply KVL:

$$Q3\_VC = Q1\_VBE1 + Q3\_VBE$$

$$= 2 \times (VBE)$$

Q4_VE = Q2_VC:

$$Q2\_VC = (2 \times VBE) - VBE = VBE$$

There is an additional current branch from Q5 to Q6. Similar to Q1 and Q2's VBEs, Q6's collector is one VBE. These constant voltages at collectors Q1, Q2, and Q6 keep them out of saturations. This is a useful feature when transistors are used as current sources. All design solutions come with trade-offs; the cascode current mirrors are no exception. What you lose is the head room. Head room is the voltages across the collector and emitter. From the IC-versus-VCE curves, it indicated that having large VCE is desirable in order to keep the transistor out of saturation. By adding a row of cascode devices, Q3, Q4, and Q5 transistor's head room would be reduced (reduced head room). This is particularly apparent in low voltage applications. In general, cascodes are not designed for extremely low voltage design due to the head room issue.

Darlington Pair

The Darlington pair configuration is a popular circuit. This circuit was invented by Mr. Sidney Darlington in 1953 when he worked at Bell Lab. It's still used today in many modern ICs. Let's use PNP devices, this time connected as Darlington shown in figure 4.71. This circuit is a differential amplifier using PNP as the input stage, NPN as active load. The Darlington pair provides two features in this circuit: **1)** maximum current gain, and **2)** input voltage conversion. On point 1, current is increased from Q1 to Q2 using the transistor beta rule as follows: Assume transistors' beta (β) are equal,

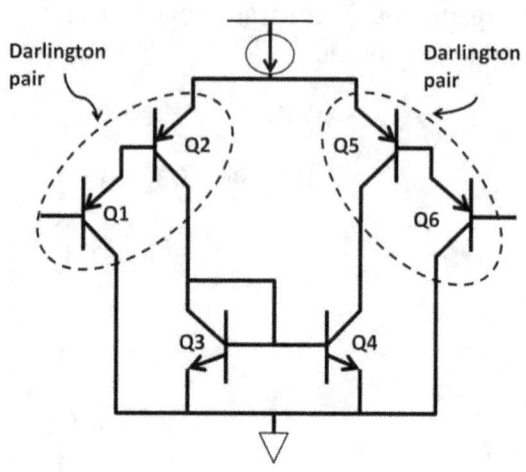

Figure 4.71: PNP Darlington pair

$$\beta = \frac{Q1\_IC}{Q1\_IB}$$

$$Q1\_IC = (Q1\_IB) \times \beta$$

Q1_IE = Q1_IB + Q1_IC:

$$Q1\_IE = Q1\_IB + (Q1\_IB \times \beta) = Q1\_IB (1 + \beta)$$

Q2_IB = Q1_IE:

$$\beta = \frac{Q2\_IC}{Q2\_IB}$$

$$\beta = \frac{Q2\_IC}{Q2\_IE}$$

$$\beta \times (Q1\_IE) = Q2\_IC$$

Q1_IE = Q1_IB X (1 + β):

$$Q2\_IC = \beta \times Q1\_IB (1 + \beta)$$

β >> 1,

$$Q2\_IC = \beta \times Q1\_IB \times \beta$$

$$Q2\_IC = \beta^2 \times Q1\_IB$$

This shows that output current IE (Q2_IE) is much larger than the input current (Q1_IB) by β^2. For example, if **Q1_IB = 10 uA**, beta are all 100. Q2's emitter current:

$$Q2\_IE = 100^2 \times 10 \text{ uA} = 100 \text{ mA, 10,000 times larger}$$

On point 2, the input voltage at the Q1 base is "lifted" up two VBEs at Q2's emitter (Q2_VE). If the input is 2 V, VBE is 1 V, and Q2_VB is at **(2 V + 1 V) = 3 V**. Adding one more VBE gives 4 V at Q2's emitter, keeping Q2 and Q3 out of saturation. This input voltage conversion is likely needed especially when the input voltage is relatively low. For designs that require low input voltage, the Darlington pair becomes an ideal choice as an input stage. Imagine using the same circuit without the Darlington in figure 4.72. With 1 V input at Q1 and 1 V VBE, emitter voltage at Q1 = **1 V + 1 V = 2 V**. Q3 collector stands at 1 V from Q3's VBE. VEC across Q1 is now **2 V – 1 V = 1 V**. For a particular bipolar process, 1 V VEC may be too low, forcing Q1 into saturation. Saturation should be avoided at all costs because it takes time for the transistor to recover from saturation during switching, hurting timing performance.

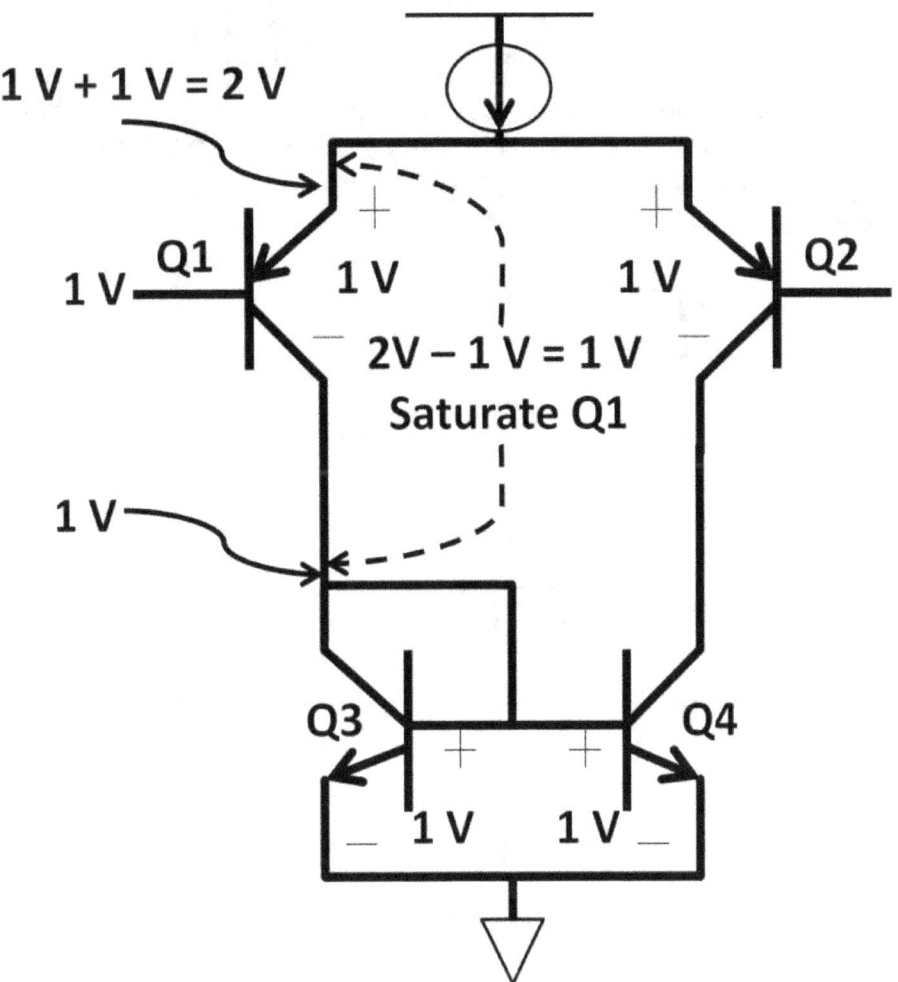

Figure 4.72: PNP differential pair with low input voltages

CMOS Cacosde

Cascode can also apply to CMOS transistors. A CMOS cascode circuit is shown in figure 4.73. Q1 and Q2 gates and Q2 drain are tied to each other. This makes Q1 gate-to-source voltage (Q1_VGS) equal to Q2 drain voltage. Gate-to-source voltage of Q2 is Q2_VGS. Apply KVL, source voltage of Q2, **Q2_VS = (Q1_VGS – Q2_VGS)**. Plug some numbers into the circuit. You will gain some real insights into how it works. For example, if **Q1_VGS = Q2_VGS = 1 V** for a given transistor size, VT, and temperature, then according to the Q2's source voltage equation, it is equal to: **Q2_VS = (1 V – 1 V) = 0 V**. This makes Q1 drain-to-source voltage (Q1_VDS) zero volt cutting off Q1. This circuit does not operate properly. To fix it, the device size needs to be changed; use the drain current equation:

$$Id = \left(\frac{uCox}{2}\right) \times \left(\frac{W}{L}\right)(VGS - VT)^2$$

By changing the transistor size, VGS could be modified for a given VT and drain current. In this case, we would like to increase the Q1 VGS to be larger than Q2's. We double Q1's width to change Q1_VGS from 1 V to 2 V. Q2_VS is now: **2 V – 1 V = 1 V**. For low voltage CMOS process, 1 V is possibly enough to keep Q1 from cut-off.

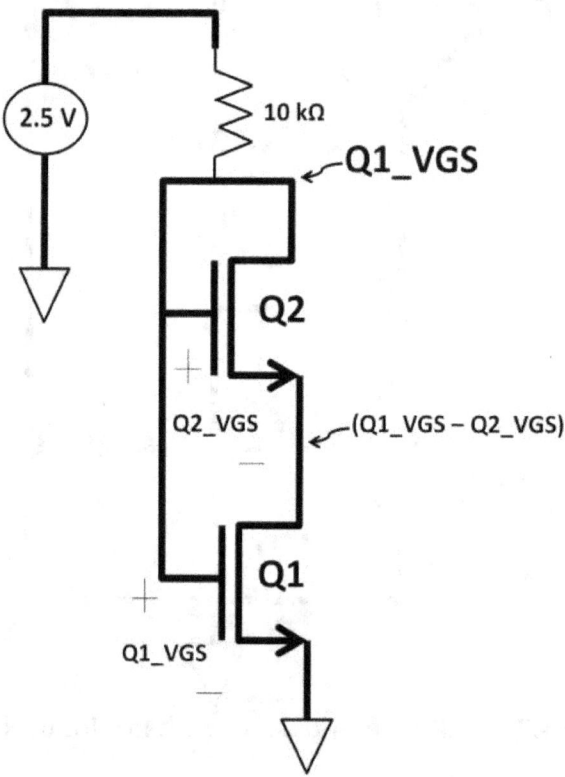

Figure 4.73: CMOS cascode circuit

Buffer (Voltage Follower)

Let's now go over some op-amp circuits to reinforce what we learned. A very common op-amp usage is a buffer. Its purpose is to provide high input and low output impedances (voltage divider concept) to maximize signal levels. By definition, the buffer output is the same as the input. An op-amp can be connected as a buffer, shown in figure 4.74. It's called voltage follower (unity gain amplifier) because the output "follows" the input with gain of one. Comparing to source and emitter followers (single-ended), a voltage follower is superior because the output is the same as the input without any voltage drop (recall source and emitter follower output is slightly less than the input).

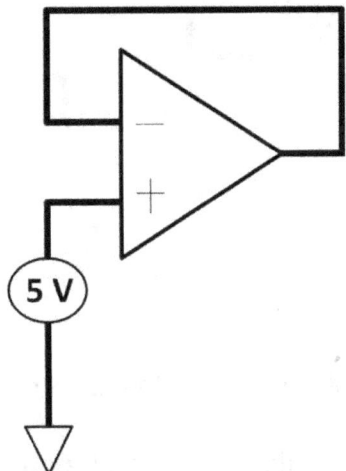

Figure 4.74: Voltage follower

This op-amp configuration above is a non-inverting amplifier where input voltage connects to the positive terminal. Using the voltage gain equation developed earlier:

$$\frac{Vout}{Vin} = \frac{Rf + Ri}{Ri} = 1 + \frac{Rf}{Ri}$$

Because Ri and Rf do not exist, this makes voltage gain transfer function to be 1, i.e., **Vout = Vin.** In figure 4.74, the op-amp output is 5 V, which is equal to the input.

$$\text{Voltage gain} = \frac{Vout}{Vin} = 1 + \frac{0}{0} = 1$$

$$\frac{Vout}{Vin} = 1$$

$$Vout = Vin$$

Summing Amplifier

The next circuit in figure 4.75 is called summing amplifier. It is an inverting amplifier with multiple inputs connecting to the negative terminal.

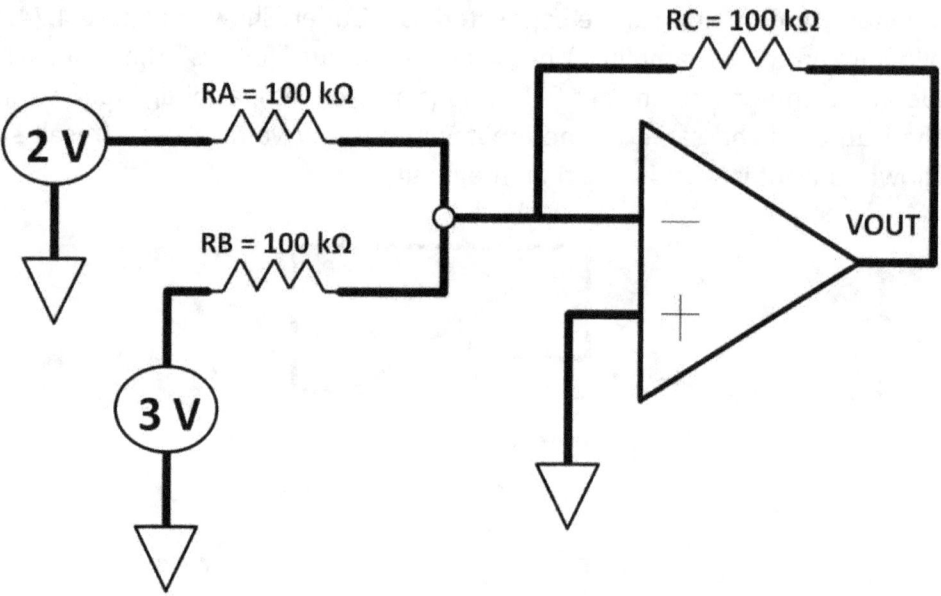

Figure 4.75: Multiple-inputs op-amp

Apply inverting amplifier and KCL rules, and virtual ground is established in the following circuit (see figure 4.75a) and modified circuit in figure 4.76 on the next page.

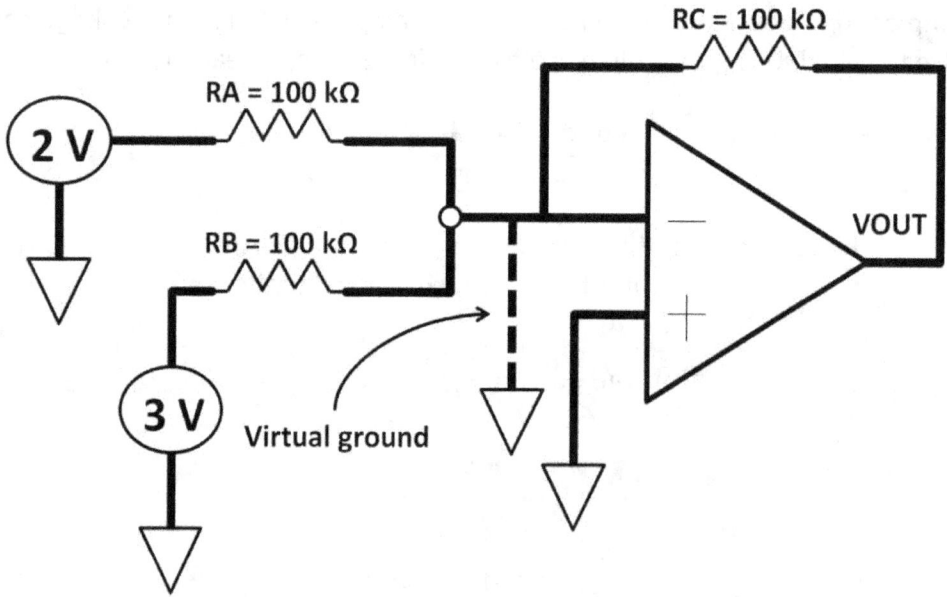

Figure 4.75a: Virtual ground at negative terminal

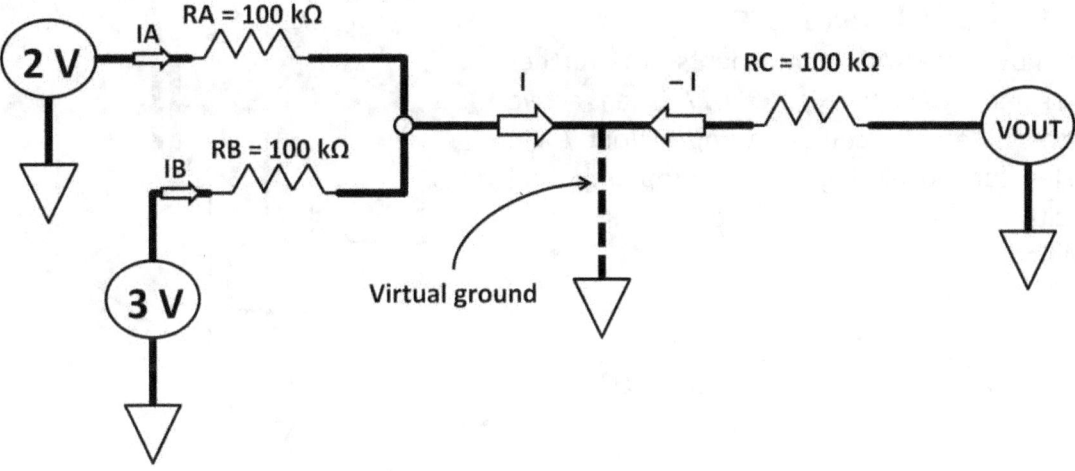

Figure 4.76: Modified multiple-input circuit

Apply KCL, **IA + IB = I**

$$IA = \frac{2\text{ V}}{100\text{ k}\Omega}$$

$$IB = \frac{3\text{ V}}{100\text{ k}\Omega}$$

$$I = IA + IB$$

$$I = \frac{2\text{ V}}{100\text{ k}\Omega} + \frac{3\text{ V}}{100\text{ k}\Omega}$$

I and − I are equal but flow in opposite directions:

$$-I = \frac{\text{VOUT}}{100\text{ k}\Omega}$$

$$I = -\frac{\text{VOUT}}{100\text{ k}\Omega} = \frac{2\text{ V}}{100\text{ k}\Omega} + \frac{3\text{ V}}{100\text{ k}\Omega}$$

$$\text{Vout} = -(2\text{ V} + 3\text{ V}) = -5\text{ V}$$

This circuit is a summing amplifier circuit with an inverted output. It adds all input voltages together. The result of the sum arrives at Vout is phase-shifted by 180 degrees.

Active Low-Pass Filter

Let's now use AC components to further understand op-amps. Figure 4.77 is an active low-pass filter. We could develop a Vout / Vin transfer function using standard op-amp and capacitive reactance rules. For an inverting amplifier:

$$\frac{Vout}{Vin} = -\frac{Xc}{Ri} = -\frac{1}{(2\pi f)(Ri)}$$

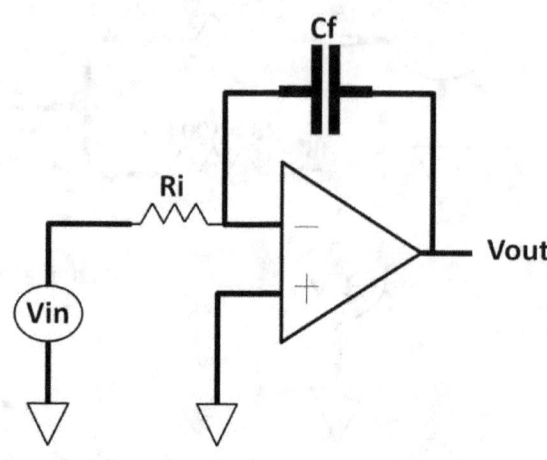

Figure 4.77: Active low pass filter using op-amp

This is a low-pass filter with high input and low output impedances by the op-amp (active device). In some cases, you may want to maintain finite gain in high frequency. A simple change to the circuit (adds Rf) in figure 4.78 achieves that. Revised transfer function:

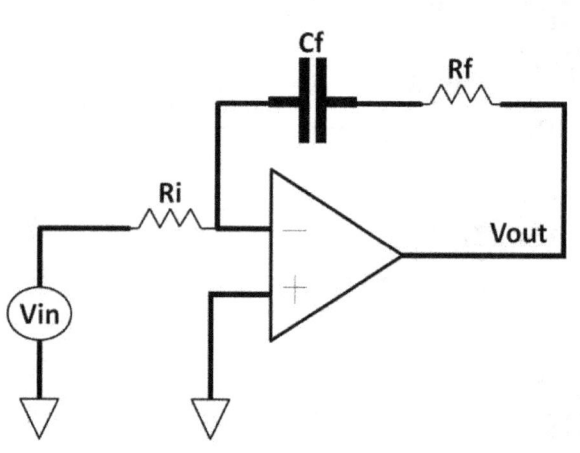

$$\frac{Vout}{Vin} = -\frac{Xc + Rf}{Ri} = -\frac{\frac{1}{2\pi fC} + Rf}{Ri}$$

$$= \frac{\frac{1 + (2\pi fC)(Rf)}{2\pi fC}}{Ri}$$

$$= \frac{1 + (2\pi fC)(Rf)}{(2\pi fC)(Ri)}$$

Figure 4.78: Add Rf in active low-pass filter

Based on this transfer function, starting at low frequency, the denominator is close to zero. Vout / Vin is large. Input frequency starts to increase, and Vout / Vin starts to fall at 20 dB / decade rate. At extremely high frequency, voltage gain remains roughly constant because $2\pi f C$ cancel out each other:

$$\frac{Vout}{Vin} = -\frac{Rf}{Ri}$$

The transfer function is best described by a bode plot (see figure 4.79).

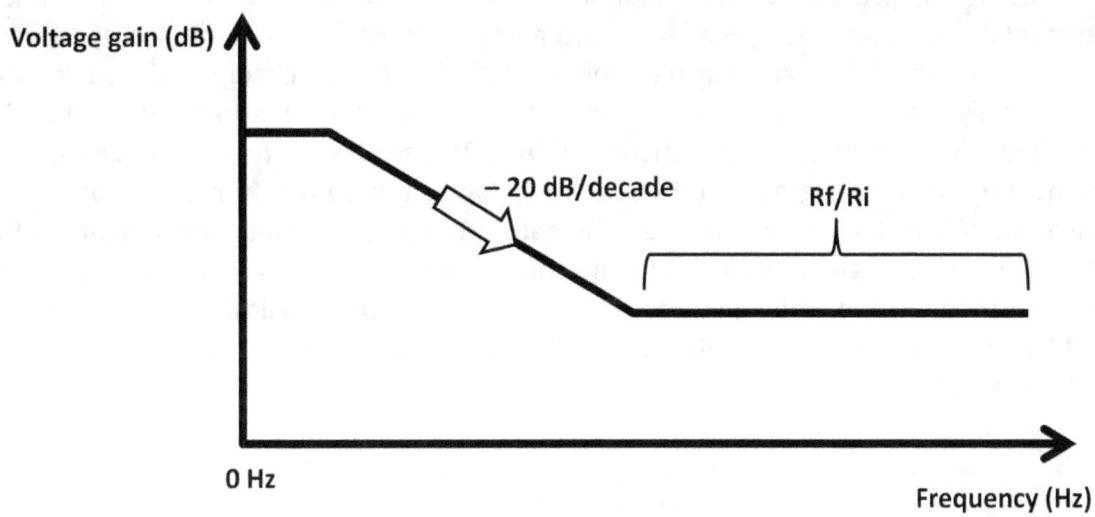

Figure 4.79: Bode plot

Again from Vout / Vout transfer function:

$$\frac{Vout}{Vin} = -\frac{Rf}{Ri}$$

At high frequency, voltage gain remains constant and holds steady by Rf to Ri's ratio. The bode plot above is an excellent method to verify circuit behaviors and performances. With the use of capacitors and inductors in feedback circuits, you need to take phase shift into consideration because it could potentially cause oscillations. Recall R C, voltage, and current (lead, lag) characteristics in chapter 3, AC. Feedback signal arriving back at the input may either lead or lag output signals. These L, C components could cause circuits to behave erratically (circuit oscillation). Unwanted oscillations create noise and unstable output in the system. They should be prevented at all costs. The criteria of oscillation depend on phase shift that exceeds 360 degrees when gain is above unity. In circuit design analysis, gain and phase margins are often used to determine oscillation criteria. We will look a closer at these circuit design criteria later in the positive feedback section.

Circuit Simulator

On circuit design process, circuit simulation software like Multisim (made by National Instruments) is popular among academia. Often used in electronics course labs by students, Multisim constructs (schematic entry) analog and digital electronic circuits at the device level. You can easily place schematic symbols and connect them by wires in software. Multisim offers simulation capability (DC, transient, and AC). The simulation results can be displayed on computer monitors in graphs and waveforms. You can place test probes on nodes (nets) to measure V and I anywhere in the schematic. Adding electronics instruments (DMM, oscilloscope, function generator, etc.) is convenient with a few mouse clicks. It's a great way to confirm theories and verify applications before building the physical circuits. Figure 4.80 and 4.81 show an op-amp simulation bench, component selection window, and scope waveform window.

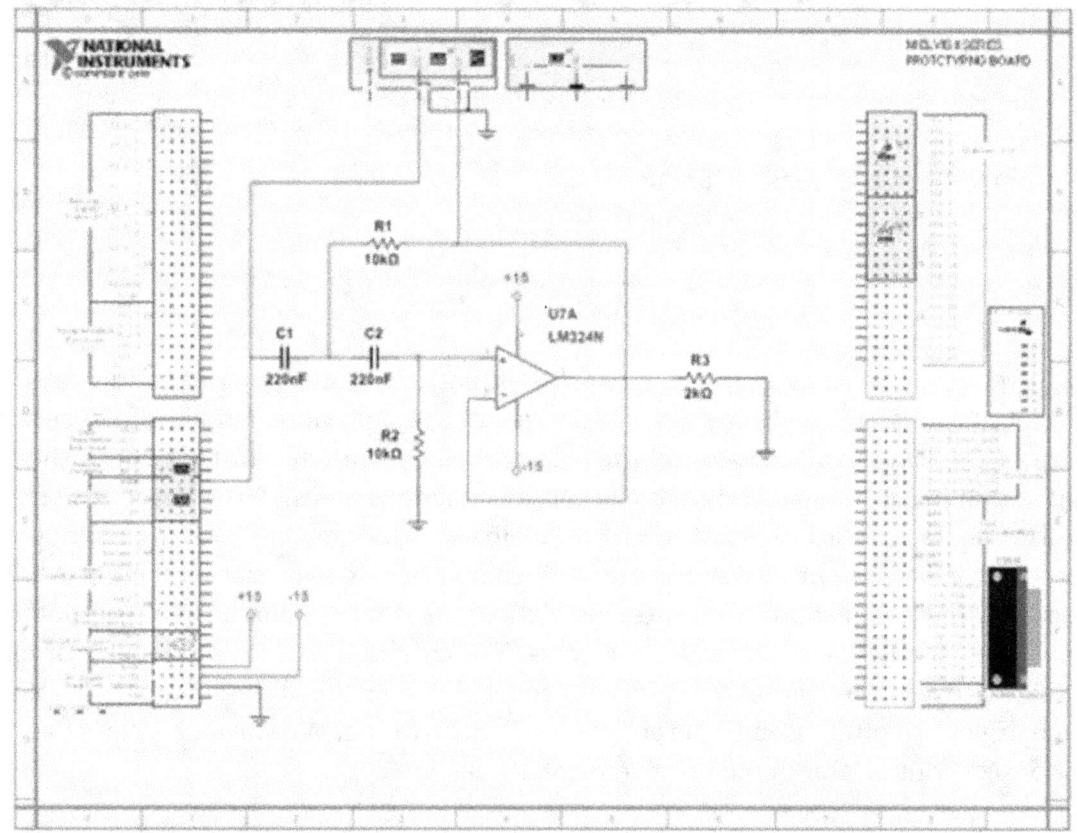

Figure 4.80: Multisim schematic capture (Courtesy of National Instruments)

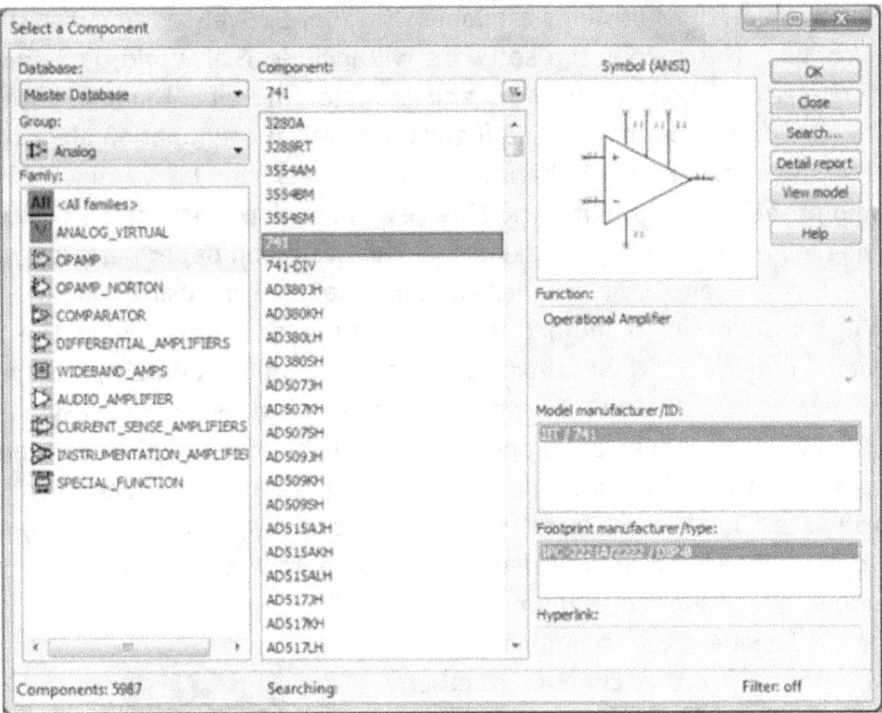

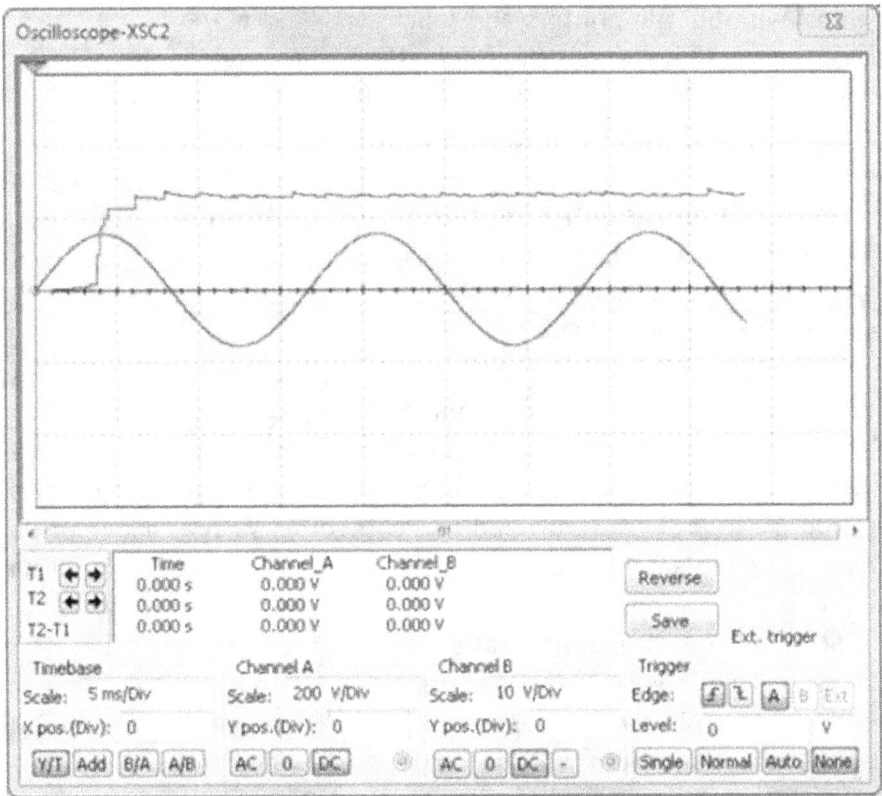

Figure 4.81: Multisim component selection and oscilloscope simulation waveform

The commercial version of Multisim is available with more advanced features such as device model modifications. This means the software will include real world parasitic parameters into transistor, resistor, diode, capacitor, and inductor models. The computer simulation software then uses these parameters and feeds them into the simulation algorithm to reflect what could be the realistic circuit behavior. The results can be verified using graphical waveforms and probes in the schematics. This design-check process offers tremendous cost- and time-saving benefits in terms of making sure the design on paper performs closely to the final hardware. Being able to verify the design to a certain degree using computer simulations before running through the manufacturing saves time and money. On the other hand, simulations could only mimic the real-world scenario to limited degrees, depending on how accurate the models are. In spite of model imperfections, new chip design (first silicon) coming out of fab usually meet basic specifications. Beside National Instruments, Cadence Design Systems, Mentor Graphics and Synopsis are market leaders in IC design and simulation software. On the test front, in addition to DMM, power supplies, and function generators, oscilloscopes are standard equipments to measure AC circuits. Oscilloscopes (scopes) are time-measuring test equipment taking input from an AC signal. They come in a wide variety differentiated by the channel numbers, resolutions, and speed. Many high-end scopes are capable of measuring in gigahertz (GHz) or gigabits per second (Gbps) with built-in printers and touch screen displays. Scopes have connectors (plugs) that allow Bayonet Neill-Concelman (BNC) cables to be connected to it. At the other end of the cable would be the AC signal being measured. The scope displays X-axis as time, Y-axis as either the current or voltage. Users can zoom in and out of the waveform using voltage and time scales knobs. Figure 4.82 shows an Agilent DSO5012A Series Oscilloscope with dual channel, 100 MHz, 2G sample/s. The voltage probe in the figure connects electronic circuits and oscilloscopes. One end of

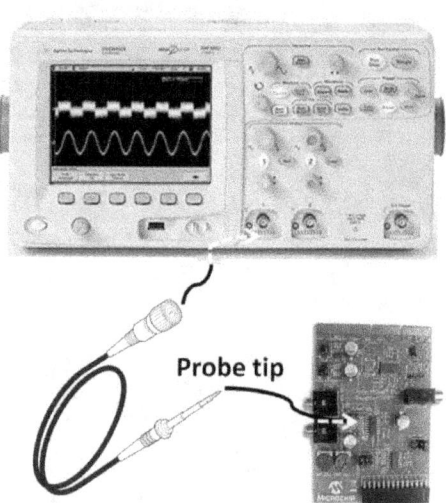

Figure 4.82: Agilent DSO5012A Series Oscilloscope with scope probe

the scope probe connects to the scope connector. The probe tip on the other end connects to the circuit of interest. Probes are divided into categories such as active or passive. Active types contain amplifiers to amplify signals. Passive ones are less expensive with resistors and capacitors built into them. Many probes come with switchable attenuation settings, e.g., 1X, 10X. The X represents the attenuation ratio. For 1X, the signal at the test pin to scope connector is 1:1 (no attenuation). 10X means the signal arriving at the scope is reduced by 10 times relative to the test pin signal. Probe datasheets list probe parameters including input resistance, capacitance, bandwidth, voltage range, etc. 1X and 10X probes' parameters may differ greatly. The 10X setting offers lower capacitance (<20 pF) with much wider bandwidth.

Getting familiar with these parameters helps engineers and technicians select the right probe type for a specific test or measurement task. Another popular electronic apparatus is the function generator. It generates AC signals driving to a load as an AC signal source. Most function generators are capable of producing signals such as sine, square, or triangular waves. Frequency adjustment, offset dial knobs, and output connectors can be found in function generators. Tektronix and Agilent are leading function generator suppliers. Figure 4.83 shows a Tektronix AFG2000 Function Generator with 20 MHz bandwidth, 14-bit resolution, and 250 MS/s sample rate.

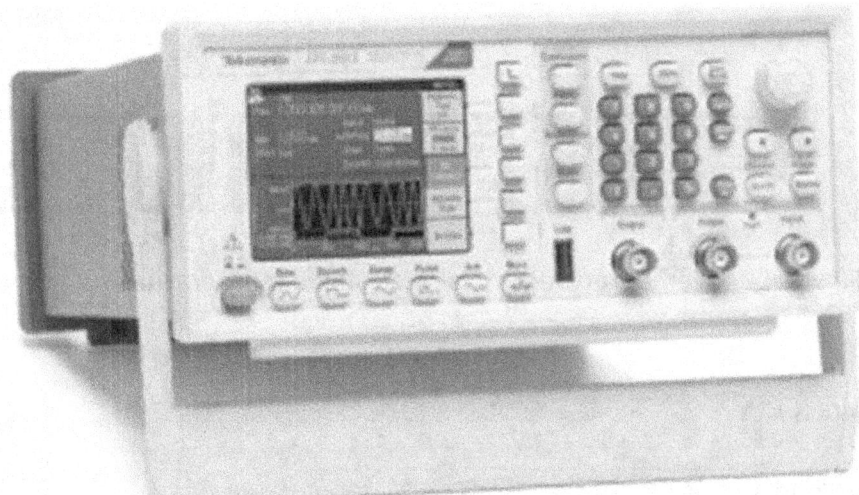

Figure 4.83: Tektronix AFG2000 Function Generator (Courtesy of Tektronix)

Hysteresis

Test equipment and electronic systems require the use of hysteresis to reduce false trigger caused by system glitches. An example is household air-conditioning (A/C) and heating systems using a thermostat. Figure 4.84 shows the temperature profile of a room over time. When the temperature rises above a 27°C set point, the A/C system turns on to bring the temperature down. Meanwhile, when the temperature falls below the set point, the heating system turns on to bring the temperature back up. The single temperature set point triggers many on-off pulses (false trigger denoted by the dotted circles in figure 4.84). This increases the wear and tear of the system over time. To prevent that, a hysteresis zone can be implemented. In figure 4.85, the hysteresis zone consists of two thresholds (upper and lower). The A/C system only turns on when the temperature goes above the upper threshold. If it falls below the upper limit, the system ignores it and the output remains high. When it crosses the lower threshold, the heating system turns on to bring the temperature up. The detailed implementation of hysteresis will be discussed in the next section (positive feedback).

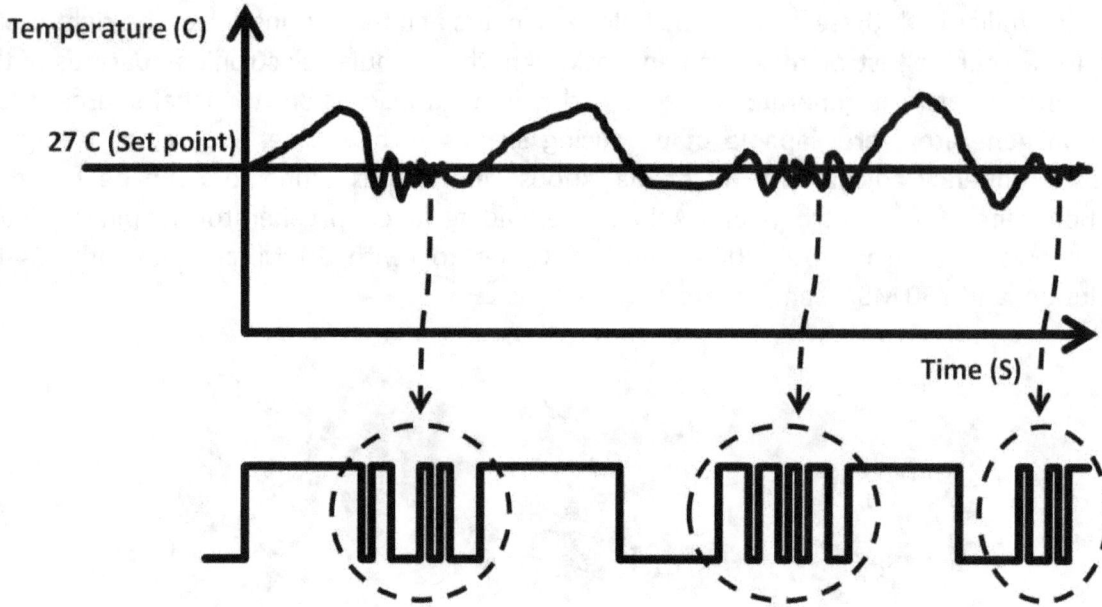

Figure 4.84: Temperature control with single temperature set point

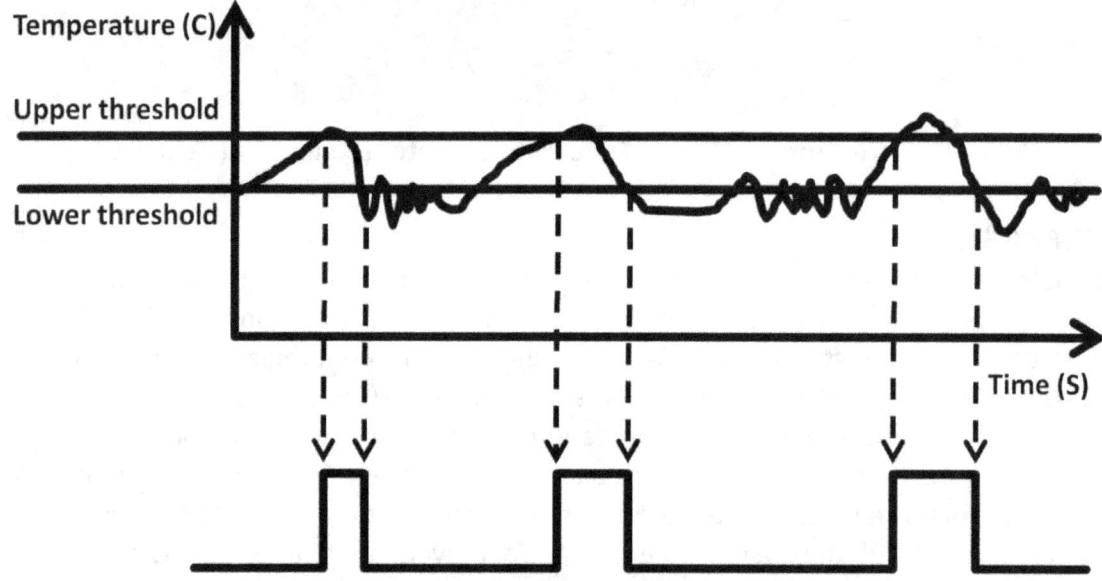

Figure 4.85: Temperature control with hysteresis

Using positive feedback in an op-amp can implement the hysteresis technique. Figure 4.86a shows a sampled op-amp hysteresis circuit with Vin and Vout waveforms. This op-amp is an inverting comparator with Vin connecting to the negative terminal (V−) while the positive terminal (V+) ties to the midpoint of the voltage divider (R1 and R2) forming a positive feedback network. V+ becomes the upper and lower thresholds of the comparator set by the R1, R2 voltage divider. When Vin starts at 0 V (Low) and rises, Vout flips to the positive rail 5 V saturating the op-amp output stage **(V− < V+, inverting amplifier)**, as shown in figure 4.86b. The upper threshold is now set at 2.5 V by the voltage divider. While Vin continues to increase from 0 V

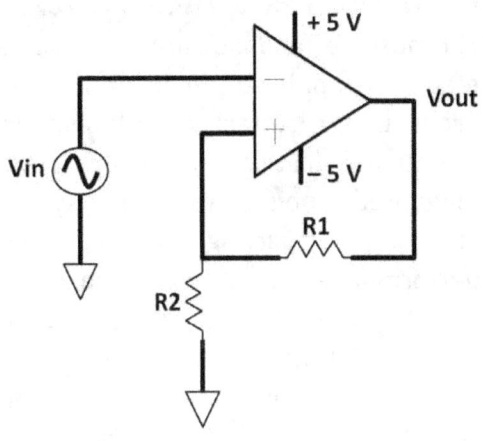

Figure 4.86a: Op-amp with hysteresis, Vin

(before 2.5 V), Vout stays at 5 V due to the comparator's high gain. Once Vin rises slightly above 2.5 V, Vout flips to the − 5 V rail **(V− > V+)**. Now, the comparator's threshold V+ is at − 2.5 V (set by the voltage divider). Vin continues to increase above 2.5 V while Vout remains at − 5 V **(V− > V+)**. As Vin (V−) starts to fall from its peak just below 2.5 V, it continues to stay low. V− remains less than V+. Once Vin (V−) falls below − 2.5 V **(V− < V+)**, Vout flips to the positive rail. The same mechanism repeats to the next cycle. The upper and lower thresholds can be easily set by varying the sizes of R1 and R2.

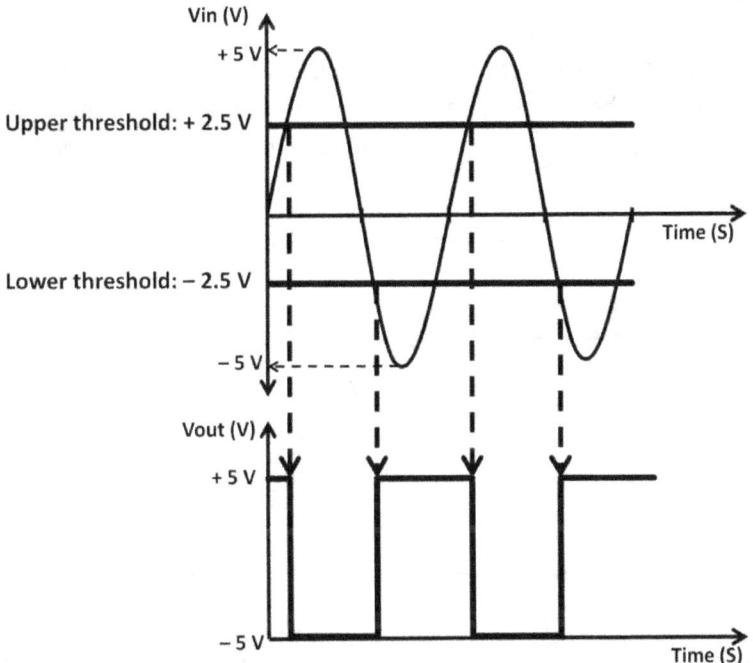

Figure 4.86b: Hysteresis waveform

Positive Feedback (Oscillation)

The positive feedback in the previous configuration is intentional. However, in any op-amp feedback topologies where the feedback resistor network is used, unwanted oscillation can occur under certain conditions. Circuit oscillations are periodic signals, which can be intentional (oscillator) or unintentional. If they are unintentional, oscillations become unwanted noise to the system, adversely impacting overall circuit performance. You need to make sure circuit oscillation does not occur unless it is intended. To understand oscillation, we first need to understand why and how oscillation occurs; then we examine how to prevent it from happening. The cause of circuit oscillation is due to positive feedback. For example, assume an op-amp's input is an AC source. The op-amp employs resistor feedback architecture (see figure 4.87). In this example, an inverting amplifier is used. There are capacitors inside and outside of the op-amp. The internal capacitors can be by design or parasitic. The external capacitor is the capacitive load (CLoad). The capacitors' present poles in the signal chain are imposing phase shift on the signal (chapter 3, AC). Capacitor voltage is lagging resistor voltage by 90 degrees. This phase shift becomes the criteria for circuit oscillation. The second effect from the pole is signal attenuation (low-pass filter), where signal is reduced by – 20 dB per decade. Each time a signal passes through a pole (capacitor), it attenuates by another – 20 dB. The larger the capacitor, the more attenuation is asserted. Both phase shift and signal reduction form the basis of oscillation. This phenomenon is explained by the bode plot shown in figure 4.88.

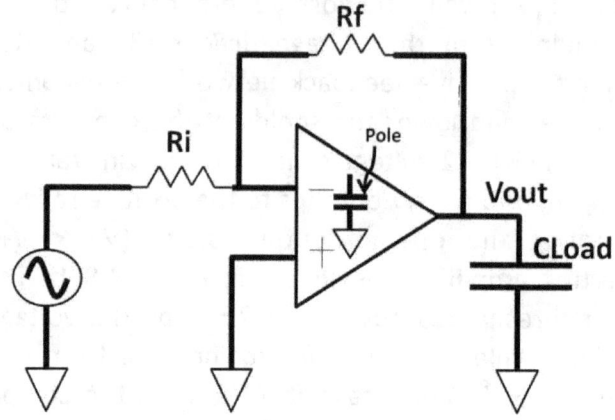

Figure 4.87: Non-inverting amplifier, poles

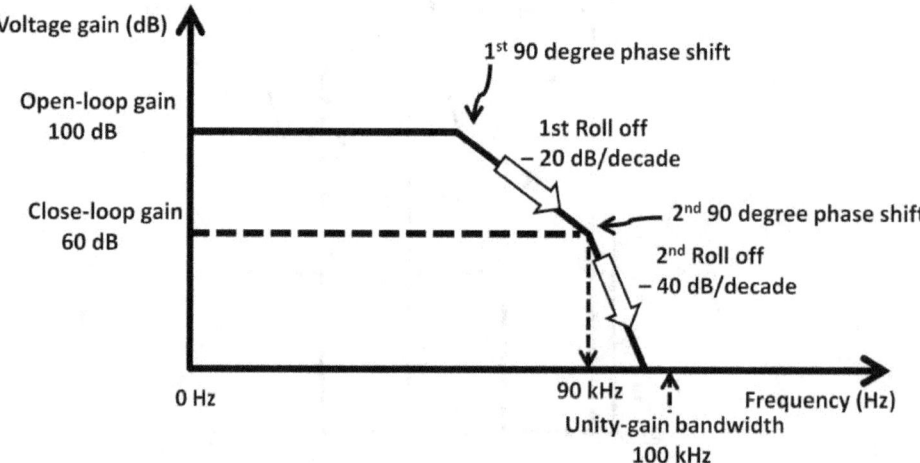

Figure 4.88: Gain, phase bode plot

This amplifier presumes to have 100 dB of open-loop gain and 60 dB closed-loop gain at 90 kHz. There are two poles. The first pole (CLoad) rolls off the amplifier gain by – 20 dB and causes a 90-degree phase shift. The second pole gives yet another – 20 dB totaling – 40 dB roll off. This second pole contributes to another 90-degree shift that gives a total 180 degrees. Because capacitors lag resistor voltage by 90 degrees, two poles (90 + 90) plus the capacitor lag voltages (90 + 90) yield a total of 360-degree phase shift. Both feedback and the output signal are now superimposed with each other (positive feedback). With the gain at the second roll off still above unity gain (0 dB), this circuit is now in unstable condition (oscillation).

To ensure stability, phase shift needs to be less than 180 degrees for any gain larger than unity (0 dB). In other words, amplifier gain at a 180-degree phase shift needs to be less than 0 dB gain. These conditions become the stability criteria. To achieve this, we add a dominant pole (external or internal) at the low frequency, deliberately moving the gain curve to the left so that by the time it rolls off to unity, phase shift is less than 180 degrees. This technique is called dominant pole compensation (see figure 4.89). The downside to this frequency compensation technique is that desired gain is now at a lower frequency, reducing amplifier bandwidth. Essentially, you are trading gain for bandwidth by adding a dominant pole. There are other types of compensation schemes to ensure amplifier stability such as lead, lag, and feed-forward compensations. The details of these techniques are beyond the scope of this book and will be discussed in other publications by the author.

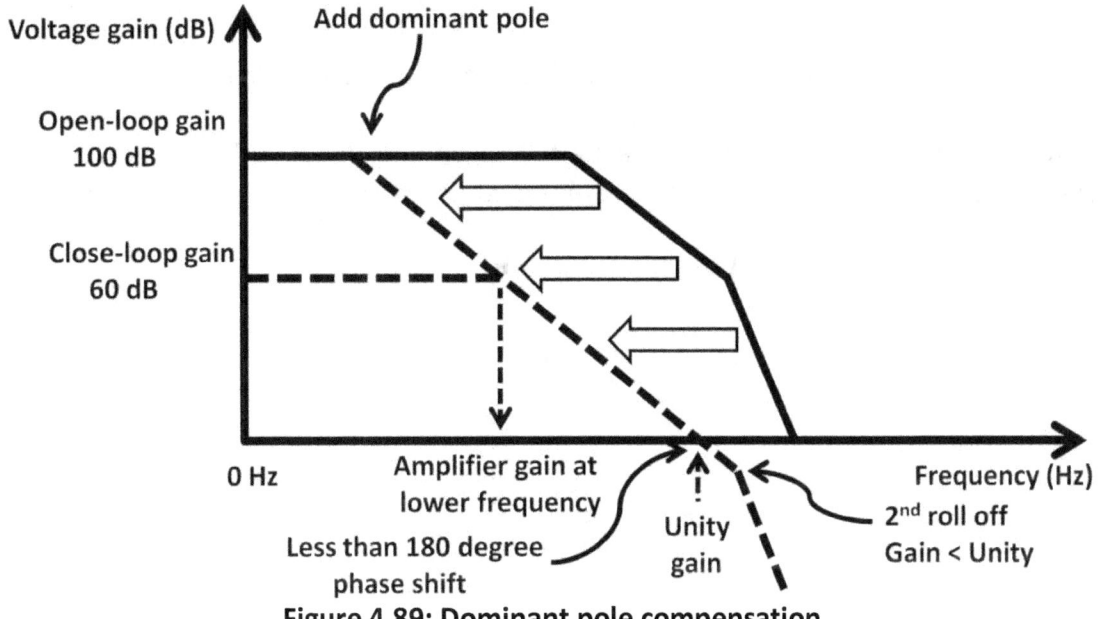

Figure 4.89: Dominant pole compensation

Instrumentation Amplifier

So far, to control gains, all op-amps were constructed with external feedback. An instrumentation amplifier (INA) allows gain control with external feedback while maintaining high input impedance. The primarily use of INA is to offer high differential gain and reject common mode signal originating from noise. INAs come in many forms. One of the most popular one is the two op-amp INA shown in figure 4.90.

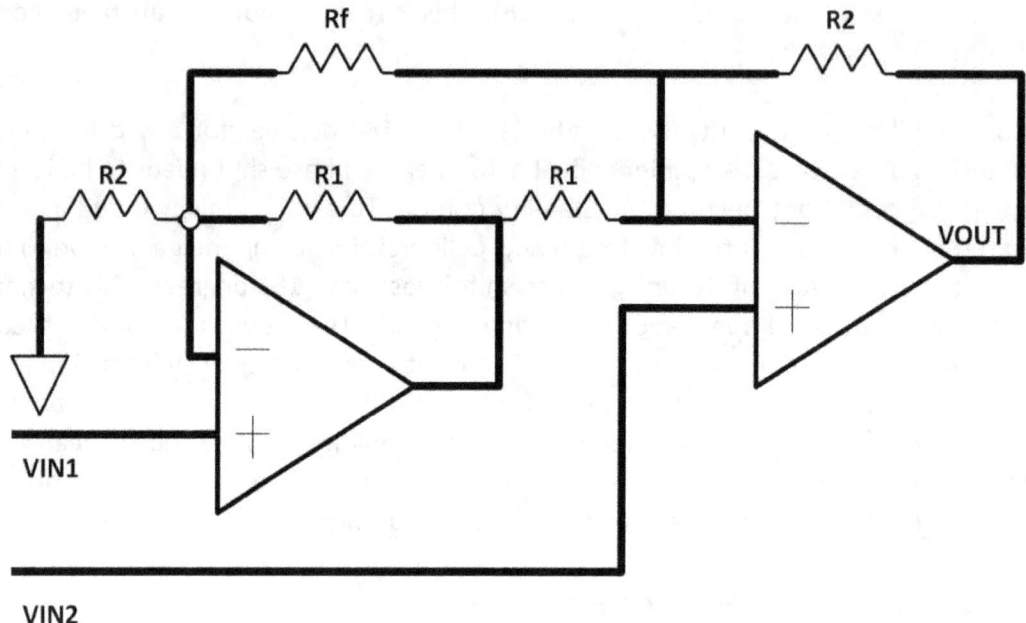

Figure 4.90: Two op-amp instrumentation amplifier

Both inputs VIN1 and VIN2 connect directly to the op-amp inputs offering extremely high input impedance. The differential gain transfer function of the above INA is as follows:

$$\frac{Vout}{Vin1 - Vin2} = 1 + \frac{R2}{R1} + \frac{2 \times (R2)}{Rf}$$

Linear Regulator

As previously discussed in chapter 2, Diodes, and 3, AC, a zener diode is a linear regulator. The circuit from figure 2.12 is shown again in figure 4.91. A zener regulator comes with deficiencies: zener's cathode (node Z) is high impedance. Unless a load's impedance is extremely high, output degrades substantially (voltage divider). This problem can be solved by using low-output impedance of emitter or source followers as buffer shown in figure 4.92. The dotted rectangle represents the impedance transformation model from high- to low-output impedance (upper right of figure 4.92).

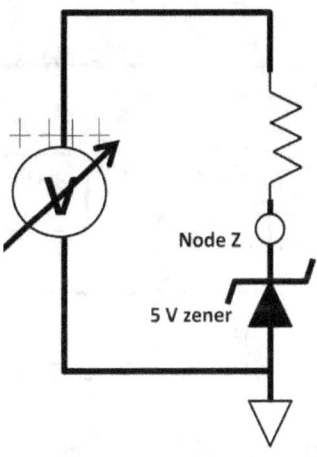

Figure 4.91: Zener regulator

There are two voltage dividers. The zener's cathode is high impedance (RZ + Rgate). After buffering it with an NFET, the source is now the output offering low impedance (RS). Rs forms yet the second voltage divider with RLoad. The second divider retains as much VZ as possible. The trade-off of this design is the loss of one VGS. Assume **VGS = 1 V** for a given NFET size:

$$VOUT = 5\text{ V} - 1\text{ V} = 4\text{ V}$$

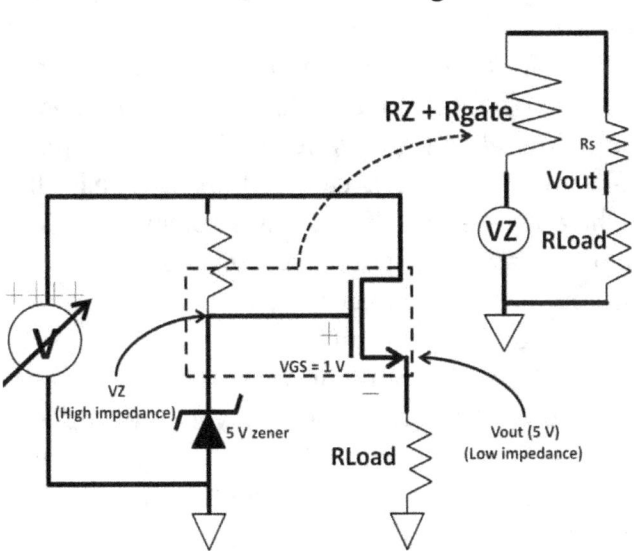

Figure 4.92: Buffered zener regulator

If sizing the NFET properly, VGS and drain current optimize the output voltage while meeting output current requirements. In mixed-signal IC design, it's desirable to have multiple internal voltage sources. The motivation is to isolate noise (high-speed digital circuits) coupling to the analog circuitries and vice versa. With superb transistor-matching capabilities in microelectronics, high accuracy internal-voltage regulators are possible. Figure 4.93 shows an implementation example. VDD_A is the internal supply to analog circuits; VDD_D power the digital circuits.

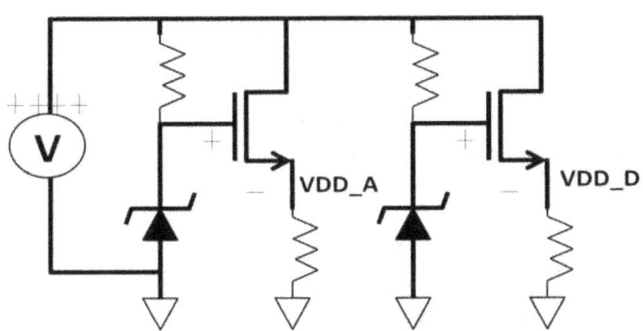

Figure 4.93: Multiple internal zener supplies with NFET buffer

Low Drop-out (LDO) Regulator

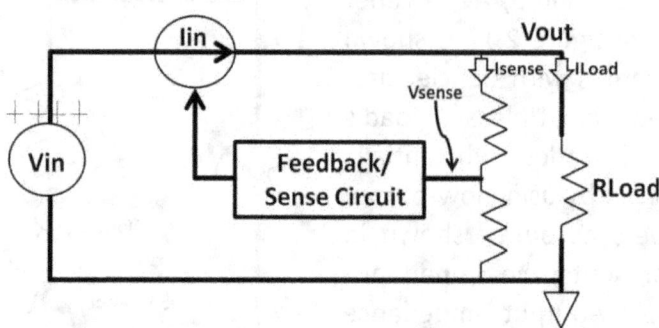

Figure 4.94: LDO functional block diagram

A low drop-out regulator is a linear regulator operating by feedback network with sensing circuits. Figure 4.94 shows a functional block diagram of LDO. RLoad could change regularly. For example, fan speed varies causing its load current to fluctuate. These changes result in load current change, ultimately changing Vout. LDO has a feedback network that senses the output voltage (voltage divider). This voltage (error voltage) is then used to adjust the input current **(Iin = Isense + ILoad)** accordingly to keep Vout at its desired value. It's merely a negative feedback system where the input current modulates from the results of the sense circuit. If the Vout falls, Iin increases to bring Vout back up. It also works the opposite way. If Vout goes up, Iin decreases to bring Vout down. It's called "low drop-out" because transistors are used as a current source in LDO. By forcing a transistor into saturation, Vout can get fairly close to Vin before dropping out of regulation. This is advantageous from a power efficiency standpoint. As a transistor goes into saturation, it dissipates the least power amount increasing efficiency. For this reason, LDO is suitable for battery-powered applications where low power consumption is desirable. The trade-off of LDO is the need for compensation to keep the negative feedback loop stable. Figure 4.95 shows an LDO example. It uses a PNP transistor called the pass device as a switch. The on-chip error amplifier senses the output by the voltage divider R1 and R2 at Vsense. This feedback voltage feeds into the negative terminal of an op-amp (error amp). Vsense is constantly comparing against the reference voltage (Vref). The error amplifier will do whatever it takes to make the two input terminals equal (op-amp rule).

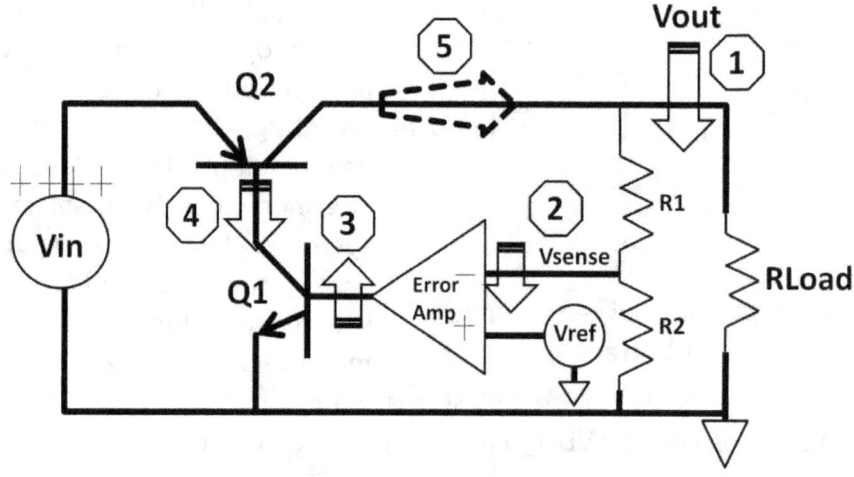

Figure 4.95: LDO example

For example, the lead-acid battery, a popular battery type for portable devices such as rechargeable radios and lamps, is used as Vin. If the battery operates at 6 V nominally, Vout regulates at 5 V by the LDO. As RLoad changes, Vout falls below 5 V (Step 1). Vsense is now lower than Vref (Step 2). The error amp is an inverting amplifier. When input goes low, output rises. The error amp effectively captures a sample of the error, lifting its output (Step 3). The op-amp output then raises Q1 base turning it on more (Step 3), pulling its collector down (Step 4). Because Q1's collector ties to Q2's PNP base, PNP now turns on more as the base gets pulled down. As a result, Q2, now supplies more current (Step 5) to RLoad, bringing Vout back up until **Vbe = Vref** again. The same concept applies when Vout goes higher making **Vsense > Vref**. To set the output voltage, a simple divider rule is used. For example, if 2.5 V is the desired output voltage, **Vref = 1.25 V, R1 = 1 kΩ**, R2 would be:

$$\text{Vsense} = \text{Vout} \times \frac{R1}{R1 + R2}$$

$$1.25\,V = (2.5\,V) \times \frac{1\,k\Omega}{1\,k\Omega + 1\,k\Omega}$$

R1 = 1 kΩ, R2 = 1 kΩ

The voltage where the Vout starts to fall out of regulation is called the drop-out voltage. It's a critical LDO design parameter. The drop-out voltage is the minimum voltage across the collector and emitter. The lowest drop-out voltage of this example is the VCE_{sat} (saturation voltage between collector and emitter). This is the voltage at which LDO is still able to maintain regulation. The smaller this voltage, the better the LDO is because the it utilizes the most available Vin before falling out of regulation. In this design, the PNP VCE_{sat} can be as low as 0.7 V. Drop-out voltage relates strongly with load current. For low load current, VCEsat can be as low as 50 mV. Such low drop-out voltage has propelled LDO applications to portable, handheld devices in recent years. In addition to drop-out voltage, transient response is also a design parameter. Output could change quickly. It takes time for LDO to respond. This time delay is an important consideration, especially in timing-critical applications. The type of Vin is another design consideration where Vin could be rectified AC or pure DC. Most LDOs are able to regulate Vout to as close as +/− 5% of the nominal value. LDO by itself draws current even though the RLoad is disabled or idle. This quiescent current becomes the dominating factor of draining input battery. Many modern LDOs integrate special features including thermal shut down and current limit capabilities to prevent damage from excessive temperature and current to the LDO ICs. For example, load could suddenly drop significantly, overloading the output. This excessive current could damage the pass device if current limit capability does not exist. Excessive current can also be caused by the input voltage (inrush current). The detailed design implementation of thermal shutdown and current limit is beyond the scope of this book. However, the functional block diagram of these features is shown in figure 4.96.

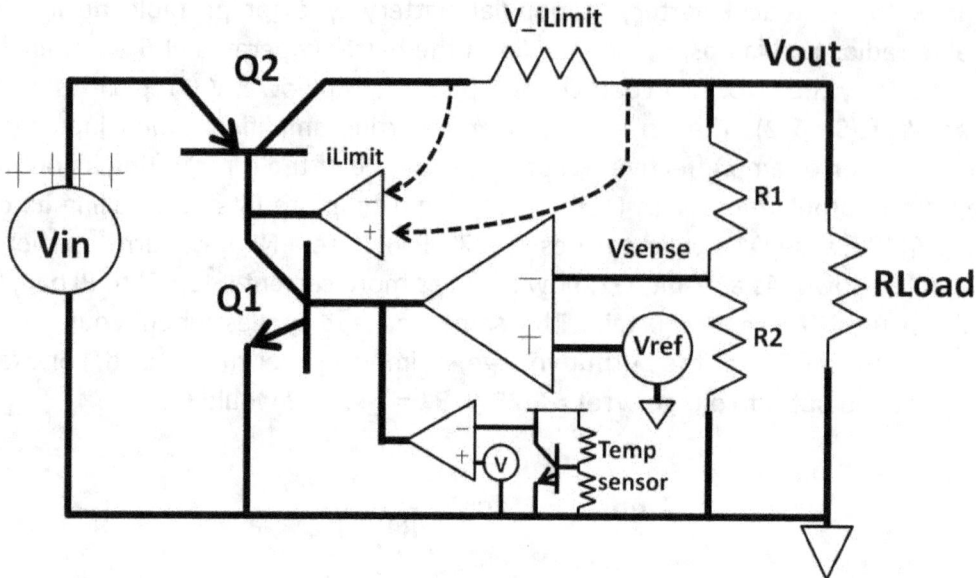

Figure 4.96: LDO with current limit and thermal shutdown features

In this example, a current limit resistor, V_iLimit (between Q2 collector and Vout) is added to the LDO. The size of the resistor determines the current limit threshold. The internal current limit comparator (iLimit) controls Q2. If over current is detected by the voltage drop across the V_iLimit resistor, for example, Vout suddenly shorts to ground. Q2's base will then pull up, shutting itself off. As a result, no current will flow to the load without damaging the pass device (Q2). The thermal shutdown circuit uses the positive temperature coefficient of the resistor to combine with the negative temperature coefficient of VBE diode. A temperature transfer function and threshold can be developed. Once the temperature goes above the designed trip point, the temperature sensor's collector pulls up, yanking Q1's base down. Q1 collector pulls up turning off Q2. LDO is then disabled. Both features prevent current flow to the load reducing the possibility of damaging the pass device.

Summary

Analog electronics interface, transform, and process many analog quantities in all kinds of applications. Analog electronics should be treated as an extension of DC, diodes, and AC, because bipolar transistors are made of two diodes. Without a complete understanding of diodes, it's difficult to get a good grasp of transistors. Full analog electronics understanding leads us to advanced digital signal processing (DSP) and more complex electronic systems. The building blocks of analog electronics are transistors. Transistors come in many shapes and forms. Bipolar and CMOS are the most popular types. Switches and amplifiers are common applications built by transistors. Transconductance small-signal models are suitable for finding out the exact voltage, current, and power gains of an amplifier depending upon the amplifier topologies. The op-amp is by far the most widely used electronic device that is implemented in a large number of designs. This chapter only covers a few op-amp circuits. It's up to the reader to further explore other circuit implementations as well as design techniques and tradeoffs. With a solid understanding of transistors and op-amps, complex circuits can be easily built, tested, and analyzed.

Quiz

1) Design a simple current source. Use one diode, one resistor, and one voltage source. Your design target is 10 uA from a 5 V supply. Assume **VBE = 1 V**. Hint: Short the NPN base and the collector together to form a diode.

2) An amplifier has the following open-loop frequency response (see figure 4.97). From DC to 10 kHz, gain is at 100 dB. Estimate unity-gain frequency.

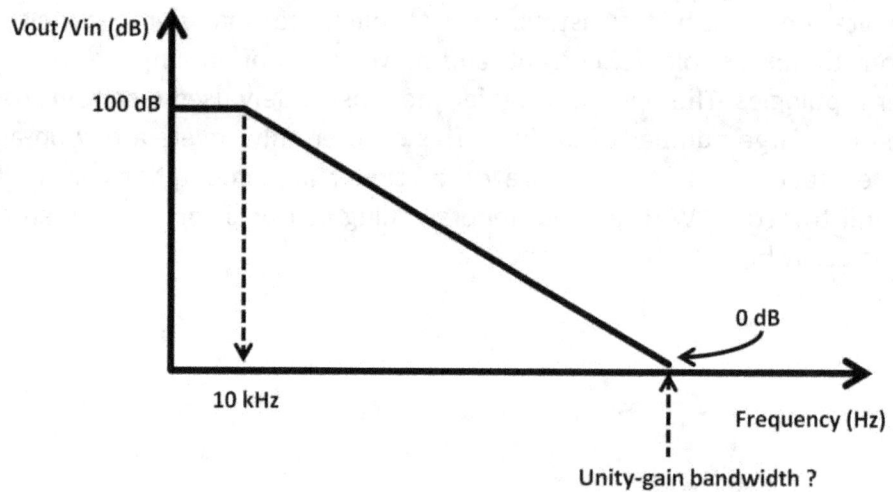

Figure 4.97: Amplifier frequency response

3) **Vin = 5 V, R = 1 kΩ**. Calculate **1)** Output current (Iout), **2)** NPN base voltage. **VBE = 1 V**. Hint: IB = 0 A. Op-amp is connected as a voltage follower (see figure 4.98).

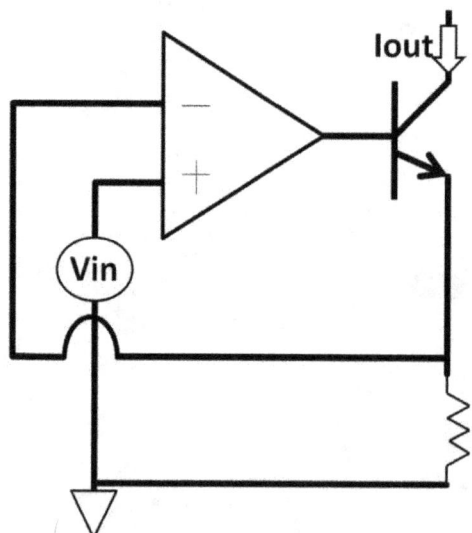

Figure 4.98: Op-amp current source

4) Many analog applications involve measuring temperature. A thermocouple is often used to measure temperature and produce an analog voltage. Thermocouple devices consist of two pieces of wire (conductor) made of different kinds of materials. The first conductor generates a voltage change from temperature change. The second conductor type would generate a voltage giving a different temperature gradient change. The transfer function of temperature per degree depends on the thermocouple type. A K-type thermocouple gives about 40 uV per °C while an S-type would give roughly 7 uV per °C. Figure 4.99 below shows a typical thermocouple application.

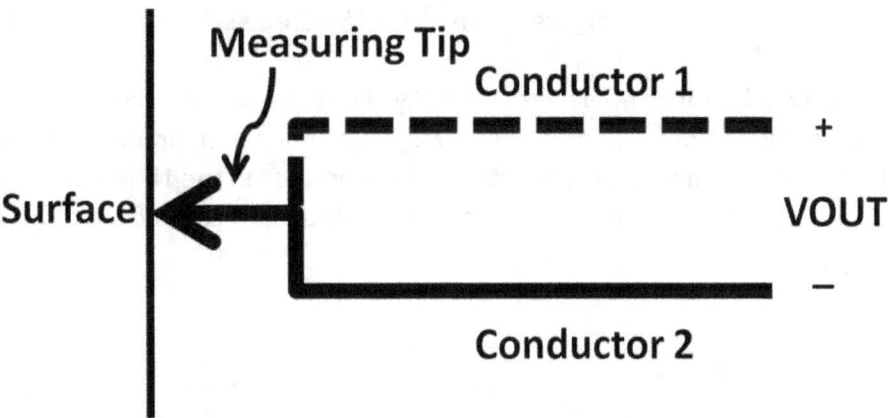

Figure 4.99: Thermocouple application

Thermocouples are generally small with fast response time. The major issues with thermocouples are small output impedance. A signal conditioning circuit may be required before driving the next stage. If the thermocouple's output impedance is 50 kΩ, it connects to an analog-to-digital converter (ADC) that has 1 MΩ input impedance. The VOUT measured by the thermocouple is 450 mV. What is the voltage that appears at the ADC input? If the minimum ADC input voltage requires 99% of the thermocouple output voltage, what do you need to add in the system to meet the ADC input requirement?

5) A Schottky diode is a special diode that features low forward bias voltage (150 mV to 450 mV) and fast response time (100 ps to 10 ns). Figure 4.100 shows a Schottky diode schematic symbol.

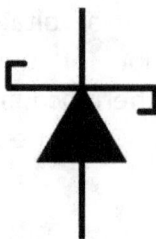

Figure 4.100: Schottky diode schematic symbol

These two features made the Schottky diode a good candidate in the switch mode buck regulator described in chapter 3, AC. One other practical use of the Schottky diode is to avoid bipolar transistor saturation. Recall the NPN symbol from figure 4.7. Figure 4.101 below shows a base-collector diode and common emitter amplifier.

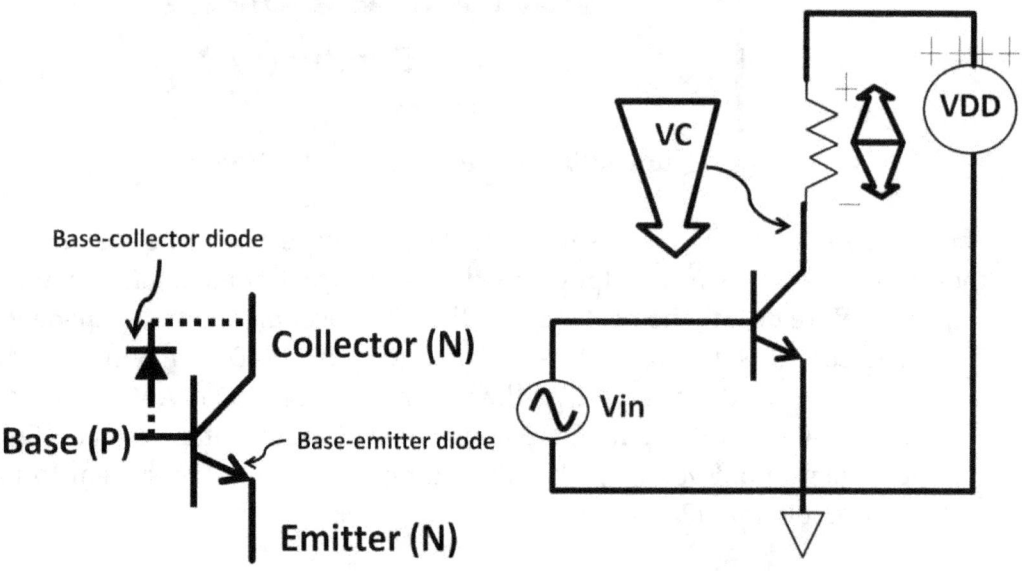

Figure 4.101: Common emitter amplifier

As Vin goes up, voltage across the collector resistor increases, causing VC to decrease. Excessive Vin increase could cause VC to go too low forward biasing the base-collector diode. How do we utilize the Schottky diode to avoid NPN saturation knowing that the Schottky diode offers low forward voltage drop?

6) A popular circuit technique is the open-collector (bipolar) or open-drain (CMOS), shown in figure 4.102. One of the applications of this circuit technique is I2C (i square c) communication protocol for clock and data lines. The drain in this circuit connects to an external R1 (pull-up resistor). It's called pull-up because when Q1 is off, the external pin pulls up to the rail generating a logic "1" (true) signal and vice versa. By knowing the on-resistance of Q1 and R1 sizes and the precise voltage, current consumption can be obtained. Assume the rail voltage is 5 V, Q1 on-resistance is 200 mΩ and R1 is 4.7 kΩ (the typical size of I2C implementations). What is the voltage at the external pin when the control signal goes high?

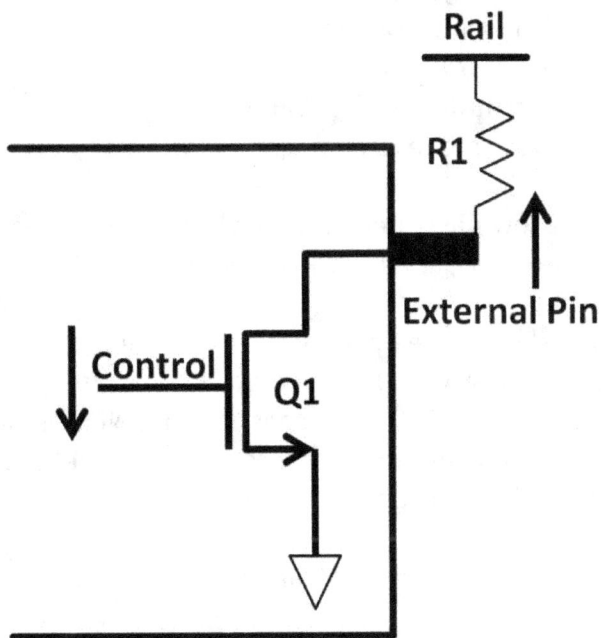

Figure 4.102: Open-drain circuit

7) Figure 4.103 shows an open-loop inverting comparator circuit using a CMOS-based input op-amp. Reference voltage (Vref) is assumed to be 2 V. Vin is a sinusoidal voltage input with **Vpeak–peak = 0 V to + 4 V**. Positive and negative rail voltages are 5 V, − 5 V respectively. Draw a Vout waveform. (Hint: As soon as Vin goes above Vref, Vout flips to the negative rail and vice versa.) The purpose of Ri is to reduce dynamic gate current due to the CMOS transistor gate being prone to damage.

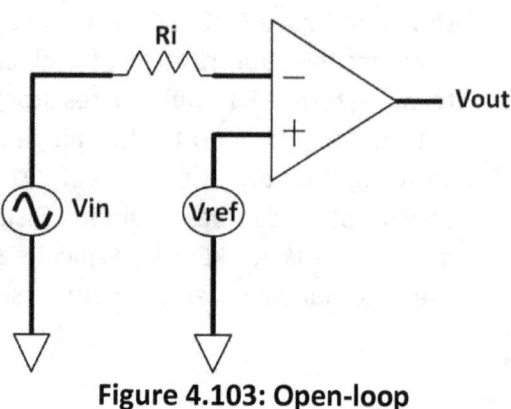

Figure 4.103: Open-loop Op-amp comparator

8) Design an active low-pass filter with f −3db at 10 kHz and a fixed gain of 10 starting at 1 MHz (see figure 4.77), assuming **Ri = 100 kΩ**.

9) Figure 4.104 shows the package of a standard MOSFET, 2N7002 by NXP Semiconductor. This MOSFET is spec at 60 V, 300 mA NFET. 60 V is the maximum drain-to-source voltage. 300 mA means that this NFET is capable of supplying 300 mA drain current (ID) at a specific VGS. Use the datasheet below and find the VGS value so that 2N7002's drain current is 300 mA.

http://www.nxp.com/documents/data_sheet/2N7002.pdf

Figure 4.104 2N7002 MOSFET (Courtesy of NXP Semiconductors)

10) On-resistance (RDSon) is non-zero in real transistors. What is the RDSon and drain current (ID) of 2N7002 if VGS = 5 V?

11) Use PFET to design a Wilson current mirror. The current mirror will produce a 10 uA reference current assuming ID is 1 uA when VGS = 1 V.

Chapter 5: Digital Electronics

Digital electronics are found in all kinds of electronic systems. Digital signals differ from analog signals in that analog quantities are non-discrete with a limitless number of possibilities, potentially leading to unwanted noise. Digital signals deal only with two, simple, well-defined, discrete levels: low (false) and high (true). The binary number system is used to describe these two levels: digit 0 (low) and digit 1 (high). The timing diagram describes the digital signals in figure 5.1.

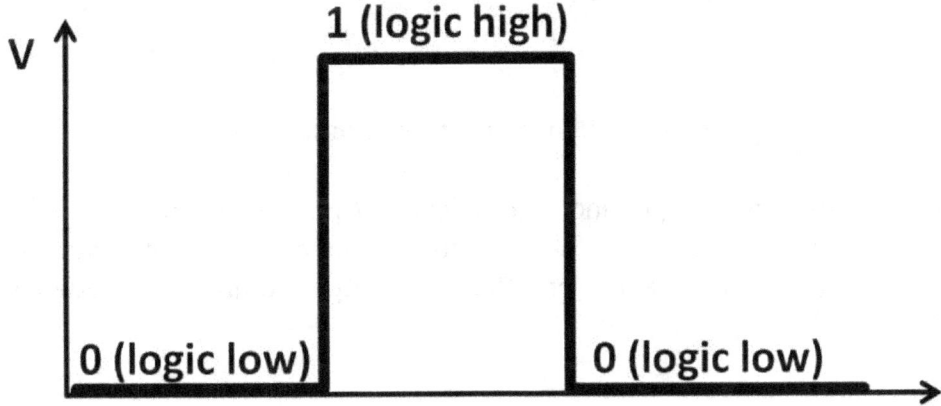

Figure 5.1: Digital signal timing diagram

With the simplicity of digital signals, they become the preferred choice to process large amounts of information (data) with high clock speed. Due to increasing demand for handling large data amounts from process-intensive applications such as high-bandwidth internet data communications, next-generation wireless technology, videos, and CPUs in computing applications, large transistor counts are needed. A CMOS transistor can be made very small and relatively inexpensively using sub-micron (less than a micrometer) manufacturing technology. Dense digital circuits like the Intel i7 Core CPU has a die (chip) size measured approximately 300 mm^2 (see figure 4.32). It contains over 1 billion transistors. Many digital circuits are called logic circuits. The exact voltage levels of the two binary digits depend on the technology. Advanced CMOS technology can have logic 1 defined as 0.5 V; logic 0 as 0 V. Basic logic circuit building blocks are collectively called logic gates. In the next few sections, we will focus on these basic logic circuits. Then we will move on to more complex digital systems. Logic gates come in wide varieties. The most basic type is the NOT-gate, discussed in next section.

1s and 0s: The Inverter

The logic NOT gate schematic symbol is shown in figure 5.2. The left-hand side of the symbol represents the input. The small circle on the right-hand side represents the output. A NOT gate can also be called an inverter.

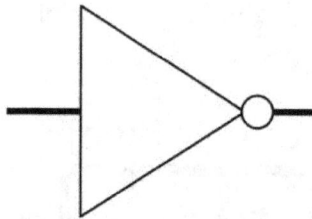

Figure 5.2: Inverter schematic symbol

An inverter is a singled-ended input and output device. When the inverter input is high, output is low, and vice versa. A truth table is often used to examine how logic gates work. A truth table lists all input, output, terminal names, and the input-output combinations. See inverter truth table (table 5-1).

Input	Output
1	0
0	1

Table 5-1: Inverter truth table

The NOT gate's truth table is divided into input and output columns. There are two possible input combinations (high or low) expressed in two rows. When the inverter input is 1, the inverter outputs 0. When the input is 0, it outputs 1. Figure 5.3 shows an AC square wave's logic input and output after being processed by the inverter.

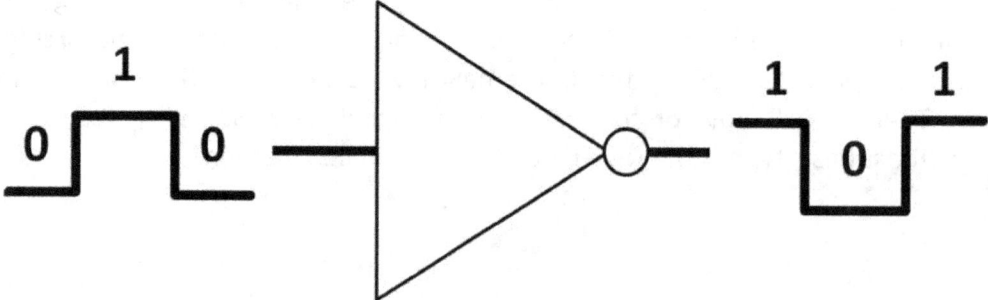

Figure 5.3: Inverter AC square wave

NMOS Inverter

In figure 4.38 from chapter 4, Analog Electronics, an NFET and resistor are used to construct an inverter. When VIN is high (e.g., 5 V logic), VOUT at the drain is low. In high density CMOS design, two CMOS complementary transistors (NFET, PFET) are used instead. The reason for that is because of power consumption. Compared to a PFET and NFET type of inverter, an NFET and a resistor inverter draw more power. This is not an ideal situation for power-sensitive applications such as high-speed CPU design. When input is high, NFET is enhanced, and current flows through the resistor; NFET and the resistor are burning $I^2 R$ power.

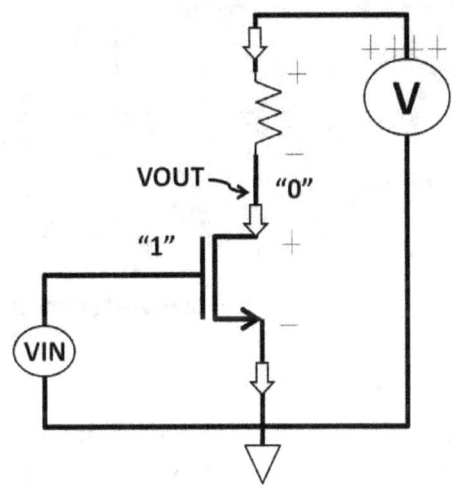

Figure 5.4: NMOS inverter

NFET and PFET Inverter

An N, PFET inverter, on the other hand, works differently. Figure 5.5 shows that the PFET connects to NFET in series. Both gates tie to each other as the input. The drains are connected together as the inverter output.

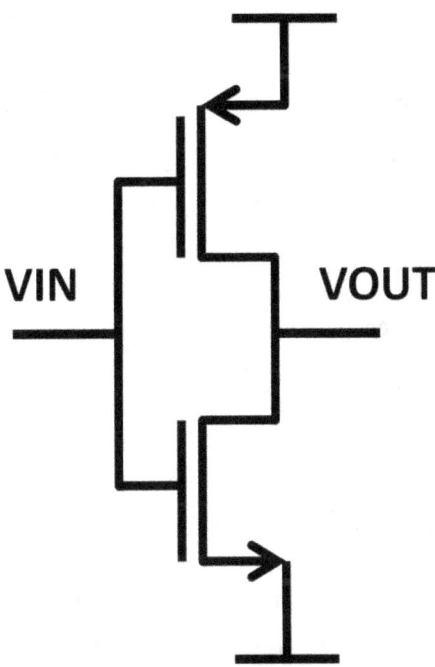

Figure 5.5: N, PFET inverter

Using table 4-4 below from chapter 4, Analog Electronics, inverter operations can be easily explained (see table 5-2).

	NFET	PFET
VGS > VT or VGS – VT > 0	ON	OFF
VDS > 0		
VSG > VT or VSG - VT > 0	OFF	ON
VSD > 0		

Table 5-2: Inverter truth table with N, PFET on, off requirements

Inverter Action

To understand NFET and PFET inverter we need to model the transistors as a switch. The following switch modes in figure 5.6 further describe these circuit operations. If the FET is on, it's enhanced (switch closed). If the FET is off, it's cut-off (switch is open). If the gate is fed high at VIN (left-hand side of figure 5.6), PFET turns off (top switch opens), NFET is enhanced, and VOUT pulls down due to a non-existing current path. When VIN is low (right-hand side of figure 5.6), NFET turns off (bottom switch open), PFET enhanced, and VOUT pulls up due to a non-existing current path.

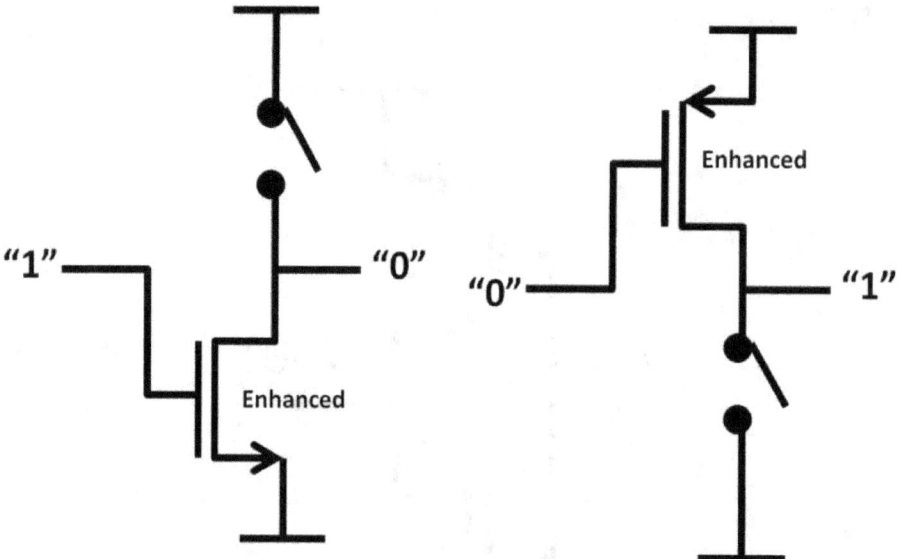

Figure 5.6: Inverter switching action

This inverter works much better compared to the NFET inverter in figure 5.4. First of all, it doesn't draw much current saving substantial power. Secondly, the size of FET can be drawn relatively smaller than a resistor (saving area, hence costs).

Shoot-Through Current

At first glance, it appears that this inverter does not draw any current at all. But if you look closely, you'll see that it draws transient current during VIN transitioning from high to low and vice versa. This current is called shoot-through current. Figure 5.7 shows the VIN transition causing the shoot-through current. Pay attention to the VIN midpoint (2.5 V). Both P and NFETs are enhanced at the midpoint of VIN causing current to flow through both transistors. The result of that is the shoot-through current occurring at each VIN transition. Figure 5.8 is a shoot-through current waveform with respect to VIN transitions.

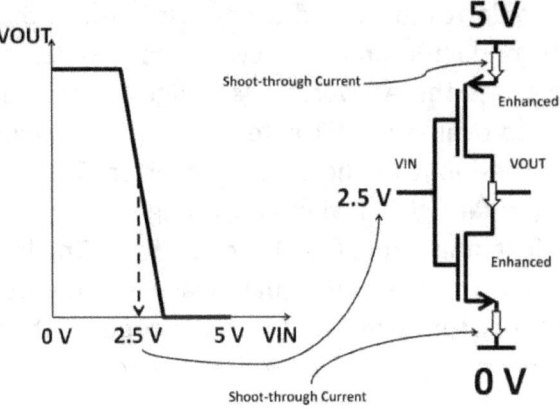

Figure 5.7: Inverter shoot-through current

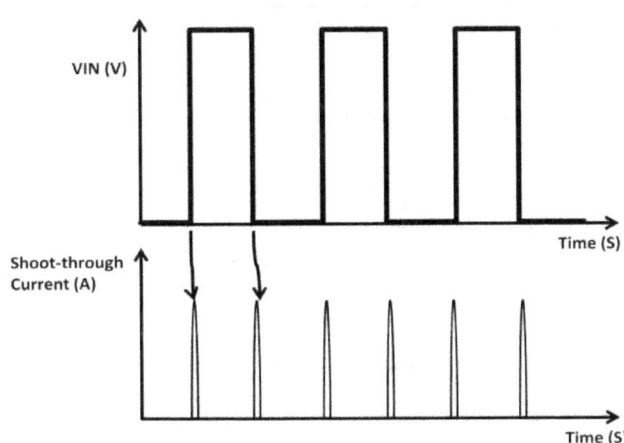

Figure 5.8: Shoot-through current waveform

Depending on the impedance of the two transistors and the components attached to them, the shoot-through current amount can be a significant source of transient noise. To avoid this, the inverter can be designed such that threshold voltage (VT) is different between N and PFETs. It means that they are no longer enhanced (on) at the same time. Varying (skewing) the width and length of the transistors could achieve just that by using the drain current equation found in chapter 4, Analog Electronics. Some refer to this technique as break-before-make or dead zone (see figure 5.8a). The shaded area is the dead zone. Within the zone, both NFET and PFET are off. This prevents shoot-through current at the expense of slower transition time. There are many other design techniques and trade-offs in designing transistor circuits. In-depth understanding of transistors is the key to design success meeting both design and tapeout target.

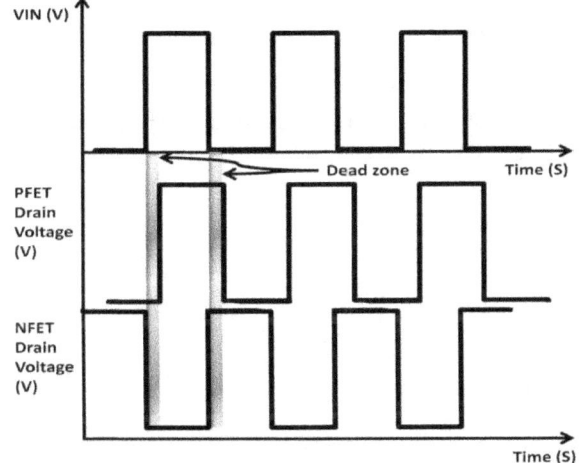

Figure 5.8a: Dead zone

Ring Oscillator

A popular circuit called a ring oscillator is made of inverters. Ring oscillators can be used in semiconductor process development to characterize device performance. A ring oscillator comprises three inverters (see figure 5.9). Suppose the input signal level (far left) is logic 0 (dotted oval), the NOT gate inverts it and yields logic 1 (first inverter output). This logic 1 feeds into the input of the second inverter. This inverter changes it to logic 0. At the third stage output (far right), it yields logic 1 again. This logic 1 (far right inverter output) resets the input of the first stage from 0 to 1 (dotted line). The logic level continues to toggle between 0 and 1. As a result, a periodic AC square wave is generated. Notice that the ring oscillator requires an odd number of inverters to function properly. If an even number of inverters were used, the ring oscillator output would be locked (latched) in one state (DC level).

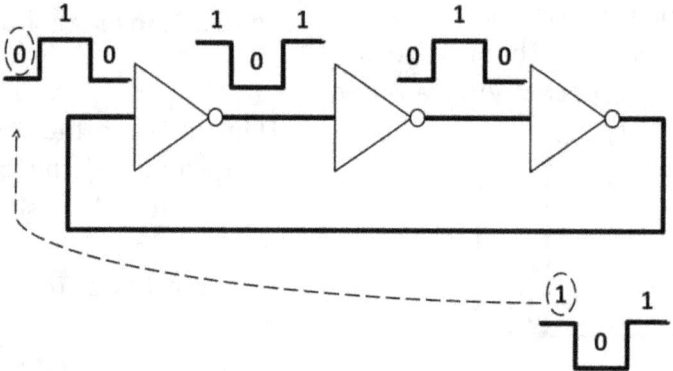

Figure 5.9: Inverter-based ring oscillator

Although the ring oscillator waveform is a square wave, it's hardly a perfect one, meaning that the rising and falling edges of the waveform are not infinitely fast. Recall that transistor gates form a capacitor between gate, oxide, and substrate. As inverter input rises from low to high, it literally charges the gate capacitor, resulting in time delay. This delay is easily explained by:

$$\Delta t = C (\Delta V) / I$$

For example, N, PFET gates' capacitance for a 3.3 V CMOS process is 10 pF. Dynamic gate current is 200 nA. The frequency of this oscillator is:

$$\Delta t = \frac{(10 \text{ pF})(3.3 \text{ V})}{200 \text{ nA}} = 0.17 \text{ ms}$$

$$T = \text{Period} = \frac{1}{\text{Frequency}}$$

$$\text{Frequency} = \frac{1}{\text{Period}} = \frac{1}{0.165 \text{ ms}} = 6,060 \text{ Hz}$$

A real inverter waveform is shown in figure 5.10.

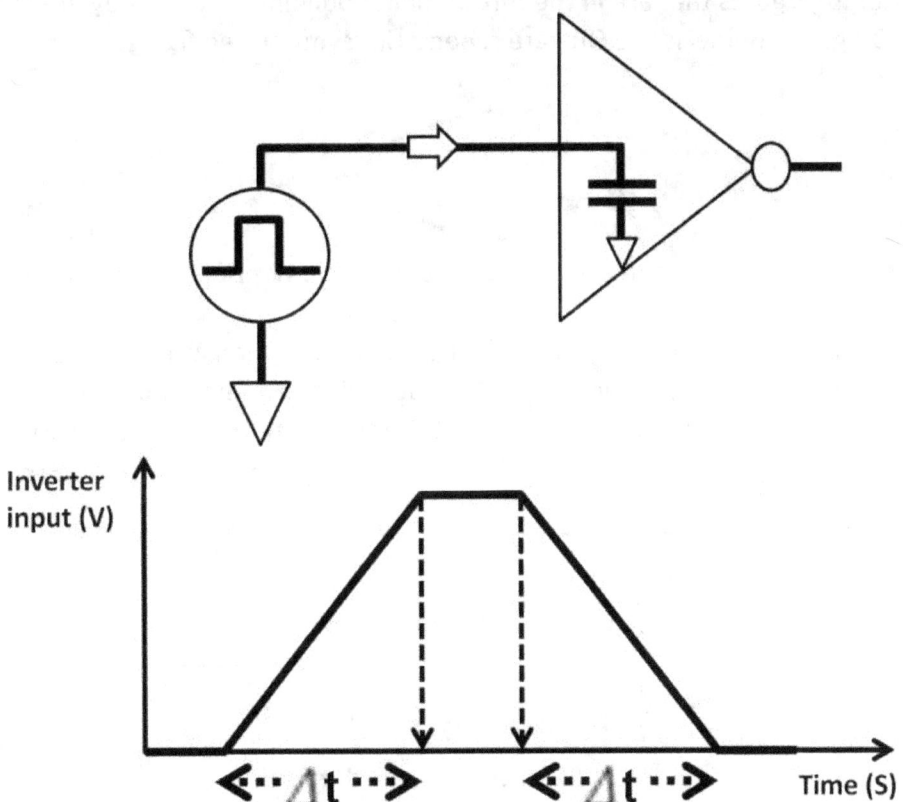

Figure 5.10: Real inverter waveform

To adjust inverter frequency, you simply increase or reduce the inverter number (increase or decrease total time delay), hence the changes in frequency. The other techniques to vary ring oscillator frequency are adjusting the width and/or length of the transistors, and adding capacitors in between inverters (see figure 5.11).

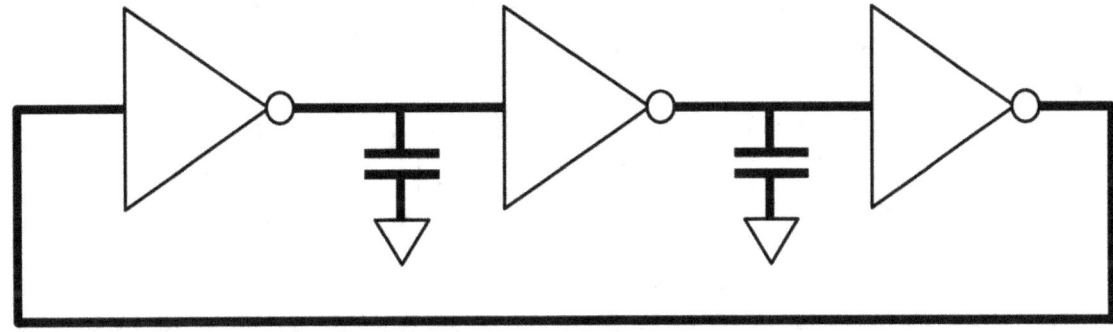

Figure 5.11: Use a capacitor to increase delay and lower frequency

OR Logic Gate

There are other logic gates that are in the mix of digital building blocks. They are OR, NOR, AND, NAND, and XOR gates. Below is the OR gate schematics symbol (see figure 5.12).

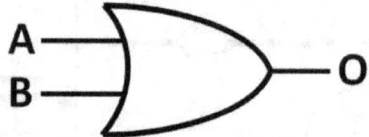

Figure 5.12: OR gate schematic symbol

There are two inputs (A, B), one output (O) in an OR logic gate. We use the binary number system to analyze digital circuits such as an OR gate. Binary numbers use the base of 2. With two inputs, the total input combinations is $2^2 = 4$. The OR gate truth table is shown below (see Table 5-3).

A	B	Output
0	0	0
0	1	1
1	0	1
1	1	1

Table 5-3: OR gate truth table

OR Gate Schematic

Several transistors are needed to construct the OR gate. Figure 5.13 shows the schematic of an OR gate. Two PFETs are connected in series. Two NFETs are connected in parallel. Recall the transistor on/off table in chapter 4, Analog Electronics. The OR gate operation is understood as such: If either A or B input is high, the NFET drain (inverter input) gets pulled down, and the inverter's output is high. The output only goes low if both A and B inputs are low. When this occurs, NFETs turn off and PFETs are enhanced, yanking the inverter input high. This results in inverter output being low. This satisfies the OR truth table. By combining N/PFETs in series and parallel form, logic gates are easily constructed.

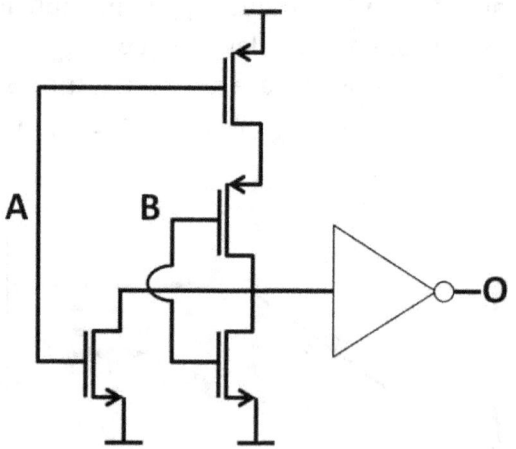

Figure 5.13: OR gate schematic

Three-Input OR Gate

An OR gate, or any other logic gate for that matter, can have more than two inputs. A three-input (A, B, C) OR gate symbol is shown in figure 5.14.

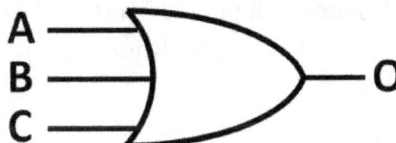

Figure 5.14: Three-input OR gate schematic symbol

With three inputs, the total number of input combinations is $2^3 = 8$. The three-input OR gate truth table is shown below (see table 5-4). There are eight input combinations starting from "000." By adding "1" to "000", it yields "001" (second row). Starting from the second row, it again increases by increments of 1. Essentially, the next row is the result of adding 1 to the previous row. This process continues until it reaches the highest value "111" (bottom row).

A	B	C	Output
0	0	0	0
0	0	1	1
0	1	0	1
0	1	1	1
1	0	0	1
1	0	1	1
1	1	0	1
1	1	1	1

Table 5-4: Three-input OR gate truth table

From the OR gate truth tables, you can devise that the output of an OR gate is high if any of the input is high. The output only goes low if all inputs are low (first row).

LSB, MSB

Among the three digits in the OR gate truth table, the number on the far right-hand side represents the least significant bit (LSB). It carries the smallest weight amount. The number on the far left-hand side is the most significant bit (MSB). It carries the largest value. This weighted approach can be explained by converting "110" back to decimal. The LSB currently has a value of "0" and has a weight of 2^0. The second digital "1" has a weight of 2^1. Finally, the MSB carries a weight of 2^2. To covert "110" back to a decimal number, multiply the corresponding binary digit by its weight, then add them up:

$$\text{MSB} \qquad \qquad \text{LSB}$$
$$\downarrow \qquad \qquad \qquad \downarrow$$
$$(2^2 \times 1) + (2^1 \times 1) + (2^0 \times 0) = 6$$

All combinations of gate input numbers yield an output of logic 1, except for an input of all 0s. The name OR gate comes from the fact that the output yields a logic 1 if one *or* all of the inputs is high. To fully understand logic gates, we can't just memorize the truth table. Instead, we need to fully understand the input and output conditions of a specific logic gate. With the understanding of these conditions, we can then come up with the truth table values.

NOR Gate

By adding a dot at the OR gate output, NOR gate is obtained (see figure 5.15).

Figure 5.15: NOR gate schematic symbol

The dot simply means that all the NOR outputs are exactly opposite (inverted) from the OR gate outputs. You can imagine there is an inverter at the output of an OR gate. The NOR gate truth table is shown below (see table 5-5).

A	B	C	Output
0	0	0	1
0	0	1	0
0	1	0	0
0	1	1	0
1	0	0	0
1	0	1	0
1	1	0	0
1	1	1	0

Table 5-5: NOR gate truth table

AND and NAND Gates

An AND gate is an important gate worth discussing. The AND symbol in figure 5.16 contains two inputs (A, B) and a single-ended output (O).

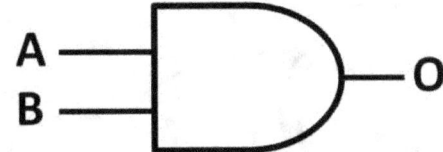

Figure 5.16: AND gate schematic symbol

The truth table below demonstrates the AND gate operations (see table 5-6).

A	B	Output
0	0	0
0	1	0
1	0	0
1	1	1

Table 5-6: AND gate truth table

The AND gate only yields high output when both A and B inputs are high (bottom row). The NAND gate is the opposite of the AND gate, where output goes low when ALL inputs are high. The NAND gate is simply an AND gate plus a NOT gate. The NAND gate symbol (solid dot at the output) and truth table are shown in figure 5.17 and table 5-7.

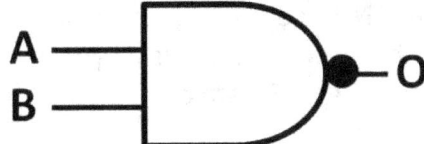

Figure 5.17: NAND gate schematic symbol

A	B	Output
0	0	1
0	1	1
1	0	1
1	1	0

Table 5-7: NAND gate truth table

XOR Gate

The last basic logic gate is the exclusive OR (XOR) gate. XOR outputs go high if the inputs are different (rows 3 and 4). If the inputs are the same, outputs stay low. The XOR symbol and operation table are shown in figure 5.17 and table 5-8.

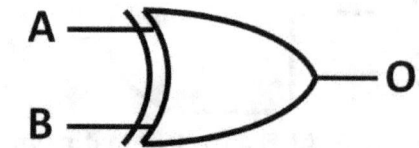

Figure 5.17: XOR gate schematic symbol

A	B	Output
0	0	0
0	1	1
1	0	1
1	1	0

Table 5-8: XOR truth table

Combinational Logic

Combing logic gates together create an endless number of possible combinations. These logic circuits are created using combinational logic. Figure 5.19 below shows a practical example. For safety reasons, automotive makers implement the windshield wiper operation in such a way that the windshield only works if three conditions are met. First, the front hood is completely closed, and both the windshield wiper switch and the ignition key are turned to ON positions. This leads to a simple three-input AND operation. Meanwhile, to make it easier for automotive technicians to work on the windshield wiper, there is a bypass switch in place to turn the wiper on regardless of the three conditions. An OR gate combined with an AND gate could accomplish that.

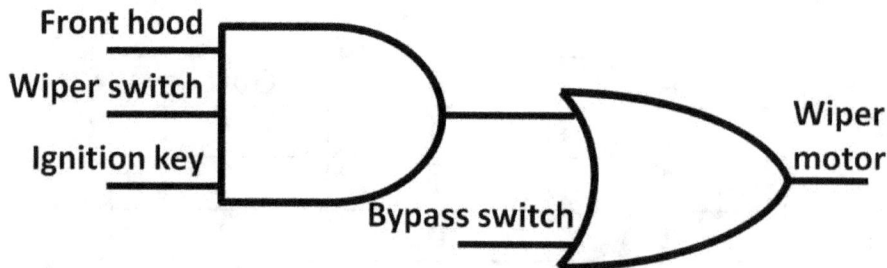

Figure 5.19: Combinational logic practical example

Boolean Algebra

To express the operations more logically, we apply Boolean algebra. For an AND gate, inputs are multiplied (solid dot) by each other. For an OR gate, inputs are added to each other. Figure 5.20 describes the logical circuits in Boolean algebra and the symbol definitions: F (front hood), WS (wiper switch), I (ignition key), B (bypass switch), and WM (windshield motor).

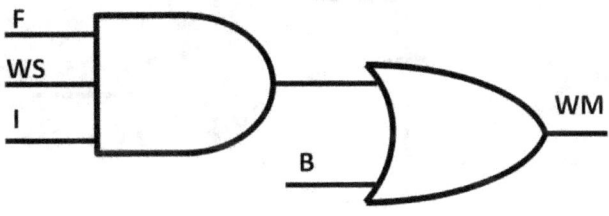

(F • WS • I) + B = WM
Figure 5.20: Logic circuits described by Boolean algebra

Using Boolean algebra, a bar on top of the letter is used to show inverted output. The inverter's Boolean equation is shown below.

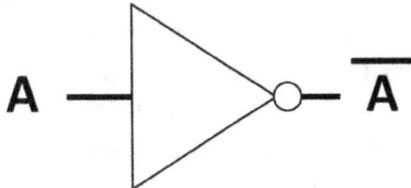

The application below shows another example of a Boolean circuit expression. Two temperature sensors are used to control a heating system. If the first or second temperature falls below a certain temperature (logic 0), the heating system turns on. The logic circuit and Boolean algebra are shown in figure 5.21.

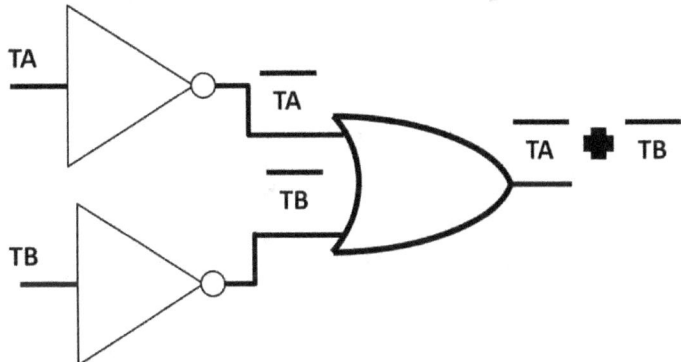

Figure 5.21: Boolean expression of a heating system application

For NAND and NOR gates, the Boolean equations can be found below (see figure 5.22).

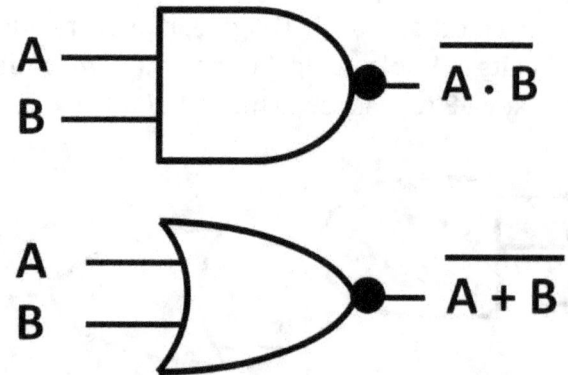

Figure 5.22: NAND and NOR gates' Boolean equations

Latch

Combinational logic output does not require any previously stored information (memory) to obtain a valid output. Many electronic systems, however, require memory to be used for desired operations. For example, when the user of a microwave oven enters the cooking time, the time is stored as memory within the microwave oven electronics. Many automobiles nowadays have memory seats. The passcode of a home security system is stored as memory within the system. Smartphone cameras store images or videos as memories. There are many more electronic applications that use memory. Digital circuits such as the latch and flip-flop are basic building blocks of digital systems and data storage elements. Digital systems combined with standard logic gates and memory are called sequential logic. The difference between a latch and a flip-flop is that a flip-flop uses a clock to determine the output states, a latch does not. A latch consists of two inputs, a set (S) and reset (R), and a differential output pair (Q, Q_bar). Figure 5.23 shows the latch schematic symbol.

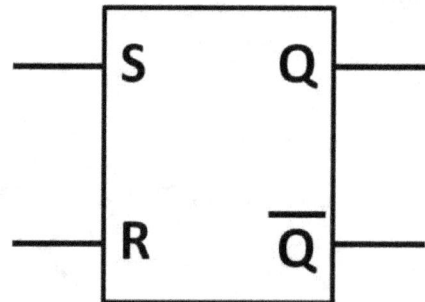

Figure 5.23: Latch schematic symbol

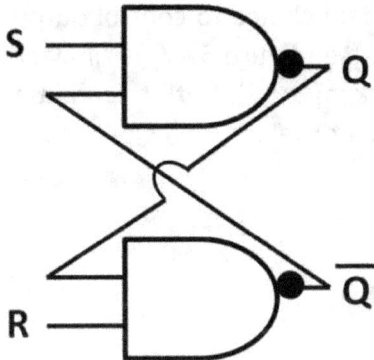

Figure 5.24: Latch made up of two NAND gates

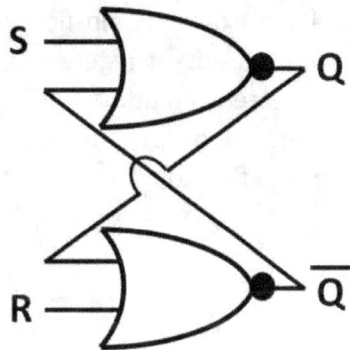

Figure 5.25: Latch made up of two NOR gates

A latch could include two NAND gates (see figure 5.24). Other than NAND gates, a latch can be constructed using NOR gates (see figure 5.25). The latch operations are described using a timing diagram below in figure 5.26. S is fed externally. When S goes high, Q goes high while R remains low. First, the rising edge of S causes Q to pull up. Q_bar is a complement of Q, i.e., 180-degrees out of phase from Q. Q continues to stay high (shaded area) even though S goes from high to low. This shaded area represents the memory is now stored. Q only goes low when R goes high, resetting the Q output. This reset occurs at the first rising edge of R. While R is purposely set high, S goes up. However, Q remains low resulting in data stored (second shaded area) on the right-hand side. Triggered by external signal, R eventually goes low while S remains high. Ultimately, the falling edge of S sets the output low on the far right.

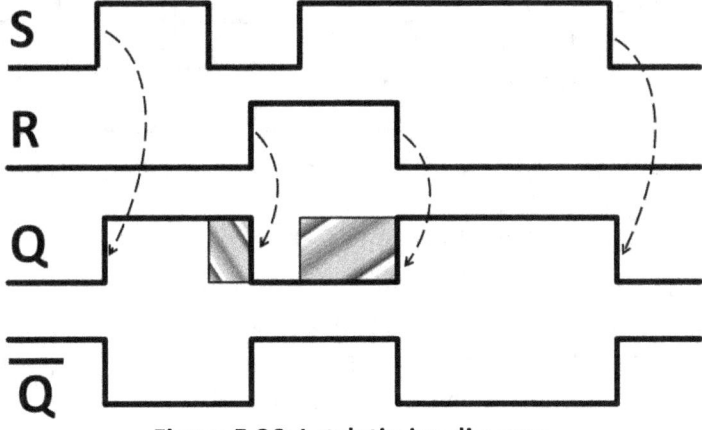

Figure 5.26: Latch timing diagram

The latch has the capability to retain information. It's free running and doesn't require any timing-specific requirement (clock) to produce a valid output. In some cases, we would like to control the output only under some particular timing constraints. This is where flip-flop comes in.

Flip-Flop

In the previous latch example, flip-flop would be an ideal choice to control output with timing requirements. The S-R edge-triggered flip-flop symbol (see figure 5.27) is similar to that of the latch except that there is an additional pin for the clock input (C). With the additional clock pin, this flip-flop triggers the output in response to the rising or falling edges of the clock, hence the name edge-triggered flip-flop.

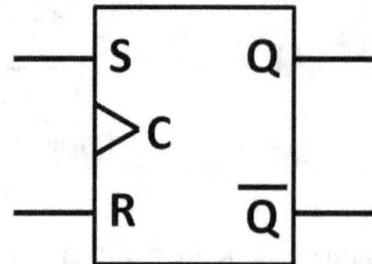

Figure 5.27: S-R edge-triggered flip-flop symbol

The operation of the S-R edge-triggered flip-flop is that the output responds only when clock is high. When clock source is low, outputs remain in their previous states. The timing diagram in figure 5.28 shows how flip-flop operates. Clock pulse C runs at a fixed frequency with a 50% duty cycle. S and R signal levels are randomly assigned. During the first rising edge of S, Q should have been set to high; instead, it stays low because the clock pulse is low. Q goes high right after the rising edge of the

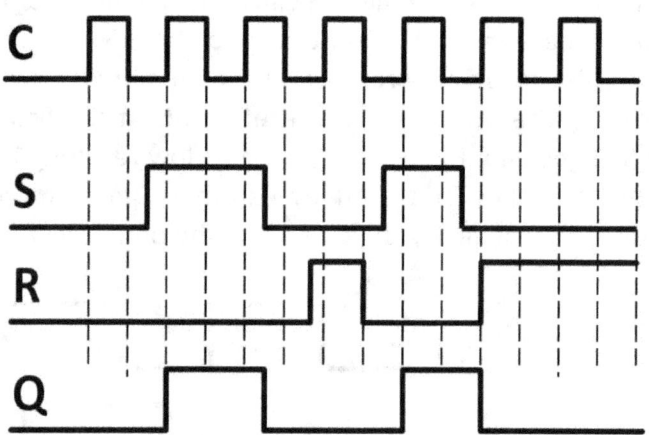

Figure 5.28: Flip-flop timing diagram

clock. Q continues to stay high while S remains high. After the first S falling edge, Q resets to low during the high clock. The first rising edge of R has no effect on Q because S is low. On the second rising edge of S, Q remains low due to clock being low. Q rises upon the next subsequent high clock. Q finally gets reset when R goes high. Some flip-flops respond to the falling edge of clock instead of a rising one. Such a flip-flop symbol is shown in figure 5.29 (dot at the C pin).

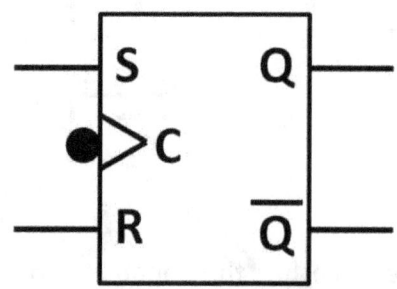

Figure 5.29: Falling edge-triggered flip-flop symbol

D and J-K Flip-Flops

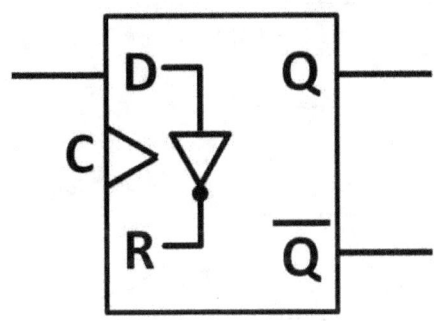

Figure 5.30: D-flip-flop symbol

Another flip-flop type is the D-flip-flop. It consists of a single-ended input (D). From a timing-function standpoint, it works exactly the same as the previous flip-flops. There are two inputs internally in the D-flip-flop. There is an internal inverter from the D input to ensure that two inputs are compliment to each other. The D-flip-flop symbol in figure 5.30 shows the internal inverter. Latches and flip-flops are just building blocks of digital circuits. J-K flip-flops, on the other hand, are a variant of edge-triggered flip-flops. They work almost exactly as S-R flip-flops except that the output toggles when the clock signal is high. The schematic symbol for the J-K flip-flop is shown below (see figure 5.31).

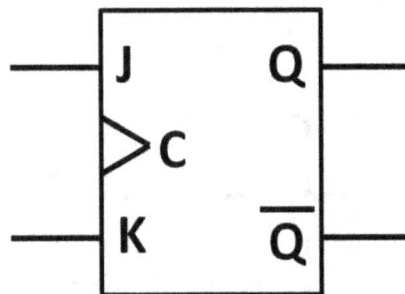

Figure 5.31: J-K flip-flop

Frequency Divider

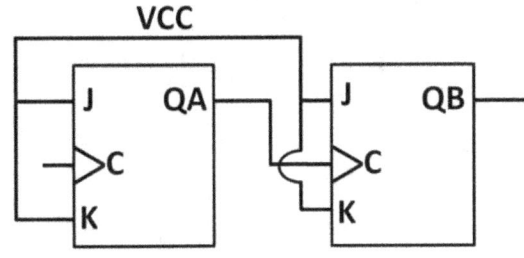

Figure 5.32: Divide-by-two frequency divider

One popular application of J-K flip-flops is the frequency divider. A divide-by-two frequency divider is shown in figure 5.32. It's a 2-bit divider. The bit is the basic unit of digital information. It's the smallest addressable unit in digital system. A bit could be assigned either "1" or "0" (transistor on or off). Digital electronics use bits and bytes to quantify memory size. For example, 8-bits of memory is equivalent to 1 byte. A bit in the frequency divider represents the number of possible combinations there are in binary system. For a 2-bit system, there are $2^2 = 4$ combinations. For a 4-bit system, there are $2^4 = 16$ combinations. Table 5-8 shows the number of possible states in decimal up to 8 bits.

Bit Number	Possible Outcomes
1	$2^1 = 2$
2	$2^2 = 4$
3	$2^3 = 8$
4	$2^4 = 16$
5	$2^5 = 32$
6	$2^6 = 64$
7	$2^7 = 128$
8	$2^8 = 256$

Table 5-8: Bit numbers and number of outcomes

Both flip-flops of the frequency divider inputs are tied to VCC (logic high). The dividing action can be seen in the timing waveform below in figure 5.33.

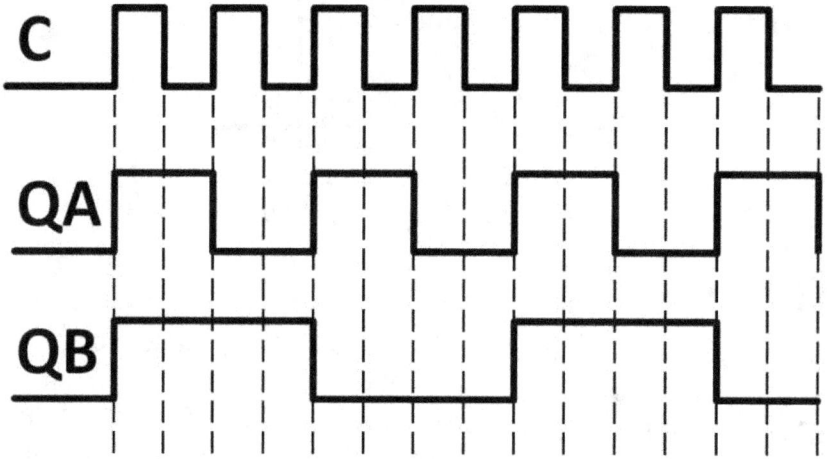

Figure 5.33: Frequency divider timing diagram

As clock goes high, QA responds by pulling up. QB goes high as well when QA is now the clock source at the second flip-flop. QA's high level is stored even after C goes low. The same goes for QB where it stays high. When C goes high again, QA now toggles back to low. QB remains high even when QA (QB's clock source) goes low. The process then continues. You can see that the clock frequency of C is divided by half through QA. QB's frequency is four times less than C. Additional dividing action can be achieved simply by adding flip-flops in series. This flip-flop utilizes the clock connected in series, i.e., each clock is independently operated. This could potentially create a timing error as one flip-flop has to wait for the output to respond before triggering the clock of the next flip-flop.

Shift Register

Flip-flop's clocks can be connected on a dedicated line making it common among all flip-flops. A well-known circuit called a shift register accepts data serially, one bit at a time on a dedicated line. The shift register output is in the exact form of the input, in this case, serially. An example of a 3-bit shift register is shown (see figure 5.34). This connection is a daisy-chain connection. Its name came from the fact that multiple devices are connected as a "chain."

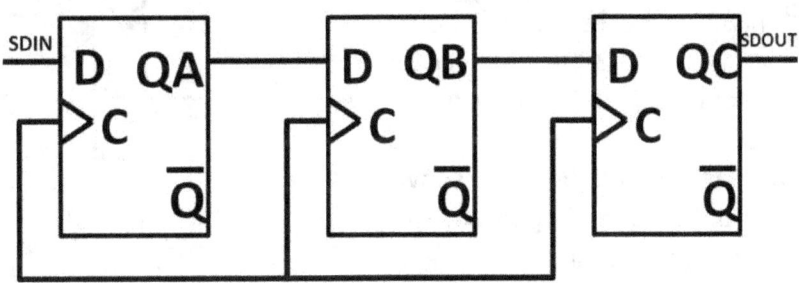

Figure 5.34: 3-bit shift register

The shift register waveform is shown below in figure 5.35.

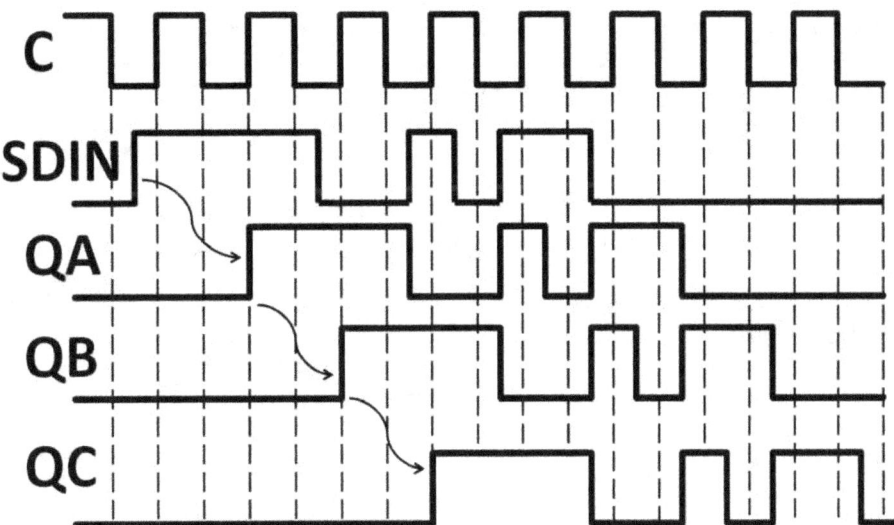

Figure 5.35: Shift register timing diagram

The first SDIN (data input) rising edge did not cause QA to rise immediately due to the clock signal being low. QA then goes high at the next rising edge. In other words, QA is delayed by one clock cycle before able to clock the data in from SDIN. Shift register is widely used in serial communications. Universal serial bus (USB) is a popular type. Others are synchronous peripheral interface (SPI), Integrated-integrated circuit (I2C) and Control Area Network (CAN). These serial transfer protocols will be discussed in chapter 7, Microcontrollers.

Parallel Data Transmission

Data can be transmitted and received via parallel communication protocols. Parallel data transmission trumps serial transmission because parallel's higher data rate with multiple data transmission can occur simultaneously. The downside to parallel data operation is the need for more transmission buses and cables, resulting in higher costs. In the 3-bit shift register example, QA, QB, and QC (SDOUT) can be retrieved in parallel while SDIN supplies data serially. A practical example in figure 5.36 shows how parallel data output gets implemented. This is a variable-gain op-amp design with gain controlled digitally by QA and QB. There are four individual gains. By clocking in SDIN serially and extracting QA and QB in parallel, four possible gain combinations can be easily selected.

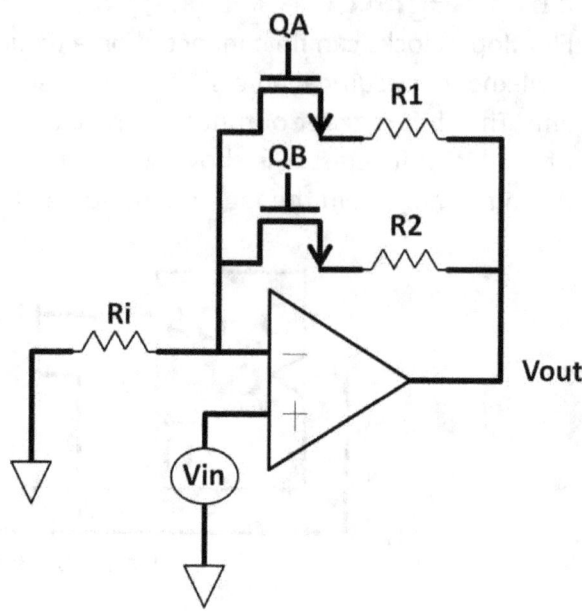

Figure 5.36: Parallel data output using op-amp

Gain 1: QA high, QB low

$$Vout = (Vin) \times \frac{R1 + Ri}{R1}$$

Gain 2: QB high, QA low

$$Vout = (Vin) \times \frac{R2 + Ri}{R2}$$

Gain 3: QA, QB both high

$$Vout = (Vin) \times \frac{R2 || R1 + Ri}{Ri}$$

Gain 4: QA, QB both low, gain of 1

$$Vout = (Vin) \times \frac{\infty + Ri}{\infty}$$

∞ >> Vin and R1:

$$Vout = Vin$$

This design example shows that electronic systems can combine both analog and digital electronics in one design. While analog output is achieved by the op-amp, low-cost and high-speed digital electronics control gain. This is a classic example of mixed-signal design.

Multiplexer

A more intuitive way to control gain is to use multiplexer (MUX). A multiplexer has multiple inputs. It selectively uses only one specific output channel depending on the control signal (CTRL). A simple MUX symbol and circuit are shown in figures 5.37 and 5.38.

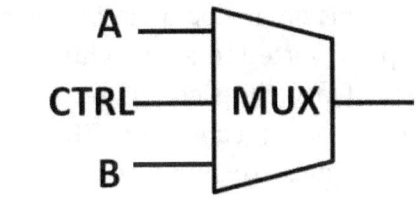

Figure 5.37: Multiplexer schematic symbol

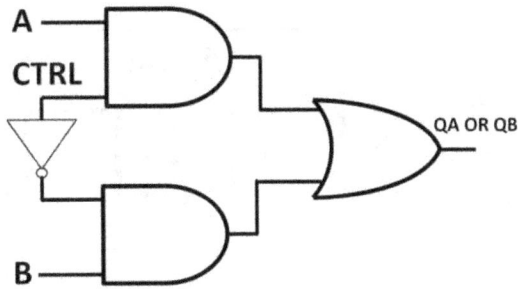

Figure 5.38: MUX made of AND and OR gates and inverter

This MUX consists of two AND gates, one OR gate, two input channels (A, B), one control pin (CTRL), and an output (either QA or QB). When the CTRL pin is high, channel A is selected while B is ignored using the AND gate logic. When CTRL is low, channel A is ignored and B channel is selected. The final gain control circuit implementation using MUX is shown below (see figure 5.39).

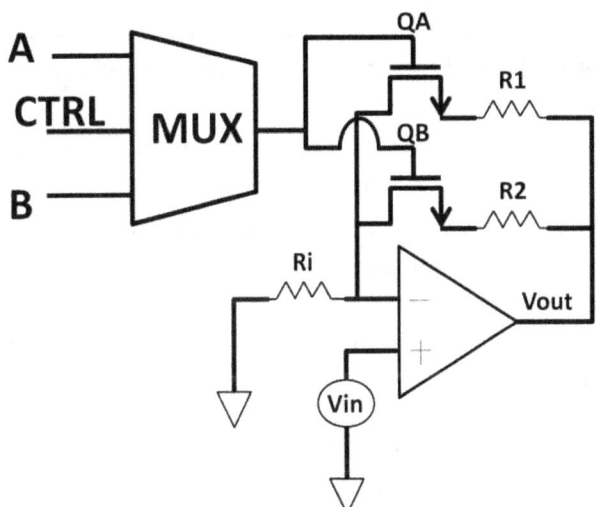

Figure 5.39: Gain control circuit using MUX and op-amp

Mixed-signal

If the op-amp on the previous page is bipolar-based, this is a mixed-signal system, meaning it combines both analog and digital circuits. Both CMOS and bipolar devices can be used in digital and/or analog designs. The trade-off comes down to power, performance, and cost. Many applications require interfacing between analog and digital quantities. For example, when you are talking on a cell phone, your voice is an analog quantity. Using analog-to-digital converters (ADCs), the voice is digitized and up-converted to a much higher frequency before transmitting as radio-frequency waves in the air. Once the signal is received by the receiving phone, the process is reversed using digital-to-analog converters (DACs) where the digital signal is converted back to sound as analog signals. This analog-to-digital, digital-to-analog concept is shown below in figure 5.40. Electronic systems such as the one below require engineers' ability to determine what type of device to use in either analog or digital systems.

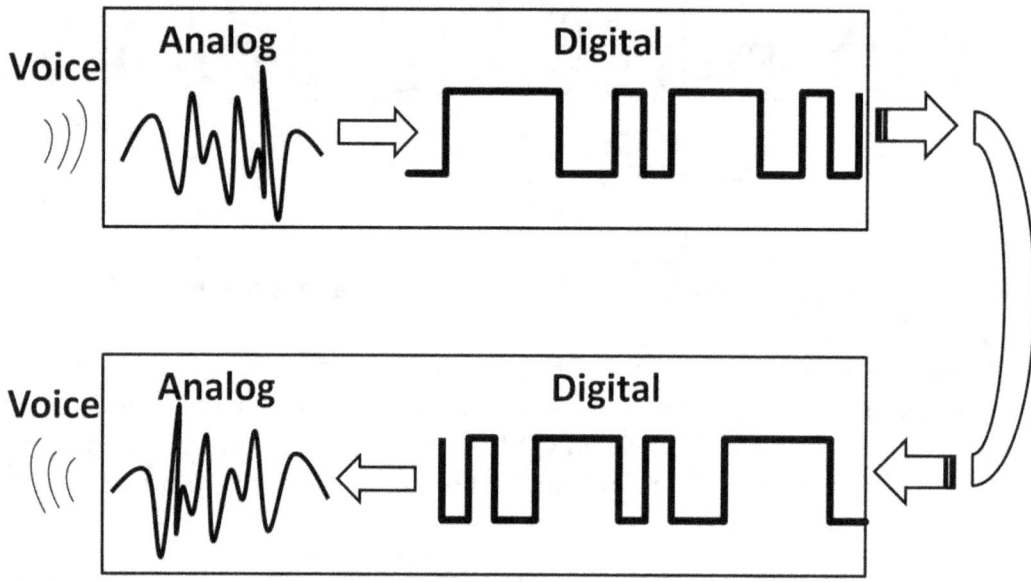

Figure 5.40: Analog-to-Digital-to-Analog concept

For industrial applications such as motor controls, you need to be cautious when integrating analog and digital designs. From a system spec standpoint, a motor takes more power and heavier load (current) to operate. CMOS devices are generally insufficient as output devices to drive a motor. Although there are special MOSFET types such as power MOSFETs that are capable of driving higher loads, bipolar devices are usually better choices when it comes to driving heavier loads. In other words, using a logic gate to drive a motor most likely would result in lack of driving capability.

Level Shifter

To resolve this load issue, a driver (level shifter) circuit can be used. A level shifter translates (shift) voltage levels from V+ to higher V++ increasing current driving capability. Figure 5.41 demonstrates this concept.

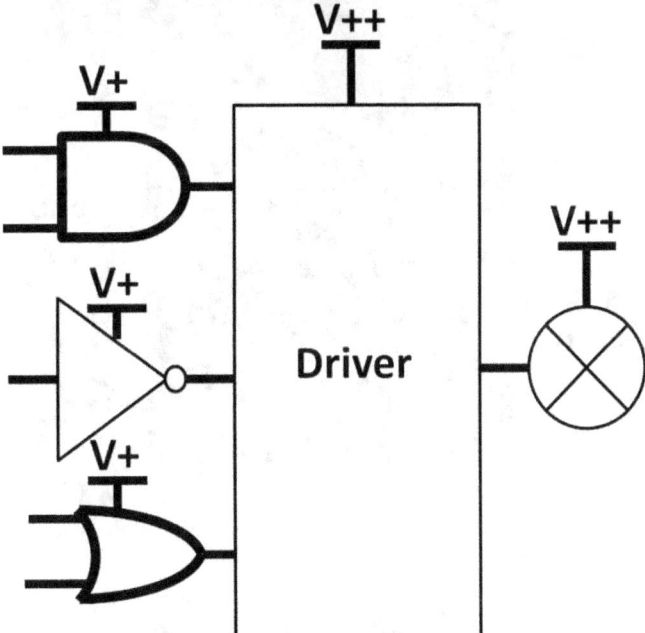

Figure 5.41: Drive as voltage level shifter

If both digital and driver circuits reside in the same system, it's a common practice to have multiple power supplies and grounds to isolate noises and minimize coupling. Undesirable, high-speed noise comes mostly from high-speed digital circuits within the systems. In microelectronic design, multiple power supplies can be generated as described in chapter 4, Analog Electronics. In addition to creating multiple supplies, digital and analog grounds can be designed to run separately so that ground currents could return to other paths. Obviously, these measures increase complexity and circuit cost with increased performance.

Multi-Layer Board

For printed circuit board design, multiple layers of power supplies and grounds are regularly implemented in printed circuit boards (PCB) with the same idea above (segregating noises). Figure 5.42 shows a student-designed circuit board (a temperature sensor application) using a microcontroller, seven-segment display, AC-DC conversion, and transformer. The bottom of figure 5.42 shows a Microchip Technology audio development board for audio applications. This board divides the power and ground into multiple layers. The input and output jacks (connectors) of this board are 3.5mm.

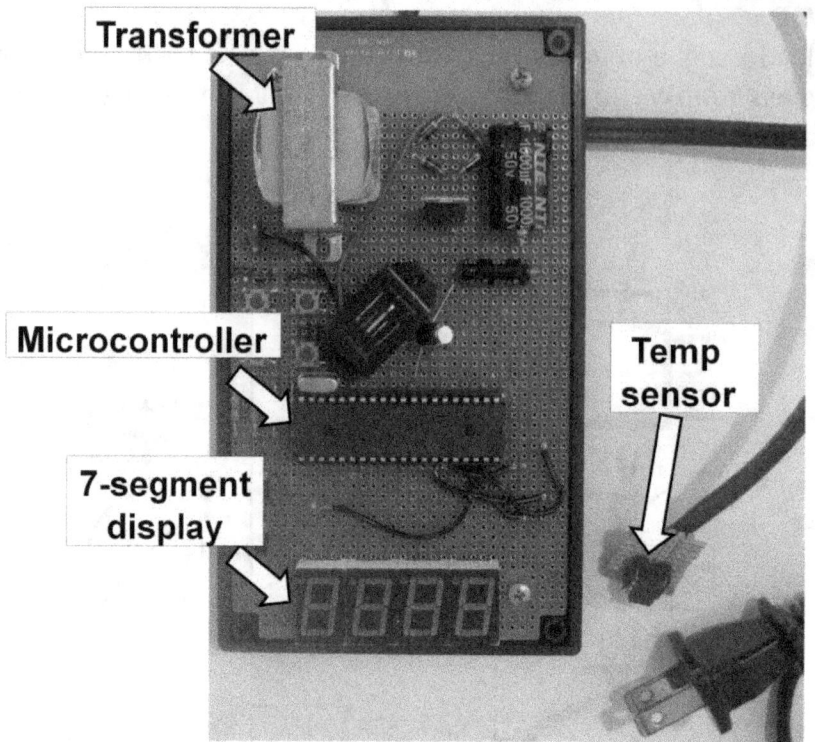

Figure 5.42: Temperature sensor PCB (Top), Audio development board (Bottom)

Many logic gates in the marketplace are grouped together into a single semiconductor package. Major IC manufacturers sell digital chips in various types. Reading device datasheets thoroughly and clearly ensures the correct chip types are used to meet your system specifications.

Digital Voltage Levels

Among digital IC specifications, you need to know the exact voltage level that defines whether it's logic 0 or 1. Transistor-Transistor Logic (TTL), CMOS, and Emitter-Coupled-Logic (ECL) are popular voltage standards found in digital designs. Among these logic families, propagation delay, toggle speed, and supply voltage are the main parameters. Table 5-9 shows the parameter comparisons of these three families. Each family has gone through multiple iterations over the years. The numbers were assembled from the latest versions.

Family	Propagation delay (ns)	Toggle speed (MHz)	Supply voltage (V)
CMOS	3	125	3.3
TTL	3	100	5
ECL	1	250	5

Table 5-9: TTL, CMOS, and ECL specifications

Analog-to-Digital Converter

Analog-to-digital converters (ADCs) and digital-to-analog converters (DACs) are found in literally all kinds of electronic products. Let's first look at ADC. The ADC schematic symbol is shown in figure 5.43. ADCs come in wide varieties and they are categorized by performance parameters such as speed (sampling rate in Hz) and resolution (number of bit) in addition to channel numbers, noise levels, temperature, voltage ranges, and

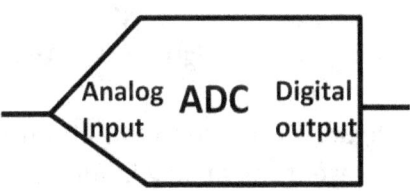

Figure 5.43: ADC schematic symbol

accuracy. Analog Devices, Texas Instruments, Linear Technology, Maxim Integrated Circuits, and Microchip Technology are among major ADC suppliers. Most offer online parametric product search such as this site from Analog Devices:

http://www.analog.com/ps/psthandler.aspx?pstid=10169&la=en

to help customers choose the right parts for their designs. Typical resolution ranges from 8-bit for low-end ADCs to high-end 24-bit ADCs. The higher the resolution, the more accurate the ADCs are. For example, an 8-bit ADC with 5 V analog reference voltage yields 256 steps, $2^8 = 256$. Each step, therefore resolves to **5 / 256 = 19.5 mV**. If it were a 24-bit ADC, a 5 V reference voltage results in 298 nV per step, a much more finer and accurate ADC. The analog-to-digital conversion of an 8-bit ADC is described in figure 5.44. The analog input signal is reproduced then converted to a digital signal as seen in the waveform. ADCs can be classified in different market segments. From industrial measurement, video, audio, and data acquisition, to high-speed instrumentation and radio-frequency applications, ADC topologies are categorized by architecture. Popular ones are sigma-delta (Σ-Δ), successive approximation (SAR), and pipeline. The differences among their architecture are characterized by resolutions and sampling rate. Sigma-delta ADCs operate high resolutions (12 to 24-bit) operating at low sampling rate (10 to 10 kHz). SARs operate in mid range performance (12 to 16-bit, 100 kHz to 10 MHz). Pipeline runs in the highest sampling rate (10 MHz to 1 GHz) with the lowest resolutions (8 to 16-bit).

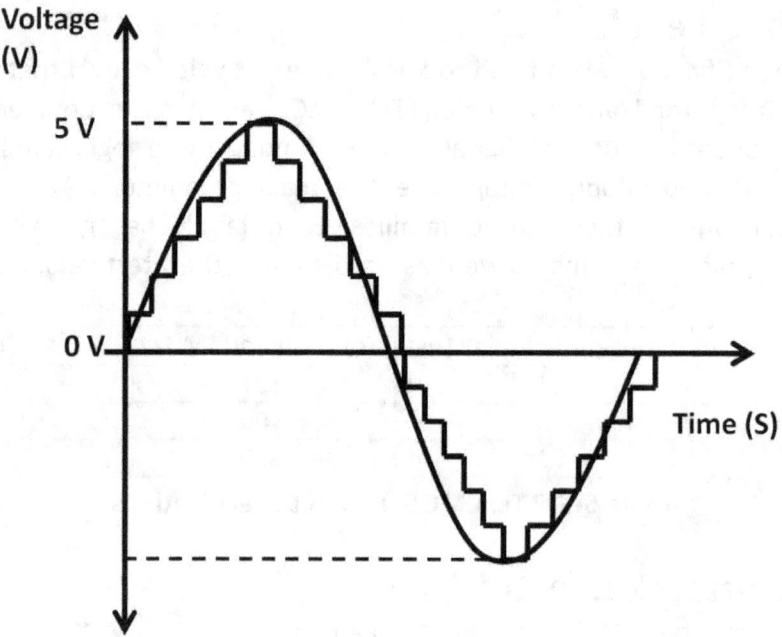

Figure 5.44: Analog-to-digital conversion of an 8-bit ADC

From figure 5.44, due to low bit number and resolutions of the 8-bit ADC, the digital output did not represent the analog input waveform quite accurately. With a 24-bit ADC, the waveform in figure 5.45 shows that the digital representation is closer to the analog input, offering much higher accuracy and a better replication of the input.

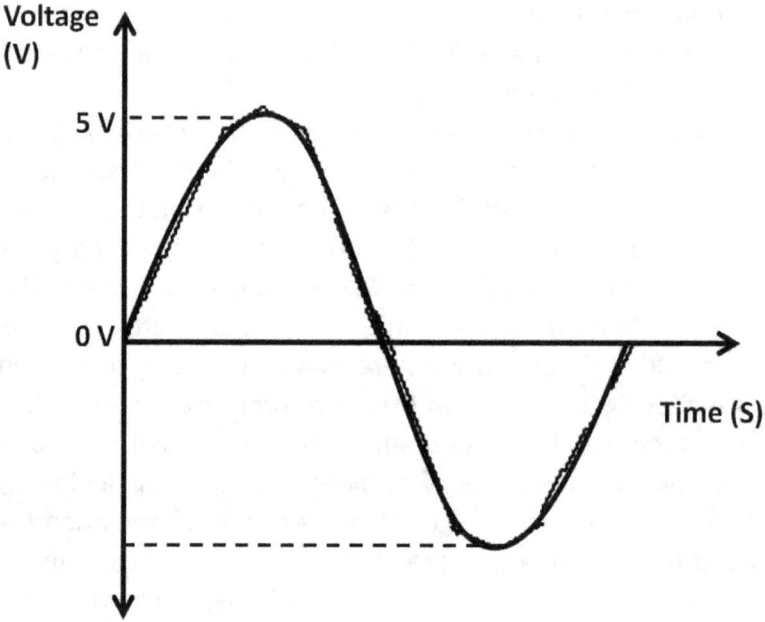

Figure 5.45: Analog-to-digital conversion of a 24-bit ADC

Nyquist Frequency

When we talk about sampling rate, it's identified as how often the ADC takes an analog signal sample. The higher the sampling rate, the more accurate the output would be. Another ADC spec is throughput rate. It's defined as mega-sample per second (MSPS). Low end, low cost ADCs run in the 100 Hz range, with high-end ones running in the 1 GHz range. The waveform below (see figure 5.46) shows that the sampling frequency is running twice as fast as the input signal. It's converting the analog-to-digital signal twice in every input signal period. The two-times sampling frequency is the Nyquist frequency. It's the minimum frequency that the sampling signal needs, i.e., at least twice as fast as the input signal (and preferably more than twice), in order to convert an analog value into a digital value with less error.

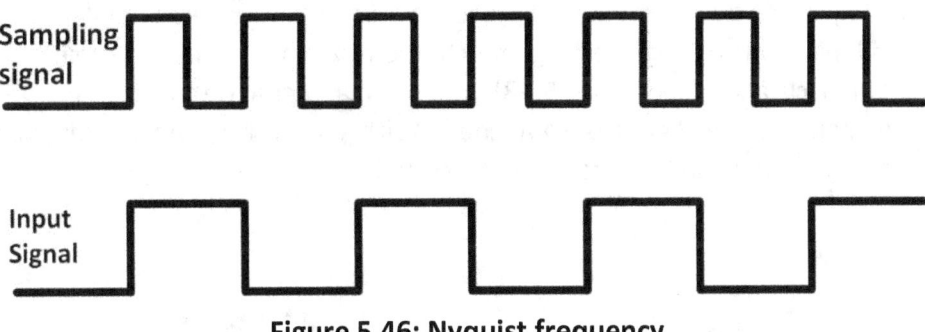

Figure 5.46: Nyquist frequency

The analog-input to digital-output transfer function of an 8-bit ADC is demonstrated in the graph below in figure 5.47, assuming the reference voltage is 8 V.

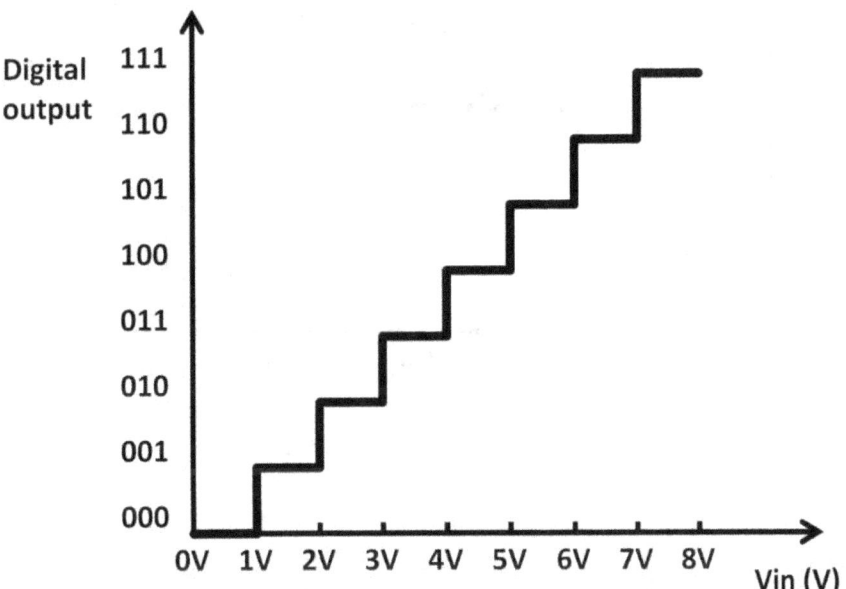

Figure 5.47: 3-bit (8 levels) analog input to digital output transfer function

The digital outputs look like ladder steps. These outputs are 3-digit binary numbers with 8 possible output combinations **($2^3 = 8$)**. Starting from "000", the value corresponds to 0 V analog input. Going up one step in the ladder, "001" will be resolved to 1 V input, so on and so forth. There are eight individual analog input ranges: 0 to 1 V, 1 V to 2 V, etc. These produce a discrete output code for each analog input. Each analog input voltage range can literally take an infinite number of values (the definition of an analog signal), causing differences between the actual analog input and the exact value of the digital output. This uncertainty is collectively called quantization error. This error ultimately leads to quantization noise with the ADC.

ADC Gain and Offset Errors

Like any other analog circuits, ADCs come with imperfections originating from design errors and the manufacturing process. Understanding these errors gives engineers knowledge about ADC's capabilities and limitations through testing and characterizations. Gain and offset errors are the main sources of inaccuracies (see figure 5.48). The original digital output is linear where analog input precisely maps to the digital output code. With gain error, the ladder step output is shifted to the right, resulting in the wrong digital code from the analog inputs.

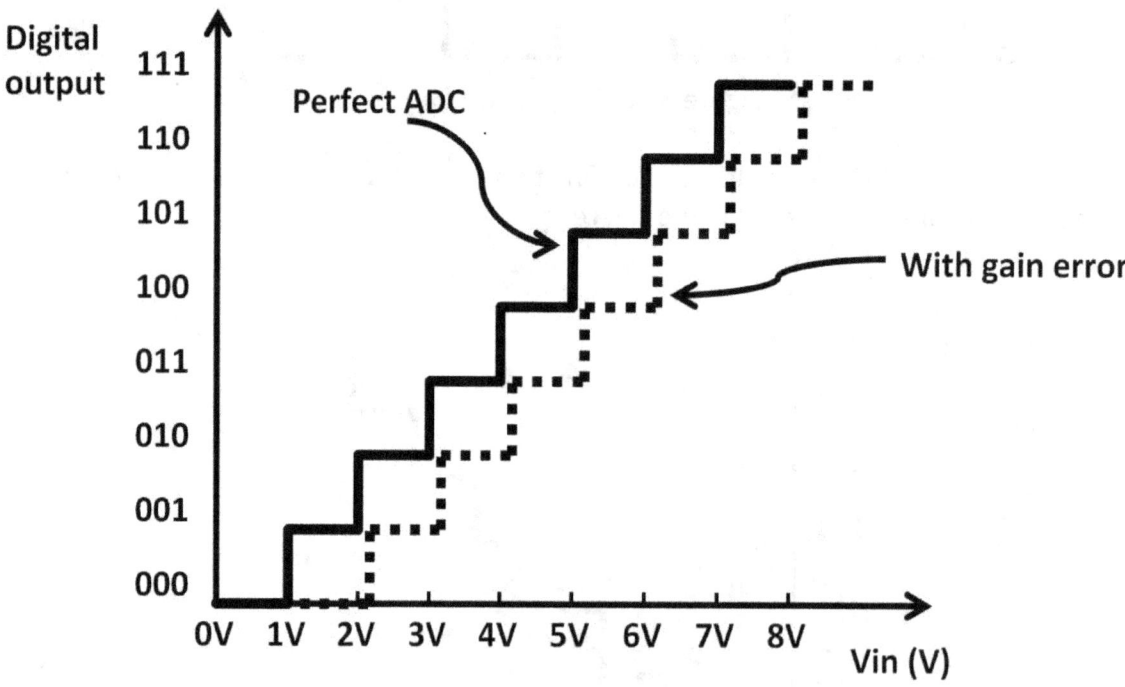

Figure 5.48: ADC gain error

Offset error, on the other, hand gives a tilted digital output as shown in figure 5.49. Both offset and gain errors are categorized as drift (changes with respect to temperature). Offset drift is measured in V / °C (voltage per degree Celsius). A 24-bit sigma-delta ADC could feature less than 5 nV / °C offset drift. Gain drift is measured in parts-per-million per °C (ppm / °C). A high resolution 24-bit ADC could have gain drift as low as 1 ppm / °C. Parts-per-million is simply a way to interpret percentage. 1 ppm means (1 / 1 million) X 100 percent. 24-bit sigma-delta ADCs are good candidates for measurement equipment applications such as temperature, pressure or weight measurements.

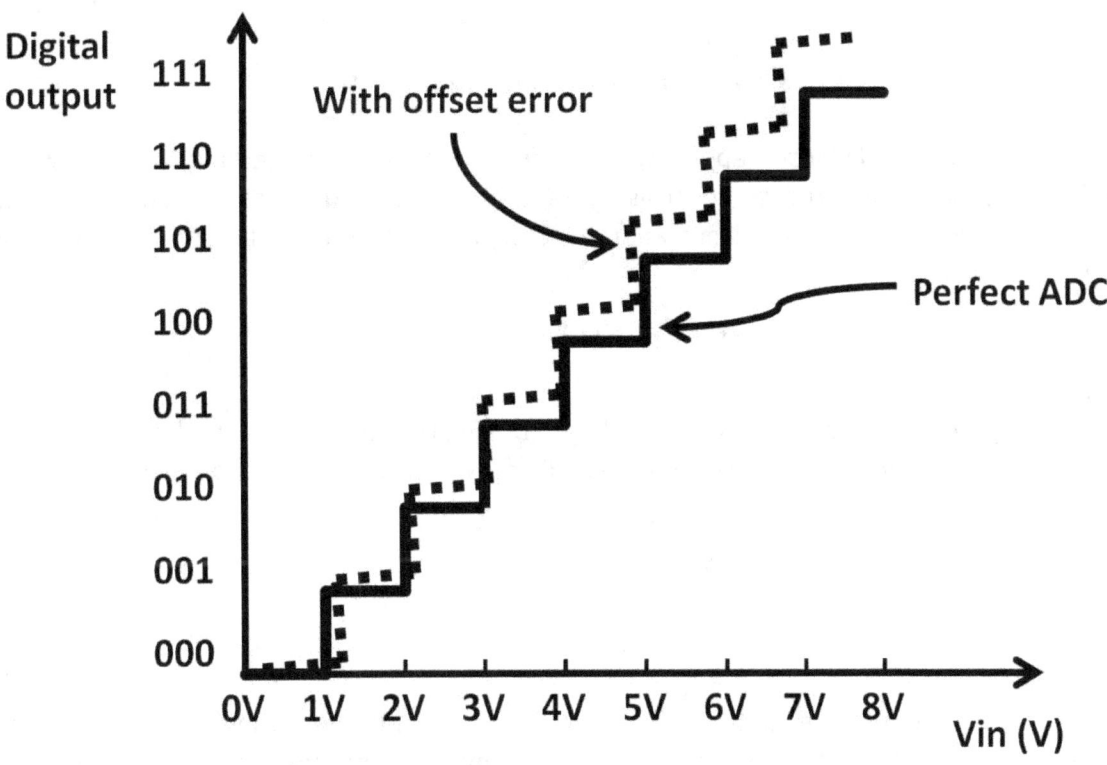

Figure 5.49: ADC offset error

Both gain and offset errors and quantization noise contribute to the non-linear ADC behavior. Other ADC specifications include signal-noise ratio (SNR) measured in dB. Ideally, SNR would be infinite if noise is zero. Other specs are power supply rejection ratio (PSRR), common mode rejection ratio (CMRR), power supply voltage ranges, phase noise in frequency domain (jitter in time domain), supply currents, clocking schemes, and interface types. Many ADCs in the market include signal conditioning circuits such as internal input and output amplifiers, buffers, and sampling clocks. The large number of ADCs makes system-level design challenging when it comes to selecting the right part for the applications.

Digital-to-Analog Converter

Digital-to-analog converters (DACs) are the reversal of ADCs, converting digital signals to analog ones. The DAC output is the proportional value of the digital inputs based on a reference voltage. The DAC schematic symbol is shown below (see figure 5.50).

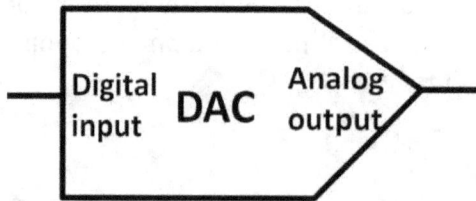

Figure 5.50: DAC schematic symbol

DACs can be found in all kinds of applications: audio, video, digital processing, wireless systems, manufacturing, motion, process controls, data acquisition, and measurement that require digital programming capabilities, just to name a few. The DAC transfer function can be derived below:

$$Vout = (Vref \times \frac{D}{(2^n - 1)})$$

Vout: Analog output; Vref: Reference voltage; D: Digital input code; n: Bit numbers. For example, a 3-bit DAC with 5 V reference voltage (Vref) with digital input code "101" results in:

$$Vout = (5 \times \frac{6}{2^3 - 1})$$

$$= 4.28 \text{ V}$$

The "101" digital inputs are first converted to a decimal number using a binary-to-decimal conversion method. Regarding DAC architecture, many academic texts cover resistive dividers and binary weighted and R-2R ladder DACs. As with ADCs, DACs' applications are widespread, from cameras, audio and video processing, and medical imaging, to wireless communications and advanced TV applications. Many end-system designs now incorporate system-on-chip (SOC) methodology where analog, digital function and circuits are integrated in one single piece of silicon, motivated by small die sizes, less board space (lower costs). Majority of high end IC suppliers design, manufacture system-on-chip ICs. One example is in the wireless industry where transceivers (transmitter and receiver combined in one design) transmit and receive radio signals. Individual circuit blocks could include ADCs, DACs, amplifiers, buffers, phase lock Loop, multiplexers, filters, voltage-controlled oscillator, voltage, current references, and other logic circuits all on one single die. To successfully design highly integrated products, engineers must understand the entire system-level specifications. Many designs involve circuit and behavioral blocks simulations to verify design functionality prior to manufacturing.

Binary-Weighted DAC

Figure 5.51 is a simple DAC example called binary-weighted DAC. It's based on a closed-loop inverting op-amp using summing amplifier topology. D0, D1, and D3 are digital inputs making it a 3-bit DAC. VOUT is the analog output. All three digital inputs will have the same voltages. Since D0 input has the largest resistor resulting in the least amount of current, it's the LSB of the DAC where D2 is the MSB. Applying the inverting amplifier gain rule from chapter 4, Analog Electronics, if all D0 to D3 are high "111" at 5 V, the VOUT is derived as below.

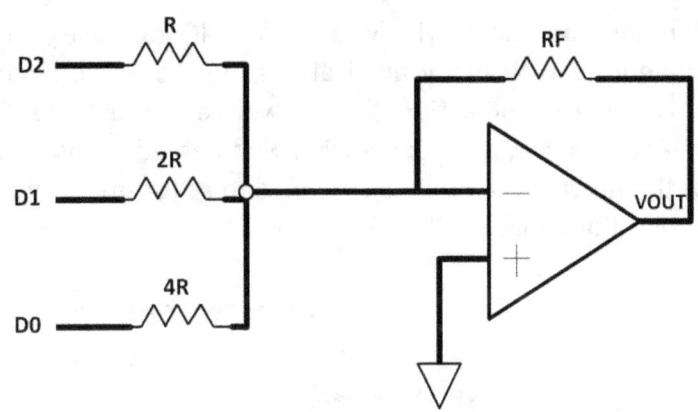

Figure 5.51: Binary-weighted DAC

$$VOUT = -\left(\frac{5}{R} + \frac{5}{2R} + \frac{5}{4R}\right) \times RF$$

For example, if **R = 10 kΩ, and RF = 5 kΩ, VOUT = − (5 / 10 kΩ + 5/ 20 kΩ + 5 / 40 kΩ) X 5 kΩ, VOUT = − 4.38 V**

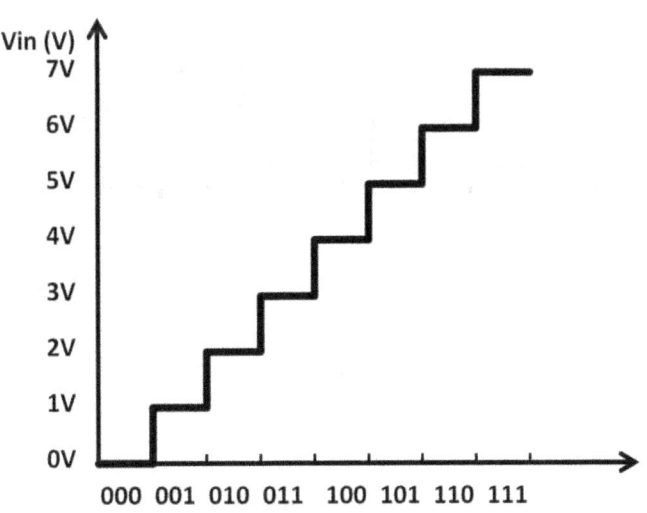

Figure 5.52: DAC transfer function

The analog output versus digital input transfer function graph is shown in figure 5.52. Many DAC design parameters are similar to those of ADCs. Gain, offset errors, PSRR, CMRR, temperature, supply voltage variations, system noise, and sampling clock rate error all affect analog output accuracy. Regardless of DAC parameters, engineers and technicians need to be concerned with the type of load the DAC is driving. In many cases, an interface device or circuit is required to provide sufficient load. Some loads require current or voltage output, hence the need of V-I or I-V conversion at the DAC output. In some cases, a separate clock or voltage reference IC is needed for clocking and providing voltage supply to the DAC or ADC, because there may not be one single data converter that is able to meet all design requirements.

555-Timer

Perhaps the most widely discussed IC in college curricula is the 555-timer. It can be implemented in many applications, e.g., precision timing, oscillation, pulse generation, and pulse width modulation (PWM) with an adjustable duty cycle. The original 555-timer was invented by Mr. Hans Camenzind who passed away in 2012 at the age of seventy-eight. It's one of the most successful ICs ever invented. It remains widely used in academics and commercial applications. Figure 5.53 shows the 555-timer block diagram and pin names.

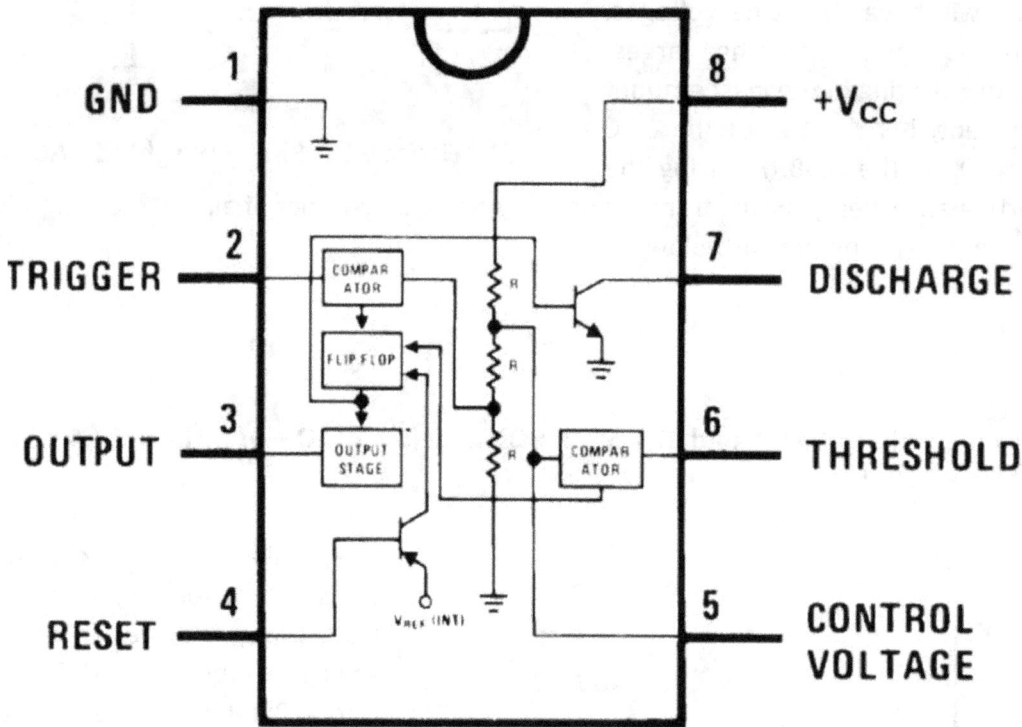

Figure 5.53: 555-timer block diagram (Courtesy of Texas Instruments)

The electrical specification of the 555-timer is shown in table 5-10 below.

Electrical Characteristics [1] [2]

($T_A = 25°C$, $V_{CC} = +5V$ to $+15V$, unless otherwise specified)

Parameter	Conditions	Limits LM555C			Units
		Min	Typ	Max	
Supply Voltage		4.5		16	V
Supply Current	$V_{CC} = 5V$, $R_L = \infty$ $V_{CC} = 15V$, $R_L = \infty$ (Low State) [3]		3 10	6 15	mA
Timing Error, Monostable					
Initial Accuracy			1		%
Drift with Temperature	$R_A = 1k$ to $100k\Omega$, $C = 0.1\mu F$, [4]		50		ppm/°C
Accuracy over Temperature			1.5		%
Drift with Supply			0.1		%/V
Timing Error, Astable					
Initial Accuracy			2.25		%
Drift with Temperature	R_A, $R_B = 1k$ to $100k\Omega$, $C = 0.1\mu F$, [4]		150		ppm/°C
Accuracy over Temperature			3.0		%
Drift with Supply			0.30		%/V
Threshold Voltage			0.667		x V_{CC}
Trigger Voltage	$V_{CC} = 15V$		5		V
	$V_{CC} = 5V$		1.67		V
Trigger Current			0.5	0.9	μA
Reset Voltage		0.4	0.5	1	V
Reset Current			0.1	0.4	mA
Threshold Current	(5)		0.1	0.25	μA
Control Voltage Level	$V_{CC} = 15V$ $V_{CC} = 5V$	9 2.6	10 3.33	11 4	V
Pin 7 Leakage Output High			1	100	nA
Pin 7 Sat [6]					
Output Low	$V_{CC} = 15V$, $I_7 = 15mA$		180		mV
Output Low	$V_{CC} = 4.5V$, $I_7 = 4.5mA$		80	200	mV
Output Voltage Drop (Low)	$V_{CC} = 15V$				
	$I_{SINK} = 10mA$		0.1	0.25	V
	$I_{SINK} = 50mA$		0.4	0.75	V
	$I_{SINK} = 100mA$		2	2.5	V
	$I_{SINK} = 200mA$		2.5		V
	$V_{CC} = 5V$				
	$I_{SINK} = 8mA$				V
	$I_{SINK} = 5mA$		0.25	0.35	V
Output Voltage Drop (High)	$I_{SOURCE} = 200mA$, $V_{CC} = 15V$		12.5		V

Table 5-10: 555-timer electrical specifications

Figure 5.54 shows the simplified internal circuit diagram of the 555-timer.

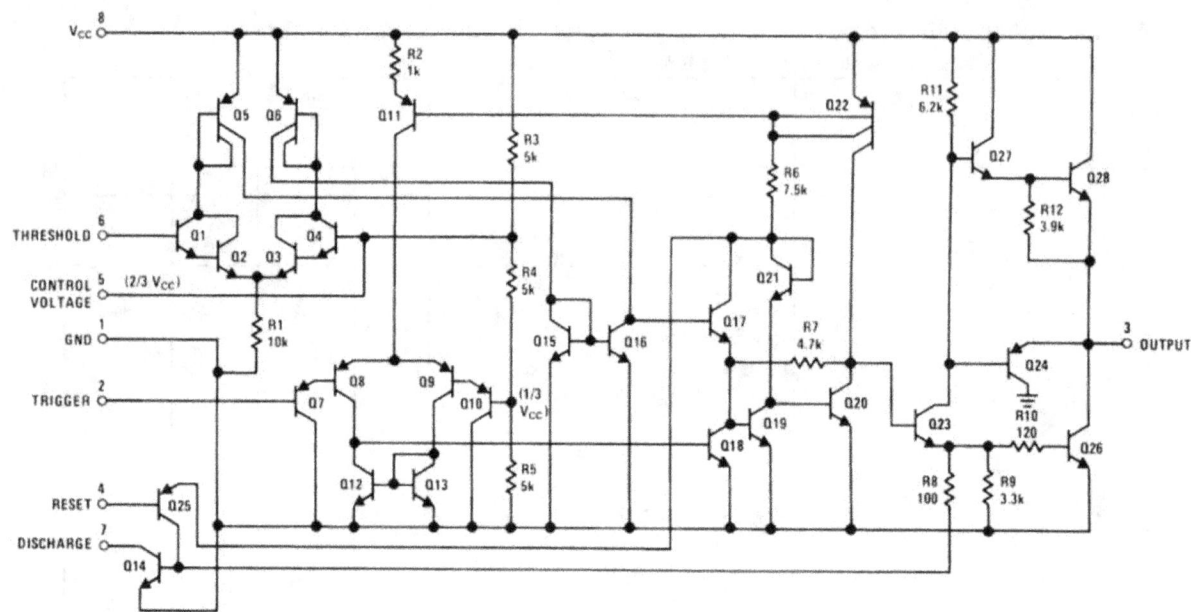

Figure 5.54: Simplified internal 555 schematic (Courtesy of Texas Instruments)

Let's take a look at a simple 555-timer monostable application (see figure 5.55) using the simplified schematic. A monostable circuit has only one stable logic state while the other state is unstable (always in transition). The presence of a trigger signal forces the 555-timer into an unstable state (R1, C1, time constant). In this example, the 555-timer functions as a one-shot timer. The reset pin connects internally to the base of the PNP (Q25 in figure 5.54)), which controls the discharge pin. Pulling the reset pin low turns on PNP. This pulls the discharge pin low, forcing output to stay low. Tying the reset pin to VCC keeps PNP off and the part out of reset state. The output pin connects to a VCC, R2, and R3 voltage divider as output load. Keep in mind, the 555-timer can source or sink only up to 200 mA to the load. A 555-timer is not suitable to drive high loads. The control voltage pin connects to an internal voltage divider (R3, R4, and R5) used as a comparator threshold. The threshold voltage is set by the internal resistor ratio (1 / 3 X VCC or 2 / 3 X VCC). The external 10 nF capacitor (C1 in figure 5.55) is mainly for noise reduction and decoupling purposes. The threshold and discharge pins are tied together upon receiving a negative pulse at the trigger pin. When the threshold and discharge pins (tied together) fall below (1 / 3 X VCC), the discharge and threshold pins charge up. The charging time depends on the R1, C1, time constant value. During this time, the internal flip-flop sets the output high. When the discharge and threshold pins rise to (2 / 3 X VCC), they trip the comparator resetting the flip-flop. This lifts the base of internal NPN, pulling the collector down and discharging C1. The output stays low (stable state) until next time there is a negative pulse at the trigger pin. This one-shot only works if the negative pulse occurs slower than the R1, C1 charge time. The trigger pulse, output, and discharge/threshold waveforms are shown in figure 5.56.

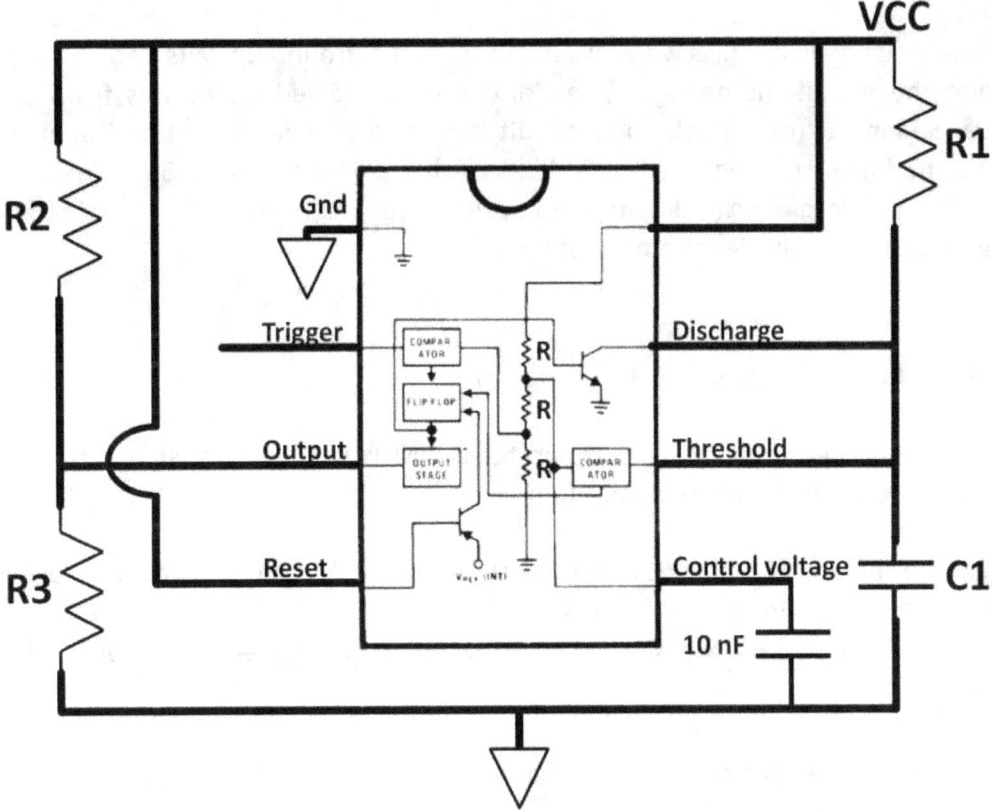

Figure 5.55: One-shot 555-timer application

Figure 5.56: One-shot 555-timer waveforms

Summary

In this chapter, digital electronics were discussed from the ground up. We started from bits "1" and "0" and the definitions of logic gates, and then explained operations from the device perspective. Spanning from simple logic circuit blocks to popular digital and analog circuits, ADCs, DACs, multiplexers, digitally controlled variable gain amplifiers, 555-timers, summing amplifiers, and other practical circuits were presented and explained in a simple manner combining real world quantities and parameters.

Quiz

1) Construct an AND gate using CMOS transistors.

2) Design a frequency divider that generates a 2 MHz square wave signal from a 16 MHz input clock. Hint: Use three J-K flip-flops.

3) Create a 1 GHz output clock from a 0.5 GHz clock source. Verify it using timing waveform. Hint: Use a two-input XOR. Separate the 0.5 GHz into two signals. Feed them to the inputs of the XOR. Make the inputs 90 degree out of phase from each other.

4) Design a variable-gain op-amp (see figure 5.36) with the following gain options: 2, 4, 8, and 16.

5) How many levels of digital outputs does an 8-bit analog-to-digital converter (ADC) have? What is the output code of the first and last levels?

6) Calculate the resolution of a 16-bit ADC if the analog reference voltage is 1.8 V.

7) Design a 555-timer application that is astable-based meaning it's unstable in both states. Draw trigger, discharge, threshold, and output waveforms Hint: Connect the trigger, and threshold pins together.

8) A 3-bit Digital-to-Analog Converter (DAC) has the following transfer function:

$$Vout = (Vref \times D) / (2^n - 1)$$

D: Digital input; Vref: Reference voltage; Vout: Analog Output voltage; n: number of bits
Calculate Vout using digital inputs below. **Vref = 2.5 V.**
a) 010
b) 111

Chapter 6: Communications

An electronic communications system's function is to transmit and receive information from one end to another and vice versa. Some communications are one-way (simplex) meaning one end can only transmit, the other can only receive. Radio and television broadcast are examples of simplex communications. Other communications techniques are occurring in both directions (bi-directional). In bi-directional systems, information can be communicated in two ways: **1)** occurring at the same time (full duplex), and **2)** one direction at a time (half duplex). Cell phones and computer networks are prime examples of full-duplex systems while walkie-talkies (two-way radios) are examples of half-duplex communications. Communication systems that are able to transmit and receive signals are called transceivers. A cell phone is a classic transceiver example. Communication systems comprise a series of analog-to-digital, digital-to-analog conversions where information is transmitted and received via a communication medium (channel). The medium could be in the form of wired or wireless (signal travels through the air). The raw material of any wired medium is typically copper. Fiber optics have gained popularity in recent years. Most wired communications are standardized as protocols by organizations such as The Institute of Electrical and Electronics Engineers (IEEE). Well-known protocols are RS-232 (computer serial port), RJ-45 (phone connector standard) and coaxial cable. Voltage levels, attenuations, impedances, and frequency ranges are clearly specified by each standard. A wireless signal goes through the air as the medium is an AC signal called a radio-frequency (RF) signal. Before transmitting through the air, signals in the communication systems are first up-converted to much higher frequencies of RF signals frequency. The RF signal frequency ranges are wide-ranging from 3 kHz to 300 GHz. This chapter primarily focuses on wireless communications. In the US, each individual frequencies region (band) hold specific purposes, from phone, radio, satellite, and television, to broadband communications. Each type occupies a specific frequency region called a frequency band (spectrum). Frequency band allocations are controlled by the Federal Communications Commission (FCC), a government agency. The picture below shows a portion of the frequency spectrum designated by the FCC. The numbers on the top represent the frequencies in Hz. Each rectangle defines the names of usage and frequency ranges. Communication systems work mostly on frequency domain.

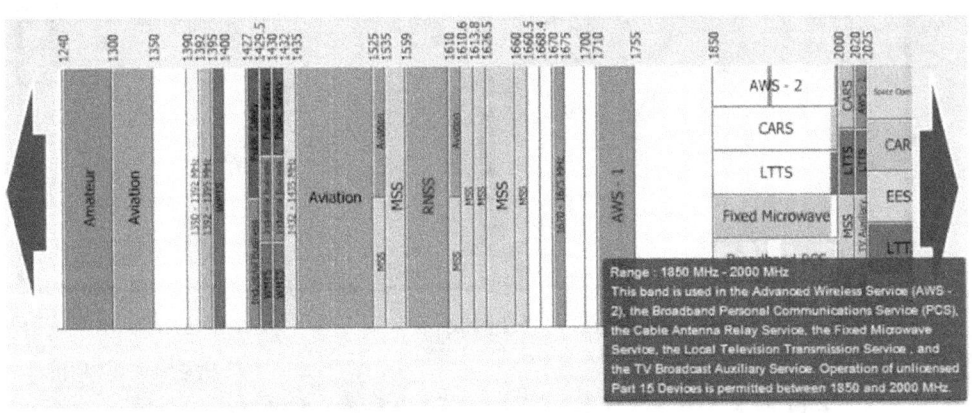

Time versus Frequency Domains

There is a strong relationship between time and frequency domains (frequency = 1 / time) as described in chapter 3, AC. A periodic sine wave running at 10 kHz with 10 V peak-to-peak is displayed in figure 6.1 as a time domain waveform (top). To express it in frequency domain, a spectrum analyzer displays voltage, current, or power as a function of frequency (bottom). The spectrum analyzer's X-axis is the frequency in Hz. The Y-axis could be voltage, current, or power.

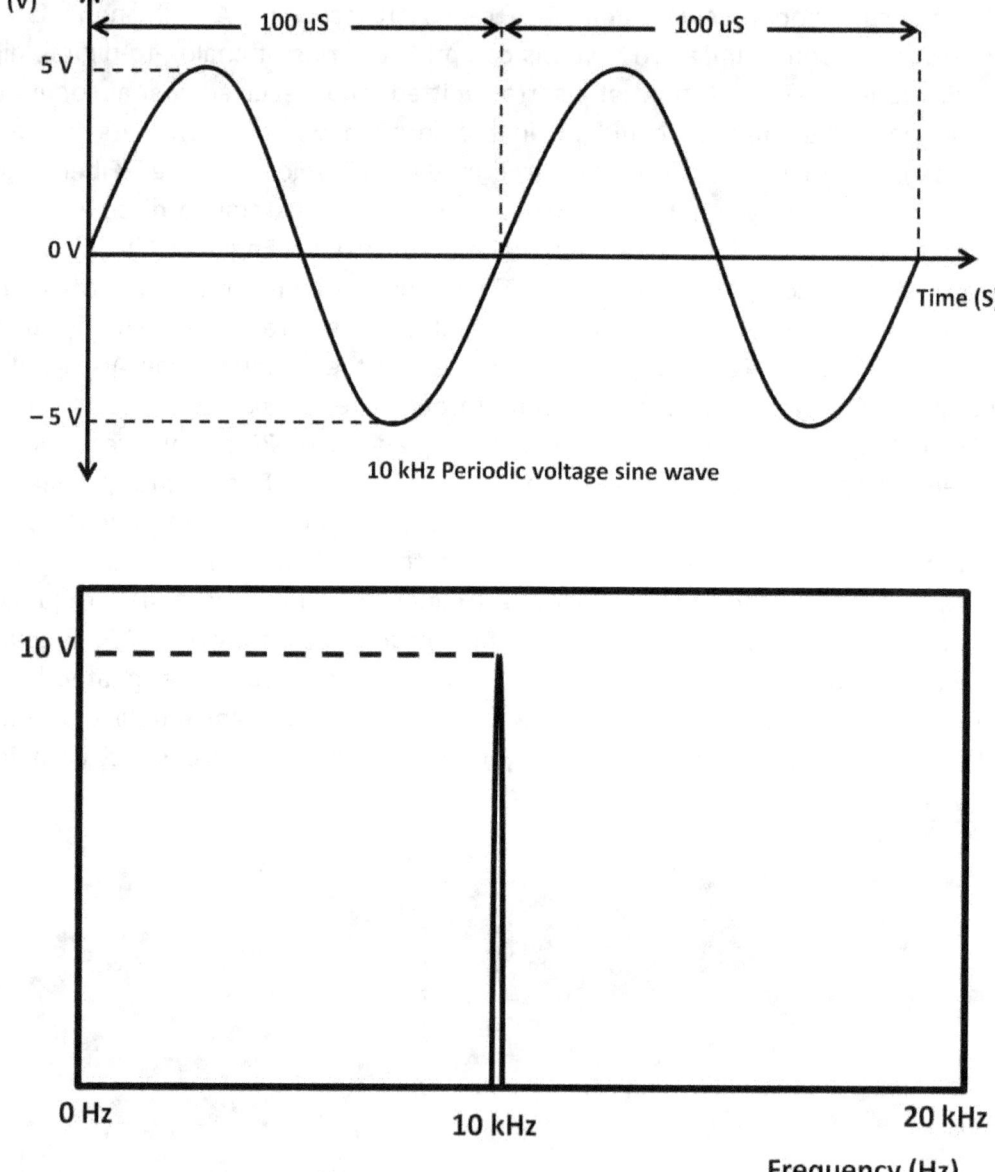

Figure 6.1: Time, frequency domain of a 10 kHz signal

In the spectrum analyzer display window, it shows that there is a sharp jump characteristic in the middle at 10 kHz. The rest of the spectrum span from 0 Hz to 20 kHz does not show any visible shapes. This demonstrates that the signal frequency is constant at 10 kHz. Recall that in chapter 3, AC, we derived resonant frequency using LC tank circuit. Using such a circuit is a good example of producing a signal with sharp frequency response similar to figure 6.1. Most radio signal transmitters implement some type of resonant circuits to generate filtered, amplified, frequency-sharp response such as series L C where maximum current occurs **(XL − Xc = 0)**, i.e., minimum impedances. This type of design is called a band-pass filter. It allows a signal to pass through only within a specified bandwidth (frequency range). Figure 6.1a shows the band-pass current and impedance in frequency domain.

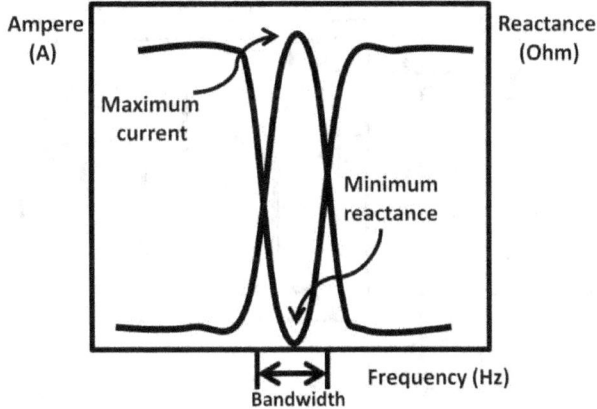

Figure 6.1a: Band-pass current, impedance frequency domain

On the receiver side, the same technique can be used to filter signals outside of a specific frequency range called a band-stop. Figure 6.1b shows the frequency response of frequency modulated (FM) bandwidth. FM will be further discussed later in the chapter. In figure 6.1b, it shows that the FM bandwidth is limited between 88 MHz to 108 MHz by band-stop filter.

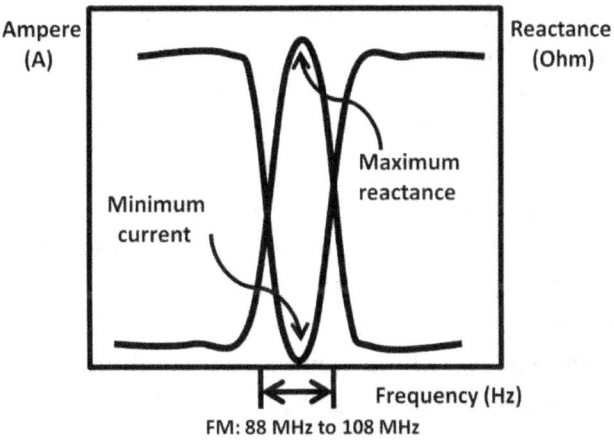

Figure 6.1b: Band-stop filter in FM receiver

The spectrum analyzer mentioned previously is the equipment of choice to test and measure band-pass and band-stop filters in frequency domain. There are several adjustments similar to the oscilloscope allowing users to zoom in and out of the frequency waveform. In figure 6.1, the display window starts at 0 Hz (far left) and ends at 20 kHz (far right). The starting, ending, and center frequencies (currently at 10 kHz) can be adjusted at any time. The Y-axis can also be scaled up and down. In the real world, it's rare to have the sharp waveform characteristic seen in figure 6.1 due to noise that exists in many places and in various forms. The noise source is usually in electrical form generated by devices in operations. The circuit (see figure 6.2) shows the noise found in the half-wave rectifier shown by the spectrum analyzer. Connecting the Vout to a spectrum analyzer, the Vout frequency waveform now shows multiple shapes.

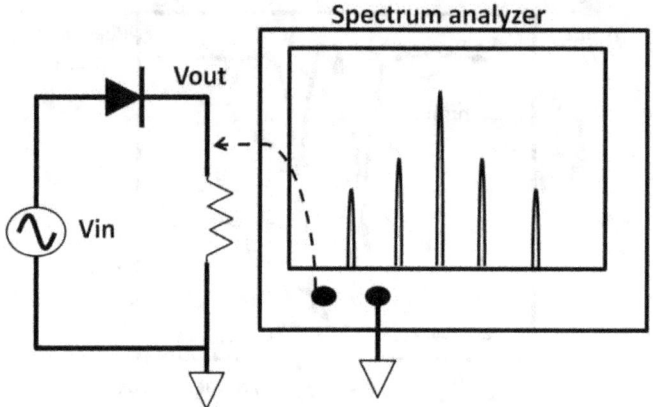

Figure 6.2: Half-wave rectifier noise

Harmonics, Distortion, and Inter-modulation

The signal at the center is called the fundamental frequency (center frequency) where the others are called harmonics. Harmonic frequency components are caused by non-linearity within the system, in this case, by the half-wave rectified waveform. The harmonics' frequency signature is constant integer multiples of the fundamental frequency. If the fundamental frequency (see figure 6.2) is 1 MHz, the first harmonic is located at **2 X 1 M = 2 MHz**, the second harmonic would be **3 X 1M = 3 MHz,** and so on. All harmonic frequencies are periodic while the harmonics amplitude is always less than the fundamental frequency. Due to the multiple-frequencies nature of harmonics, it becomes a major source of noise causing distortion to the original signal in an electronic system. Distortions are deviations or changes made to the original signal. Keep in mind that each harmonic by itself creates its own harmonics although these sub-harmonics have much less amplitude than the center frequency. Other sources of distortion in communication systems are inter-modulations, caused by the sum and difference of two frequency components. An inter-modulation products table examines the relationships between fundamental frequencies and individual products designated by order numbers (see table 6-1). Two fundamental frequencies, f1 and f2, are 100 kHz and 101 kHz, i.e., f1 and f2 are 1 kHz apart from each other.

Order number	f1 (Hz)	f2 (Hz)	Inter-modulation 1 (Hz)	Inter-modulation 2 (Hz)
First	f1	f2	100 k	101 K
Second	f1 + f2	f2 − f1	100 k + 101 k = 201 k	101 k − 100 k = 1 k
Third	2f1 − f2	2f2 − f1	200 k − 101 k = 99 k	202 k − 100 K = 102 k
Fourth	2f1 + 2f2	2f2 − 2f1	200 k + 201 k = 401 k	202 k − 200 K = 2 k
Fifth	3f1 − 2f2	3f2 − 2f1	300 k − 202 k = 98 k	303 k − 200 k = 103 k

Table 6-1: Order number, F1, F2, and Inter-modulation

From table 6-1, only odd number orders (the first, third, and fifth) are close to f1 and f2. The odd numbers become the significant noise components of the system within the spectrum. An inter-modulations spectrum is shown below (see figure 6.3).

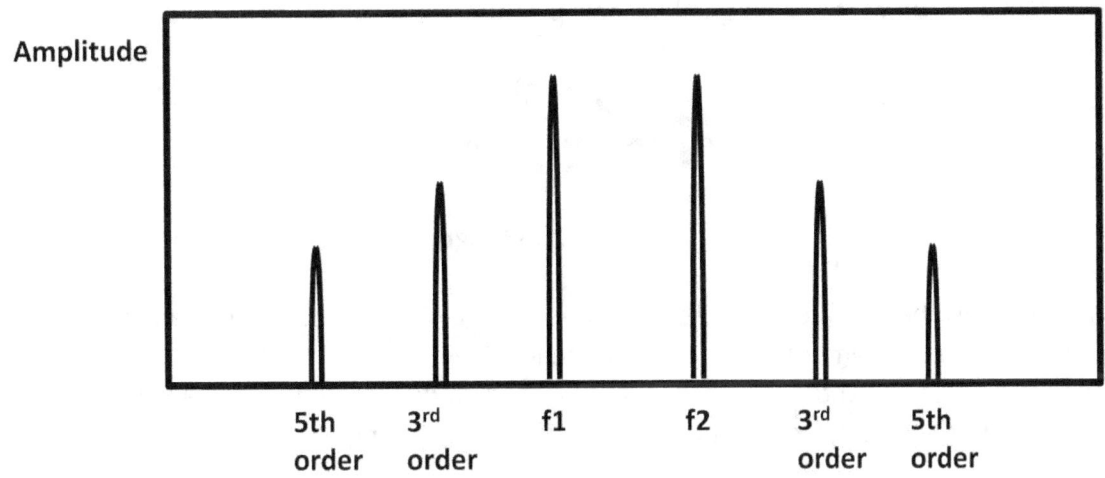

Figure 6.3: Inter-modulations spectrum

There could be numbers of harmonics and inter-modulations in non-linear systems. These components, sometimes referred to as side bands, are undesirable and need to be filtered out. Low-pass, high-pass and band-stop filter techniques can be applied. Sophisticated filter types include Butterworth, Chebyshev, and Bessel. Although the details of these filters are beyond the scope of this book, you should at least take note of their existence.

Modulation

Regardless of wired or wireless signal, most systems go through a modulation process, which is defined as combining the original information of interests with a carrier signal. A carrier frequency needs to run at a much higher frequency than the information signal. The result of this combination yields a modulated signal that includes both the original information riding along with the carrier signal. This technique squeezes more information within a certain bandwidth, raising the data rate before the signal was transmitted.

Bit Rate, USB, and Baud

In telecommunication electronics, data rate (bit rate) is quantified by the number of bits per second (bps). It a measure of how many bits are processed, transmitted, or received per one second. A popular serial data transfer protocol such as USB version 2.0 (high speed) data rate is about 48 Mbps. The newer USB 3.0 (super speed) is specified at maximum 4 Gbps. Figure 6.4 shows a USB logo commonly seen on electronic products.

Figure 6.4: USB logo

Baud rate can also be used to measure data speed. It's different from bit rate in that baud rate counts the number of symbols per second instead of the number of bits. For example, if the baud rate is 4,800 baud and each symbol represents two bits, the bit rate is 9,600 bps (4,800 X 2). Figure 6.5 demonstrates an example of 4 bauds (8 bits/second).

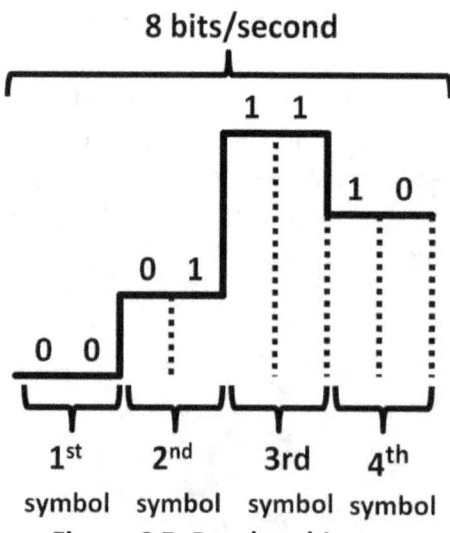

Figure 6.5: Baud vs. bit rate

Modulation is used in all kinds of transmission systems including wired and wireless internet communication (use of modems) and analog transmission such as radio transmission (amplitude modulation, frequency modulation). AM frequency bandwidth ranges from 530 kHz to 1,700 kHz. FM ranges from 88 MHz to 108 MHz. Modulation technique makes it possible by "altering" the original signal, i.e., by adding an information signal to the carrier signal creating a modulated signal.

C = F λ

To further understand why modulations are used, we need to discover the relationship between frequency, wavelength, and light speed. RF signals are simply electromagnetic waves that travel through air space at the speed of light. Wavelength's unit of measurement is the meter. It's the fundamental frequency period. The transfer function of frequency (F), wavelength (λ), and light speed (C) is defined as:

$$C = (F) \times (\lambda)$$

Speed of light (C) is a constant that is equal to **3×10^8 meter / second**. From the transfer function, if F goes up, λ needs to go down so that C remains constant. In wireless communications, antennae are used frequently. λ determines the antenna size, i.e., λ and antenna size are proportional to each other. To reduce antenna costs, it's desirable to keep the λ as small (frequency as high) as possible. The other incentive of keeping the antenna smaller in size is to prevent additional noise captured by the large antenna size. For example, the wavelength (λ) of a 90 MHz frequency modulation (FM) radio signal is,

$$C = (F) \times (\lambda)$$

$$\lambda = C / F$$

$$\lambda = 3 \times 10^8 / 90 \times 10^6 = 3.33 \text{ meters}$$

From this example, you can see that in order to keep antenna size small, frequency would need to increase with the constant speed of light (C). By using modulation technique, high frequency modulated signal can be created by adding a higher frequency carrier signal to the original signal. We will first see how amplitude modulation works in the next section.

Amplitude Modulation

Amplitude modulation (AM), often used in radio system, best describes how modulation works. In the US, the AM radio is broadcast on multiple frequency bands. The range of frequencies goes from 535 KHz to 1,705 KHz. We will figure 6.6 to further understand AM. In this example, the audio signal operates at f1; the sinusoidal carrier frequency operates at f2. We assume f1 is also a periodic sinusoidal wave for simplicity reasons. In reality, the audio signal will be in the form of random voice (analog) signals. The minimum frequency of the carrier signal (f2) needs to follow the Nyquist theorem, i.e., f2 needs to be at least twice as much as f1. The f1, f2 waveforms are shown in figure 6.6. By adding f1 and f2 together, the modulated signal can be obtained (see figure 6.7). This signal contains the original information and the carrier signal running at higher data rate than the original signal (f1). Note that the

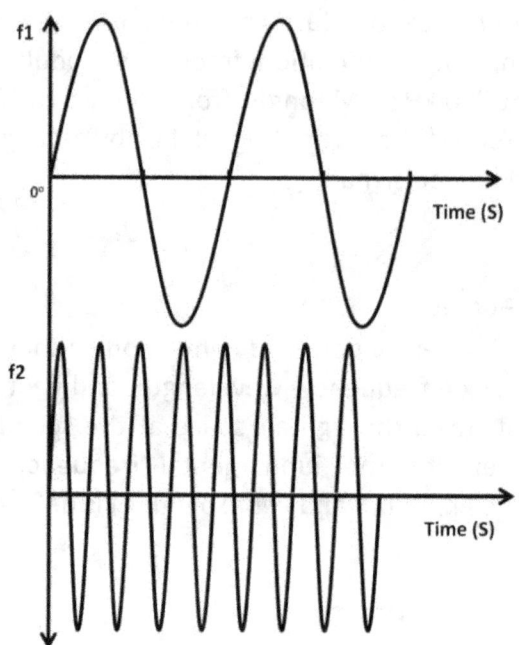

Figure 6.6: Audio and carrier signals

amplitude of the modulated AM signal changes with the f1's amplitude. The modulated signal is enclosed with a sine wave shape, the AM envelope. The AM envelope is not actually present in the modulated signal. It characterizes how well the modulated signal is created. To determine the quality of the modulated output signal, some criteria such as modulating index or the modulation factor are used. Figure 6.7a on the next page shows E_{min}, E_{max}, the minimum and maximum peak-to-peak levels.

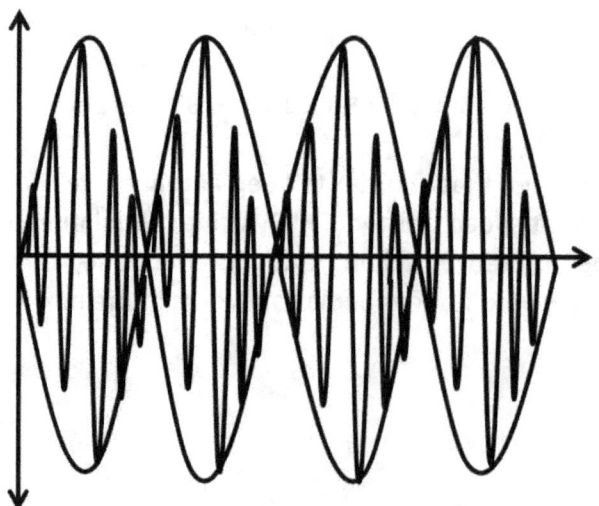

Figure 6.7: Modulated signal

Modulation Index and Bessel Chart

By definition, modulation index (M):

$$M = \frac{E_{max} - E_{min}}{E_{max} + E_{min}}$$

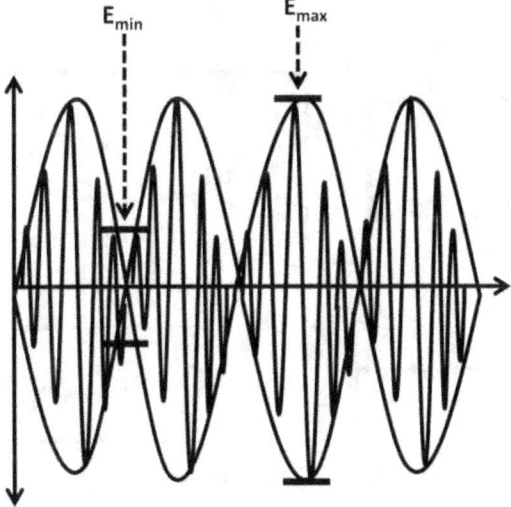

Figure 6.7a: Minimum and maximum peak-to-peak levels

Ideally, the modulation index is 1 (E_{min} is zero). E_{min} is the major error source regarding AM transmission. It represents crossover distortion where the signal is transitioning through the horizontal axis. To further quantify distortions, a Bessel chart can be used in the table 6-2 below. The table consists of the modulation factor in the far left column. The smallest factor is zero, meaning there are no harmonics, sideband components, or inter-modulation. They practically represent a DC signal. As frequency increases, so do the harmonic appearances and side bands. These cause the number of side band increases expanding to the right-hand side of the table. This chart is only showing the modulation factor up to 1.5 as an example. The modulation index could effectively go up to 10. The values in the sidebands are normalized sideband amplitude values. The side bands at the farther right-hand side would have the lowest amplitude compared to the left, e.g., 0.56, 0.23, 0.06, and 0.01 on modulation factor 1.5.

Modulation Factor	Sidebands			
	1	2	3	4
0	N/A	N/A	N/A	N/A
0.25	0.12	0.01	N/A	N/A
0.5	0.24	0.03	N/A	N/A
1	0.44	0.11	0.02	N/A
1.5	0.56	0.23	0.06	0.01

Table 6-2: Bessel Chart

AM Transmitter

The circuit below (see figure 6.8) is an AM transmitter circuit example. This circuit can be used to create the AM modulated signal in figure 6.7. It is simply a common emitter amplifier where the collector voltage is the resulting signal modulated by adding the carrier and audio signals together. By varying the collector resistors, the modulation factor can be adjusted. The LRC circuit fine-tunes the AM signal frequency using resonant frequency and band-pass techniques.

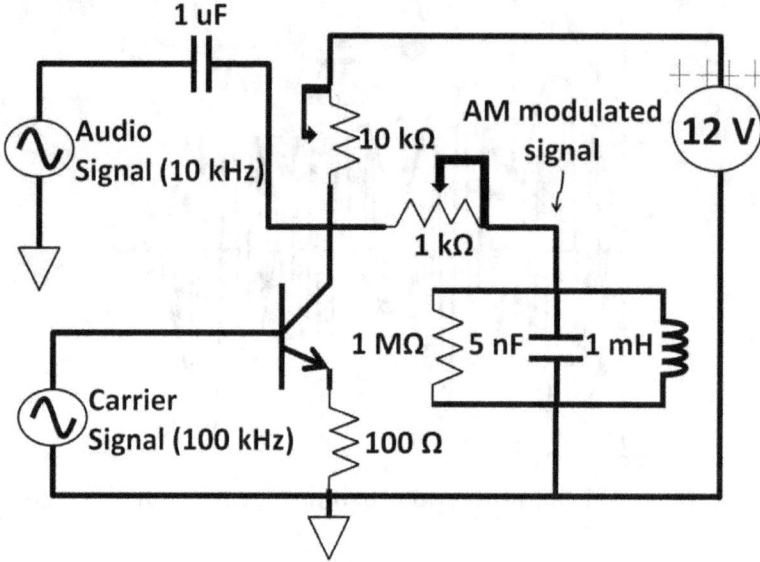

Figure 6.8: AM transmitter circuit

On the receiving side, once the AM signal is captured, it needs to be converted back to an audio signal via demodulation process. An AM detector (demodulation circuit) is needed to perform such task. A diode, resistor, and capacitor could achieve that in figure 6.8a.

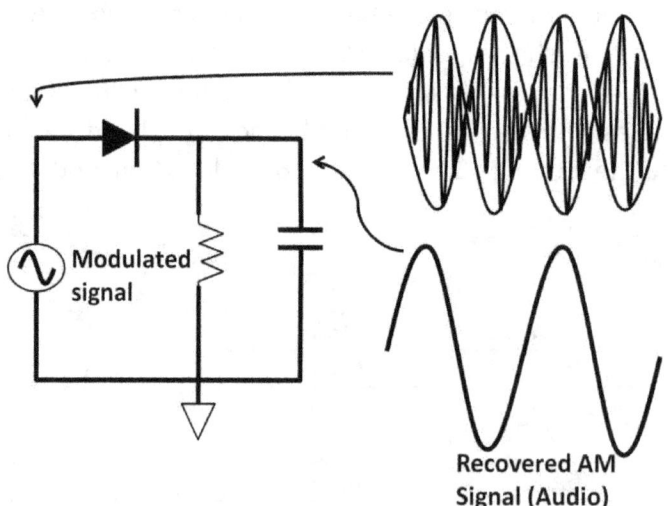

Figure 6.8a: AM demodulation circuit

Frequency Modulation

Frequency modulation (FM) works fundamentally different than AM. FM radio signal allocation in the US ranges from 88 MHz to 108 MHz. Although both AM and FM add audio and carrier signals together before transmitting via the air, unlike AM, FM's modulated signal's amplitude does not change when frequency changes with respect to the audio signal amplitude. This phenomenon was described in figure 6.9. f1 is the original audio signal adding to the carrier signal (f2). As the audio frequency (f1) reaches the peak, the frequency of the FM modulated signal is the highest. When it crosses the zero horizontal axis, it runs at the lowest frequency. It's due to this nature that FM is far superior to AM in terms of signal quality, because the AM amplitude fluctuates with the original signal. These fluctuations greatly contribute to noise. On the contrary, the FM amplitude stays roughly constant, eliminating the majority of noise components. It's for this reason radio stations use FM to broadcast higher-quality music. On the other hand, AM is used mainly for audio (talk shows) broadcast. To achieve unchanged amplitude, FM noise clipper circuit discussed in chapter 3, AC, can be used (see figure 3.44a).

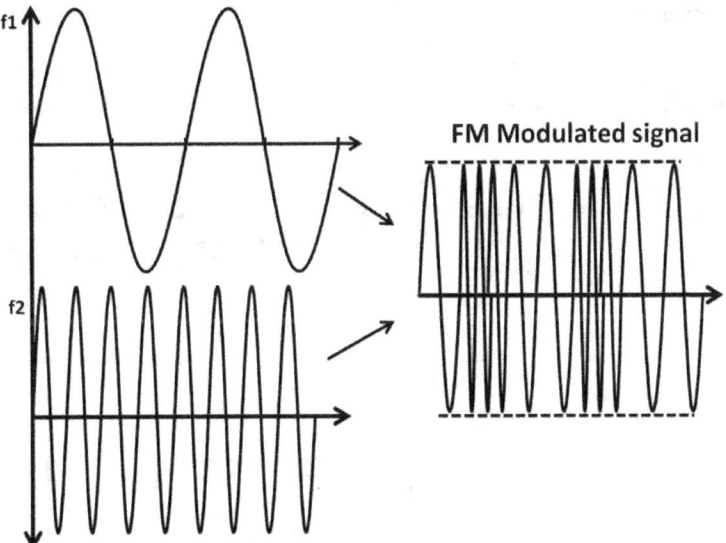

Figure 6.9: FM modulated signal

Keep in mind AM and FM transmission techniques not only apply to radio transmissions but are applicable to all other wired and wireless transmission applications including high frequency, internet, broadband, cellular, RF, and even satellite applications. Especially on RF, many mobile phones now carry multiple bands in one phone. Depending on the phone locations, it may have to switch from one band to another to receive and transmit signals. Popular cellular bands are Code Division Multiple Access (CDMA), Global System for Mobile (GSM) and Long Term Evolution (LTE). Each standard specifies a set of protocols regarding frequency band, carrier frequencies, voice encoding, decoding, and phone service security. Due to the needs of having multiple signals at other frequency ranges on one system, frequency generation or synthesis capability is needed. A popular method of doing so is phase lock loop (PLL).

Phase Lock Loop (PLL)

A typical PLL generates a very accurate clock, for example a carrier signal in an AM or FM transmission. A signal generated by a PLL would be used internally within the chip (on-chip) or supplied to other systems. PLL is a key component of radio, wireless, telecommunication, computing technology, and more. Advanced PLLs generate signals at different frequencies. Figure 6.10 shows the basic PLL building blocks including phase detector, low-pass filter, and voltage-controlled oscillator. You can see there is a feedback loop in the PLL functional block diagram. The phase detector compares the external signal to the output of the internal voltage controlled oscillator (VCO). PLL is a negative feedback system. Its task is to self-correct phase difference between the two phase detector inputs until the difference is zero. When this happens, the internal signal is "phase locked" with the external signal. At first glance, PLL does not look that practical. One might say "I could use the external signal as my clock source directly. Why do I need a PLL?" From a practical point of view, that is a correct statement and legitimate question. To answer the question, you should understand that a practical PLL is implemented with a binary divider to create signals at multiple frequencies. Figure 6.11 shows the actual implementation.

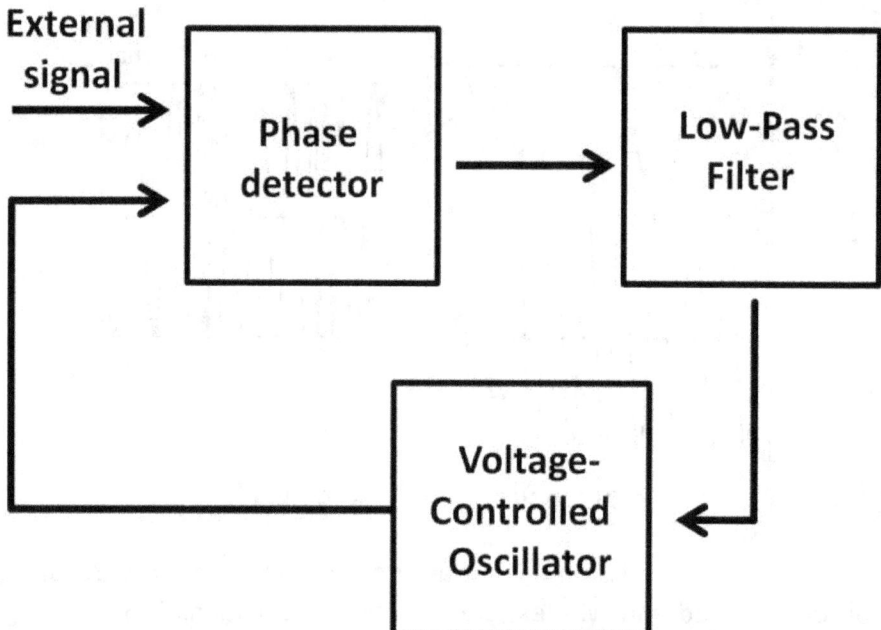

Figure 6.10: PLL functional block diagram

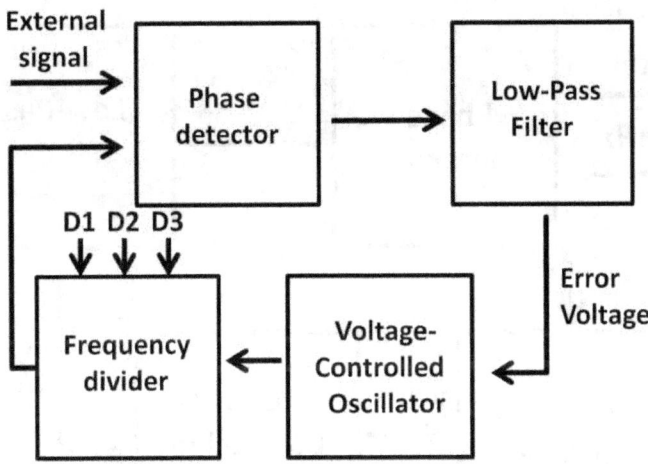

Figure 6.11: PLL implementation

D1, D2, and D3 are divider bits that can be controlled by a microcontroller. We will discuss microcontrollers shortly in chapter 7, Microcontrollers. If we assume the divider bits D1, D2, and D3 are 0, 1, and 2, the frequency divider is then able to divide incoming frequency by one, two, and four times shown in table 6-3. When the external signal arrives at the phase detector input, the VCO is running at its designed

Control bit name	Binary bit	Divider factor
D1	2^0	1
D2	2^1	2
D3	2^2	4

Table 6-3: Binary bit and divider factor

frequency. The phase detector output generates a digital pulse that represents the difference between these two frequencies in terms of phase shift. The low-pass filter converts this difference into a DC voltage called error voltage (see figure 6.11). This error voltage adjusts the VCO frequency until both phase detector inputs are the same, i.e., phase locked. At this point, the output of the phase detector is a DC voltage. It passes through the low-pass filter keeping the VCO running at the same frequency as the external signal. At any point in time, if the external signal runs differently than the locked VCO signal, the phase detector would again capture the phase difference generating a new error voltage. This voltage then skews (changes) the VCO to phase-lock the input signal again. The error voltage modulates according to the input signal variations. This topology applies the same feedback self-correction mechanism similar to what we discussed in op-amps and low drop-out regulators. Let's apply what we know to a practical scenario. Suppose the external signal comes from a 50 MHz crystal oscillator (see figure 6.12 on the next page). The divider bit, D2, is selected. The VCO output will have to be twice (100 MHz) as fast as the crystal before dividing by two to 50 MHz when the signal is phase locked. This in turn makes the PLL a frequency multiplier from the crystal. The 100 MHz signal from the VCO output can then be used to clock other circuits within the system. By using other divider bits (D1, D2, D3), multiple signals at different frequencies can be synthesized. PLL in this application becomes a frequency synthesizer, which is widely adopted in computers, high-speed digital design, microprocessors, and systems that use clock distributions.

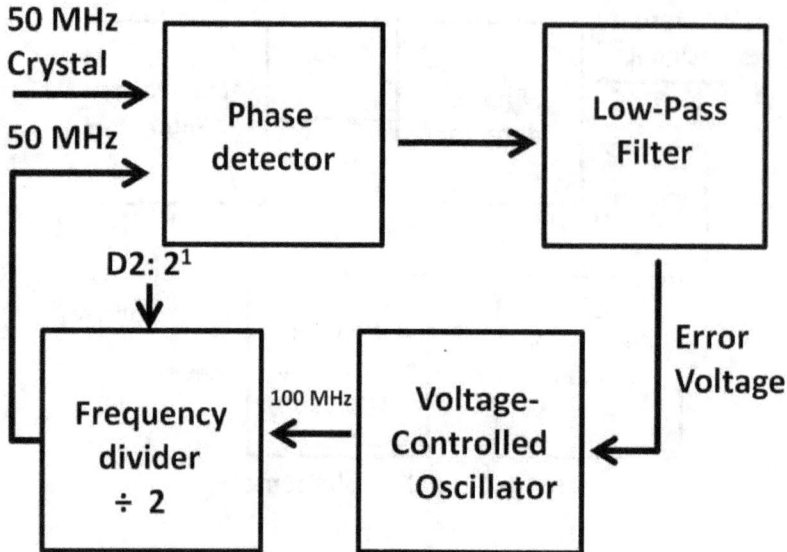

Figure 6.12: PLL frequency multiplier

The number and variety of PLL parts is staggering. Analog companies offer series of PLL chips with a variety of functions. Figure 6.13 shows an Analog Devices PLL (ADF4116/4117/4118) functional block diagram, package outline, and dimensions. According to the datasheet, "The ADF411x family of frequency synthesizers can be used to implement local oscillators (LO) in the up-conversion and down-conversion sections of wireless transceivers. They consist of a low noise digital phase frequency detector (PFD), a precision charge pump, a programmable reference divider, programmable A and B counters, and a dual-modulus prescaler (P / P + 1)". This PLL does not include VCO, which is external to the part.

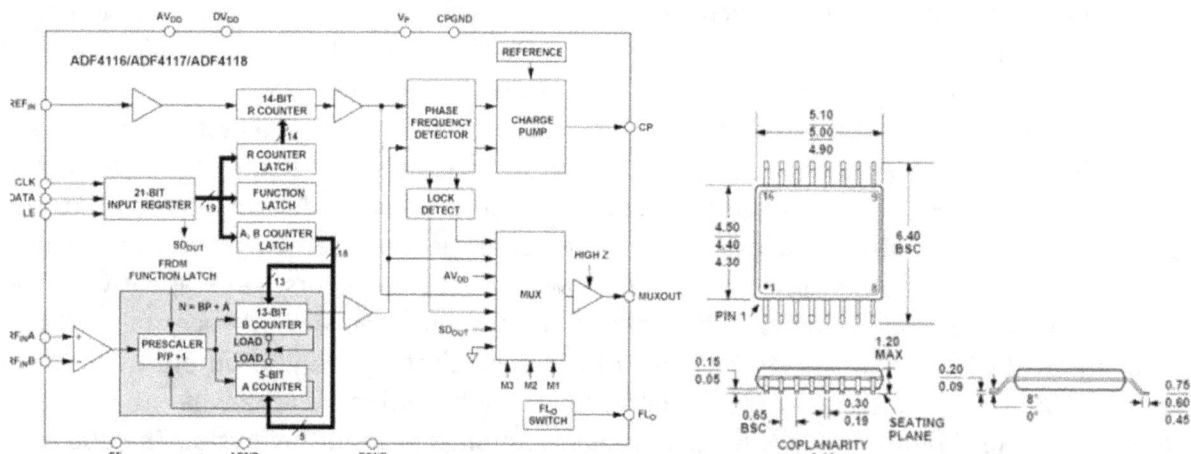

Figure 6.13: ADI PLL, ADF4116/4117/4118 block diagram, and package outline

Summary

We covered basic communication systems at the component and system levels in this chapter. Communication engineering standards, protocols, and specifications were covered emphasizing time, frequency domain, frequency, wavelength, and speed-of-light relationship. Among modulation and demodulation techniques, amplitude modulation (AM) and frequency modulation (FM) were described at the device and system levels. AM and FM circuits such as AM transmitters and receivers were reviewed. Communication system parameters such as bit rate, baud rate, harmonics, inter-modulations, modulation index, and Bessel charts were discussed. The chapter closes with phase lock loop theory and applications.

Quiz

1) From the spectrum analyzer display shown in figure 6.14, determine approximately the total bandwidth needed to transmit such a signal. Span: The difference in frequency between far left to right of the display window. Start, Center, and End are the absolute starting (far left), center (middle), and ending (far right) frequencies in the display window.

   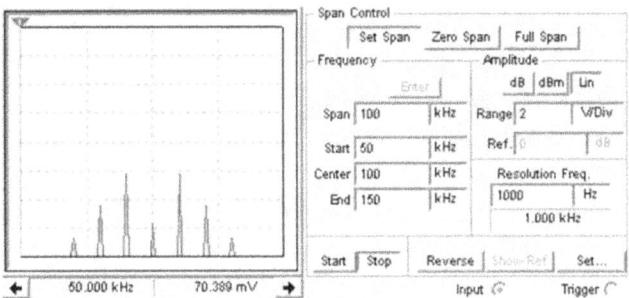

   Figure 6.14: Spectrum analyzer display window

2) If the signal frequency is 300 kHz transmitted using AM, what is the minimum frequency of the carrier signal?

3) Assuming Emin, Emax are 100 mV and 2 V, what is the modulation index, M?

4) According to figure 6.11, PLL, if the external crystal frequency is 20 MHz, and control bit D3 is high, what is the output frequency of the VCO?

5) Use the Bessel chart below in table 6-4 and determine how many significant sideband pairs a transmission would generate with a modulation factor of 0.5.

Modulation Factor	Sidebands			
	1	2	3	4
0	N/A	N/A	N/A	N/A
0.25	0.12	0.01	N/A	N/A
0.5	0.24	0.03	N/A	N/A
1	0.44	0.11	0.02	N/A
1.5	0.56	0.23	0.06	0.01

Table 6-4: Modulation factor, Bessel chart

Chapter 7: Microcontrollers

Microcontroller Units (MCUs) are silicon chips that act as the "brains" of many electronic systems. They are found in commercial, industrial, consumer, and military electronic products. Automobiles, computers, audio, video, lighting, wired/wireless network communication, LCDs (liquid crystal display), touch screens, medical devices, motor controls, temperature controls, power management, mechanical systems, children's toys, and home appliances (air-conditioners, washer, driers, microwave ovens, and refrigerators) are all controlled by MCUs. The systems containing MCUs are called embedded systems. The MCUs are "embedded" inside without direct access by the end users. The end users do not have access to the design source code (computer programs). Users only have limited numbers of programming capability. One example is a microwave oven where users "program" the cooking time by inputting the time. Users cannot change how the time is inputted (e.g., which button to use to input the time). The button locations and the beep volume and frequency are hard coded in the source programs by the embedded system designers. The source code was downloaded to the MCUs during design and manufacturing. MCUs therefore are field programmable. One MCU could have many applications as long as the source code is different, making MCUs highly configurable. Embedded system engineers use software development tools to develop and debug programs. We will discuss development environments later in the chapter. The worldwide MCU market share was US $13 billion from 2011 data. The top ten worldwide MCU vendors account for 70% of total MCU sales. They are Renesas Electronics, Freescale Semiconductor, Atmel, Microchip Technology, Infineon Technologies, Texas Instruments, Fujitsu, NXP, STMicroelectronics, and Samsung. Among major MCU markets, the automotive market accounts for almost half of the total market size. Popular programming languages used by embedded system engineers are assembly, C, and C++. In terms of MCU types, MCUs are similar to conventional microprocessors in a sense that they both have CPUs. The difference is in the peripherals (external components). Although both CPUs and MCUs communicate with peripherals through data and address communication buses, CPU peripherals are external while MCU peripherals are internal on the same chip (on-chip). With CPUs, peripherals such as volatile Random-Access Memory (RAM), non-volatile Read-Only Memory (ROM), clocks, printers, disk drives, monitors, keyboards, or mice are external devices. In MCUs, RAM (data memory) and ROM (program memory), along with other peripherals, are on-chip with the CPU. Some examples of MCU peripherals are comparators, ADCs, DACs, and timers. Depending on the type of MCU, some come with data bus interfaces including Universal Synchronous Asynchronous Receiver Transceiver (USART), Serial Peripheral Interface (SPI), Inter-Integrated Circuit (I2C), Universal Serial Bus (USB) and Pulse Width Modulation (PWM) channel. Newer MCUs come with networking protocols such as TCP/IP, Ethernet, and many other wireless network capabilities. Due to a large number of peripherals available on MCUs, they are highly configurable through software programming to control their functions. MCU datasheets that are several hundred pages are quite common. Figure 7.1 on the next page shows a simplified block diagram of a CPU and MCU.

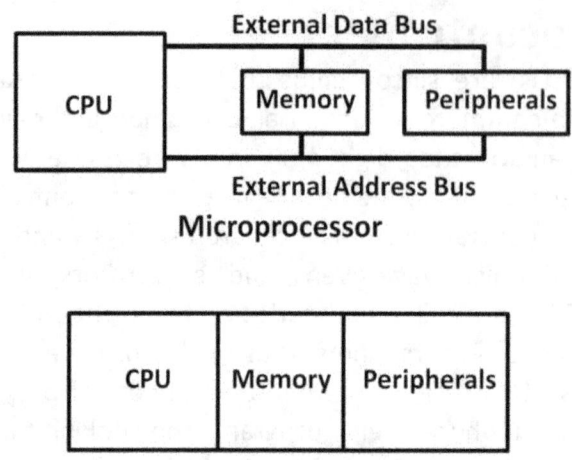

Figure 7.1: Simplified CPU and MCU block diagrams

MCU Parameters

The CPU performance within MCUs is usually lower than conventional computing ones because there is no need to design embedded systems running in multiple GHz speed. CPUs in computers often use a passive heat dissipation device called a heat sink to help disperse heat into the surrounding air due to excessive heat generated by fast clock speed. Many embedded designs involve human interactions, (e.g., by pushing a button or inputting on a touch screen). The time delay may be in the milliseconds. MHz clock is quite sufficient to meet the requirements. For this reason, a heat sink is seldom needed. An MCU with a CPU that runs above 100 MHz is considered high performance. Many MCUs' CPUs implement ARM (Advanced RISC Machines) architecture. ARM is a microprocessor family designed according to Reduced-Instruction-Set-Computing (RISC). RISC-based CPUs require a lot fewer transistors than conventional CPUs. This leads to relative slow clock speed and lower power consumption. This low power methodology ultimately benefits MCUs from a lower unit price, making it ideal for low-cost designs. This explains the large MCU application numbers in the market. Figure 7.2 shows a wireless smoke detector design reference by Microchip Technology using a conventional 9 V alkaline battery.

Figure 7.2: Wireless smoke detector (far right); smartphone, home security system (bottom)

In addition to the standard parameters such as supply voltage and temperature ranges, there are vast numbers of MCUs to choose from differentiated by types, product families, peripherals, and packages. Major MCU vendors like Microchip Technology offer close to 1,000 MCUs to customers. Table 7-1 attempts to list some MCU parameter metrics. Most MCU vendors have parametric search websites so engineers can look up parts fairly easily based on their needs.

Family	4, 8, 16, 32	Number of bits
Program memory	0 – 512	K Bytes
Data Memory	0 – 128	K Bytes
Input/Output Pins number (I/O)	4 – 80	Pins
CPU speed	4 – 120	MHz
Comparator	0 – 4	Number of comparators
ADC	2 – 32	Number of channels
USART	0 – 6	Number of channels
SPI	0 – 4	Number of channels
I2C	0 – 5	Number of channels
USB	1	Number of channels
PWM	0 – 9	Number of channels
8-bit timer	0 – 6	Number of timers
16-bit timer	0 – 9	Number of timers
32-bit timer	0 – 4	Number of timers
LCD Segment	60 – 480	Number of segments

Table 7-1: MCU parameter metrics

To select the right part for your design, you need to first know the MCU family's definitions. Popular ones are 4-bit, 8-bit, 16-bit, and 32-bit. Embedded system engineers need to understand the design trade-off versus system requirements and costs. Low-end MCUs are 4- and 8-bit families (cores) run at low frequencies for general-purpose applications. Mid- and high-end cores offer high speed and draw more power. A high-end 32-bit core offers higher performance, pin counts, power, and functionality, at a higher cost. Target application examples of high-end MCUs are accurate commercial, industrial controls, test, scientific, and medical equipment. To further understand this concept, let's take a deeper look at the MCU architecture. Most MCUs mentioned in this chapter are Microchip Technology parts. Be cautious that other MCU vendors may utilize different architectures. Engineers need to read the specific MCU datasheet for details. Microchip Technology's MCUs are named PIC® (Peripheral Interface Controller) MCUs (PIC® MCUs). Figure 7.3 shows Microchip's 8-bit product family. The graph's X-axis is the number of pins; the Y-axis is the memory size (KB). Bytes are memory units. Each byte of memory contains 8-bits of data. The bit is the basic unit of digital information (chapter 5, Digital Electronics). A bit can have a value of either "1" or "0." PIC18 is the highest performing among the 8-bit family offering the highest pin count and KB of memory.

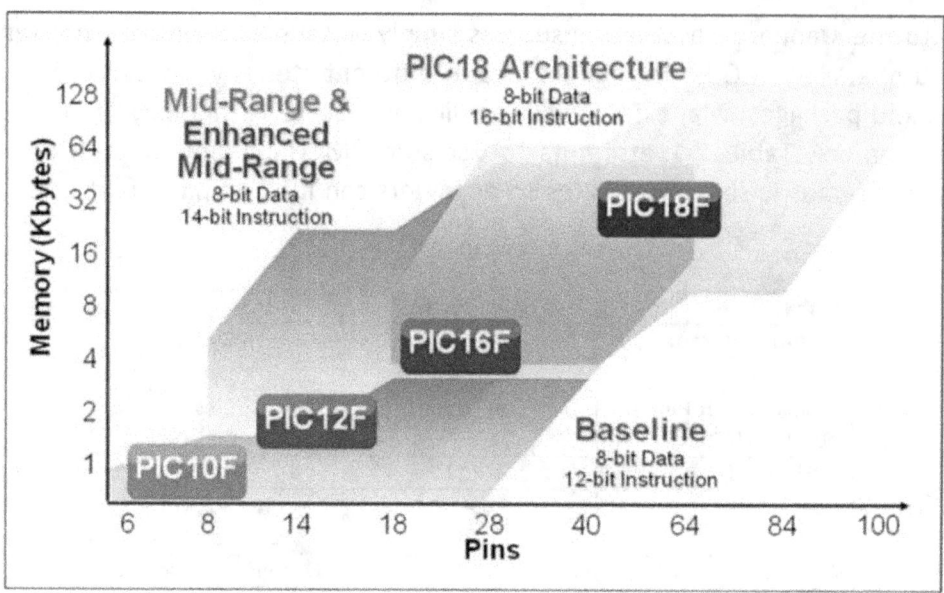

Figure 7.3: PIC 18 architecture

Figure 7.3a shows a 20-pin PIC18(L)F1XKK22 part pin package diagram and pin summary.

20-Pin DIL	20-Pin QFN	I/O	Analog	Comparator	Reference	ECCP	EUSART	MSSP	SR Latch	Timers	Interrupts	Pull-up	Basic
19	16	RA0	AN0	C1IN+	VREF-/CVREF	—	—	—	—	—	IOC/INT0	Y	PGD
18	15	RA1	AN1	C12IN0-	VREF+	—	—	—	—	—	IOC/INT1	Y	PGC
17	14	RA2	AN2	C1OUT	—	—	—	—	SRQ	T0CKI	IOC/INT2	Y	—
4	1	RA3	—	—	—	—	—	—	—	—	IOC	Y	MCLR/VPP
3	20	RA4	AN3	—	—	—	—	—	—	—	IOC	Y	OSC2/CLKOUT
2	19	RA5	—	—	—	—	—	—	—	T13CKI	IOC	Y	OSC1/CLKIN
13	10	RB4	AN10	—	—	—	—	SDI/SDA	—	—	IOC	Y	—
12	9	RB5	AN11	—	—	—	RX/DT	—	—	—	IOC	Y	—
11	8	RB6	—	—	—	—	—	SCL/SCK	—	—	IOC	Y	—
10	7	RB7	—	—	—	—	TX/CK	—	—	—	IOC	Y	—
16	13	RC0	AN4	C2IN+	—	—	—	—	—	—	—	—	—
15	12	RC1	AN5	C12IN1-	—	—	—	—	—	—	—	—	—
14	11	RC2	AN6	C12IN2-	—	P1D	—	—	—	—	—	—	—
7	4	RC3	AN7	C12IN3-	—	P1C	—	—	—	—	—	—	PGM
6	3	RC4	—	C2OUT	—	P1B	—	—	SRNQ	—	—	—	—
5	2	RC5	—	—	—	CCP1/P1A	—	—	—	—	—	—	—
8	5	RC6	AN8	—	—	—	—	SS	—	—	—	—	—
9	6	RC7	AN9	—	—	—	—	SDO	—	—	—	—	—
1	18	—	—	—	—	—	—	—	—	—	—	—	VDD
20	17	—	—	—	—	—	—	—	—	—	—	—	VSS

Figure 7.3a: PIC18(L)F1XKK22 pin definitions (Courtesy of Microchip Technology)

Harvard Architecture

PIC® implements Harvard architecture. The special feature of this architecture is the separation of program and data memory. Program memory (flash) stores user programs. The CPU fetches (retrieves) program instructions (commands) from the program memory on a dedicated bus. Data memory writes or reads data (file registers) to and from RAM and the CPU on a separate bus. The advantage of the Harvard architecture is that the CPU fetches and executes program instructions at the same time maximizing timing efficiency. These instructions perform mathematical, arithmetic, and logic operations upon interacting with the program and data memory. Figure 7.4 demonstrates the Harvard architecture.

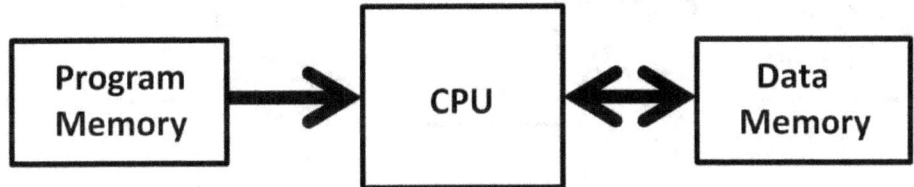

Figure 7.4: Harvard architecture conceptual view

Data and Program Memory

The advantage of Harvard architecture is that it improves operating bandwidth allowing different bus width. The PIC® bit numbers refer to the word length of the data bus. An 8-bit PIC® would have an 8-bit file register (data memory) size representing one-byte of data (contents). If, for example, the total 8-bit PIC® data memory size is 4 KB (4,096 bytes), there would be 512 file registers **(8 X 512 = 4,096)**. The last (bottom) register is 4,095 and not 4,096 because the first register's address starts with "0." Each file register occupies 8-bits (1 byte) of data. Figure 7.5 shows a simplified 8-bit PIC® data memory block diagram.

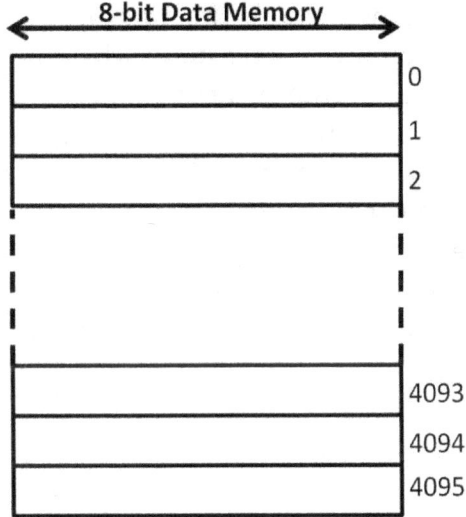

Figure 7.5: Simplified 8-bit PIC® data memory block diagram

Each register needs to have an address so that the CPU knows where to access (fetch) it. This is achieved by using the address bus between the RAM and the CPU. Once the specific register's location is known, data is then transferred between the RAM and CPU on the data bus. This addressing scheme applies to both data and program memory. To strike this point clear, the revised Harvard architecture in figure 7.6 shows the address and data bus in conjunction with the instruction bus.

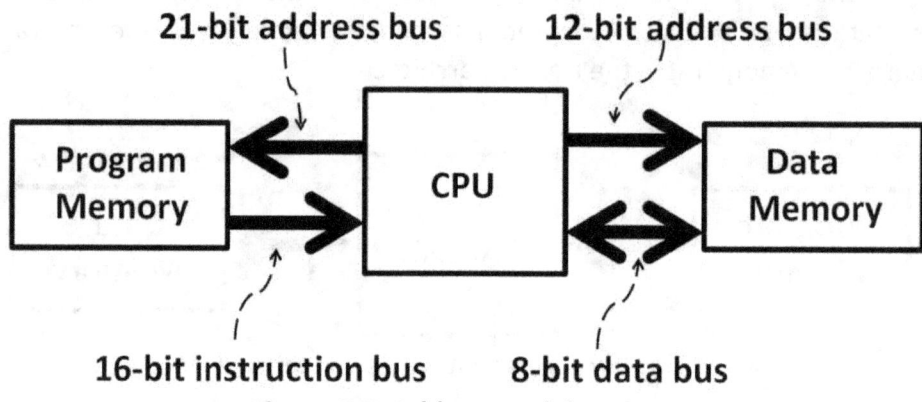

Figure 7.6: Address and data bus

As previously stated in figure 7.3, PIC18 is an 8-bit PIC® family. PIC10, 12, and 16 families all fall under the 8-bit category. The address bus for the PIC18 data memory is 12-bits wide and able to address 2^{12} = **4,096** file registers. The address bus for the program memory is 21-bits wide, capable of addressing 2 MB (2^{21}) program memory space. Each instruction therefore takes up one program memory address. Figure 7.7 on the next page describes PIC18's program memory map that shows the instruction bus is 16-bit wide. In addition to user programs, program memory contains a reset vector, an interrupt vector, an interrupt service routine (ISR), user programs, a device ID (identification), and configuration words. The reset vector is the starting point of each program execution (address 0). The interrupt vector contains the ISR's address and contents. Interrupts will be discussed later in this chapter. User memory contains source code that engineers write in their choice of programming language. Program counters and instructions work directly with the program memory. The program counter keeps track of which instruction to fetch and execute next for the CPU. Each instruction has a unique program memory address that is incremented or decremented by the program counter during code execution. After resetting the device, the program counter is clear, forcing code execution to begin at the reset vector. An External Master Clear (MCLR) pin can be used. MCLR is an active low (enabled only when low) pin that needs to pull low for a reset.

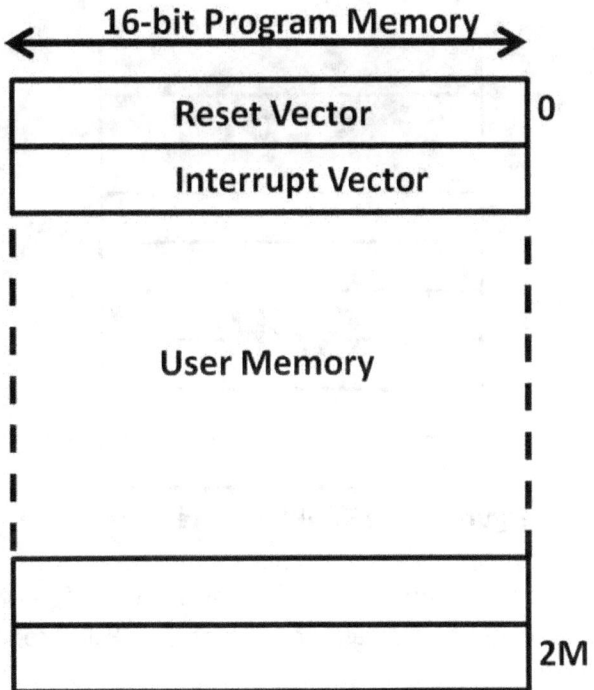

Figure 7.7: PIC18's program memory map

As mentioned previously, there are large numbers of PICs offered within a specific family. The data and program memory space is device specific, i.e., the sizes vary from one part to the next. The byte numbers shown are an example only. On PIC18 paging and banking are used in addressing memory. Program memory is divided into pages while banks divide data memory. Using the data memory map in figure 7.5 on page 251 as an example, there would be total of 16 banks with each bank occupying 256 file registers, adding up to 4,096 file registers. **256 X 16 = 4,096**. In the data memory, there are two main register types: general-purpose registers (GPRs) and special-function registers (SFRs). GPRs hold dynamic data during the execution of a program while SFRs are mainly for peripheral configurations and operations such as input and output ports (I/O), timers, ADCs, DACs, and PWMs. The SFR addresses are fixed in the data memory, and start from the lowest address (see figure 7.8 on the next page).

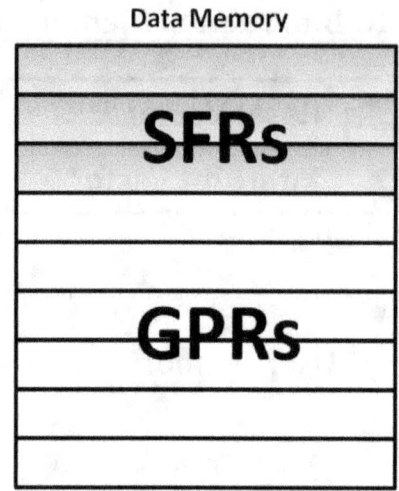

Figure 7.8: SFR, GPR in data memory

To summarize MCUs, CPUs, data memory, program memory, and peripherals, figure 7.9 shows a conceptual block diagram of PIC18F containing memory, a CPU core, and peripherals.

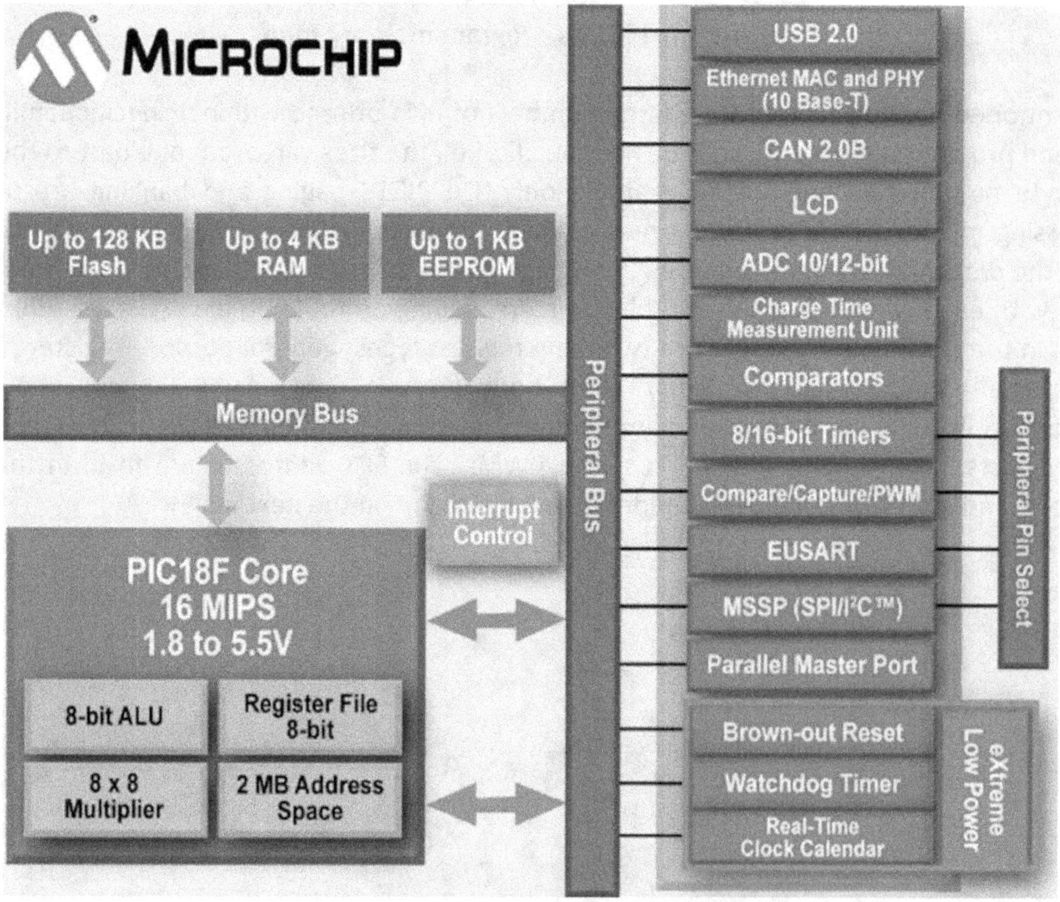

Figure 7.9: PIC18F block diagram

MCU Instructions

The MCU instruction word length varies from one PIC® family to another. The PIC18 instructions are 21-bits wide whereas the PIC16 instructions are 14-bits wide. Regardless of bit size, all instructions consist of operation code (op-code) and operand. Op-codes are the instructions that perform arithmetic and logic operations. Operands are file registers' addresses. Some instructions are byte-oriented while others are bit-oriented. Byte-oriented instructions operate on the entire register. Addition, subtraction, logic operations, data moving, and branching operations are examples of byte-oriented instructions. Bit-oriented operations perform bit operations. Bit shifting and clearing are examples of bit-oriented operations. An instruction set is a collection of all instructions. The instruction numbers in the instruction set is family-specific and may differ greatly. Table 7-2 below lists some common byte- and bit-oriented instructions in assembly language. Mnemonics in the left column represent the operands' names.

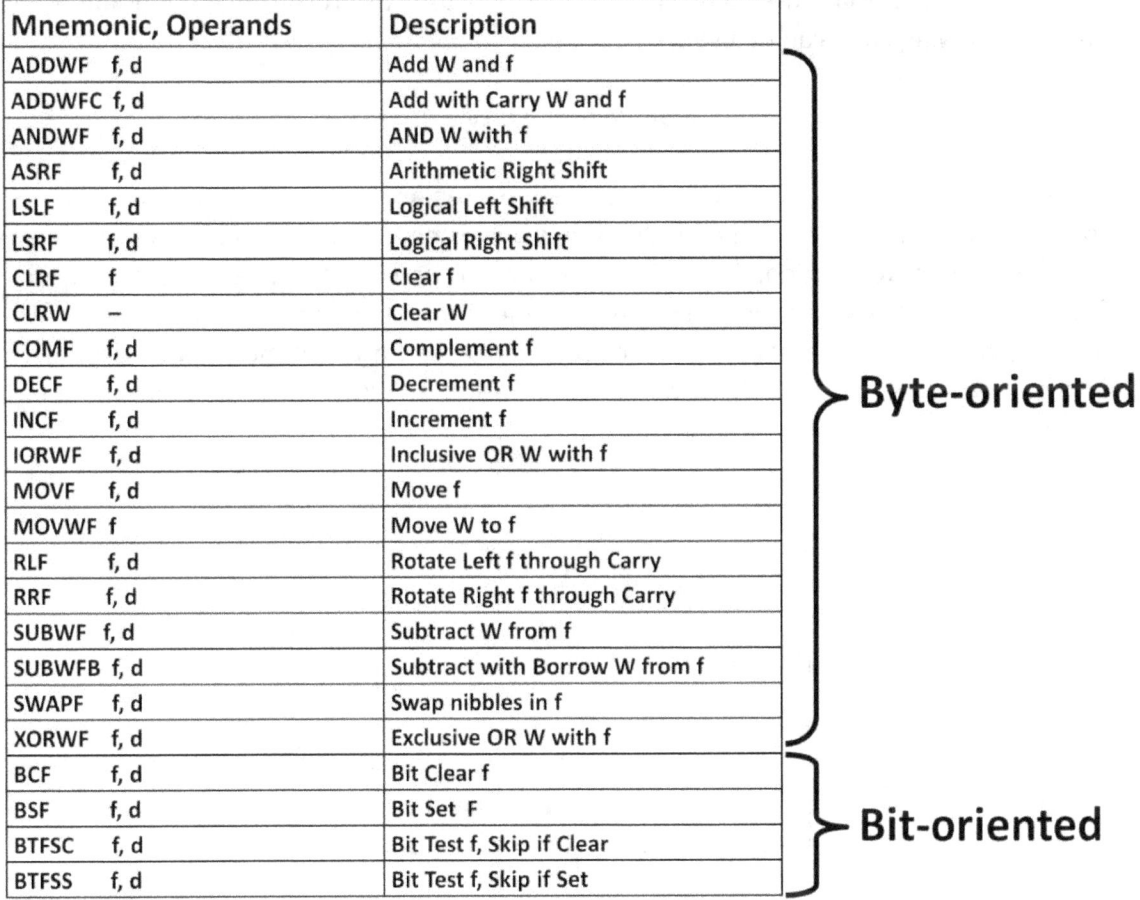

Mnemonic, Operands	Description
ADDWF f, d	Add W and f
ADDWFC f, d	Add with Carry W and f
ANDWF f, d	AND W with f
ASRF f, d	Arithmetic Right Shift
LSLF f, d	Logical Left Shift
LSRF f, d	Logical Right Shift
CLRF f	Clear f
CLRW –	Clear W
COMF f, d	Complement f
DECF f, d	Decrement f
INCF f, d	Increment f
IORWF f, d	Inclusive OR W with f
MOVF f, d	Move f
MOVWF f	Move W to f
RLF f, d	Rotate Left f through Carry
RRF f, d	Rotate Right f through Carry
SUBWF f, d	Subtract W from f
SUBWFB f, d	Subtract with Borrow W from f
SWAPF f, d	Swap nibbles in f
XORWF f, d	Exclusive OR W with f
BCF f, d	Bit Clear f
BSF f, d	Bit Set F
BTFSC f, d	Bit Test f, Skip if Clear
BTFSS f, d	Bit Test f, Skip if Set

Table 7-2: Byte- and bit-oriented instructions in assembly

F and W in the table represent the file register address and the destination. Below is an actual assembly code example:

addWF GPIO, F

This byte-oriented operation adds two 8-bit numbers together and produces an 8-bit result. The "F" on the far right designates the result destination. In this instruction, we are adding the values in the working register (W) and GPIO (file register). GPIO stands for general-purpose input output. GPIO pin functions can be changed by software to input or output pin. GPIO and W reside in the CPU core and they can store literal values (numbers, characters, or strings). Let's say the working register contains a value of "3". GPIO presumes to have a value of "2" prior to executing this add instruction. After this instruction is executed, the result (2 + 3 = 5) will be stored in the file register (F), which in this case, is the GPIO register. The value in the working register will remain the same. If you want to store the result in the working register instead, the following code can be used:

addWF GPIO, W

In this case, the value of the working register will be replaced with the addition result. GPIO remains at its original value after the instruction is performed. Using GPIO to drive a 7-segment display is a common application. A PIC18(L)F1XKK22 can be used to drive a 7-segment display (see figure 7.10). Each segment is an LED. A 7-segment display can be common anode (CA) or cathode (CC) type. All anode nodes are connected together in a common anode-type display. Turing it on requires power (5 V) to the CA node, and an RC pin to get pulled down. The resistors' functions are to limit the LED current.

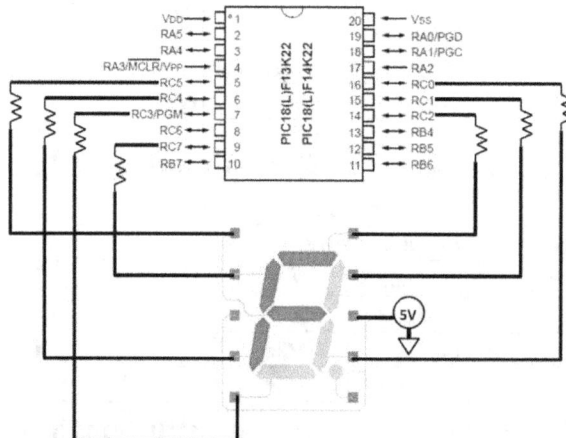

Figure 7.10: PIC18 drives a 7-segment display

On bit-oriented operations, the following assembly code clears GPIO register's bit number 3.
bcf GPIO, 2

The third type of operation is literal control. Table 7-3 shows some examples.

andlw k	AND literal with W	
call k	Call subroutine	
clrwdt -	Clear Watchdog Timer	
goto k	Go to address	
iorlw k	Inclusive OR literal with W	**Literal & Control operations**
movlw k	Move literal to W	
option -	Load OPTION Register	
retlw k	Return with literal in W	
sleep -	Go into standby mode	
tris f	Load TRIS Register	
xorlw k	Exclusive OR literal with W	

Table 7-3: Literal and control operations

movlw 0xFF

The above literal operation copies the value FF in hexadecimal (0X) to the working register (W). The control operation below takes program execution at the reset vector.

goto START

Instruction Clock

Every instruction takes time to complete. The instruction clock is derived from an external clock source that could come from two sources: **1)** mechanical resonant devices, such as crystals and ceramic resonators, or **2)** electrical phase-shift circuits such as resistors and capacitor oscillators. The trade-off between the two is that mechanical oscillator runs at much higher accuracy with low temperature drift (change). RC oscillators, in contrast, suffer from poor accuracy (more than 5%) over temperature. The precise instruction clock frequency is device specific. In the 16-bit PIC24 family, the internal instruction frequency is 2X slower than the external clock. The dsPIC30 has four external clock cycles per instruction clock. Figure 7.11 on the next page shows a PIC24's oscillator (FOSC) and instruction (FCY) frequencies. It takes two external clock cycles for one instruction period (TCY).

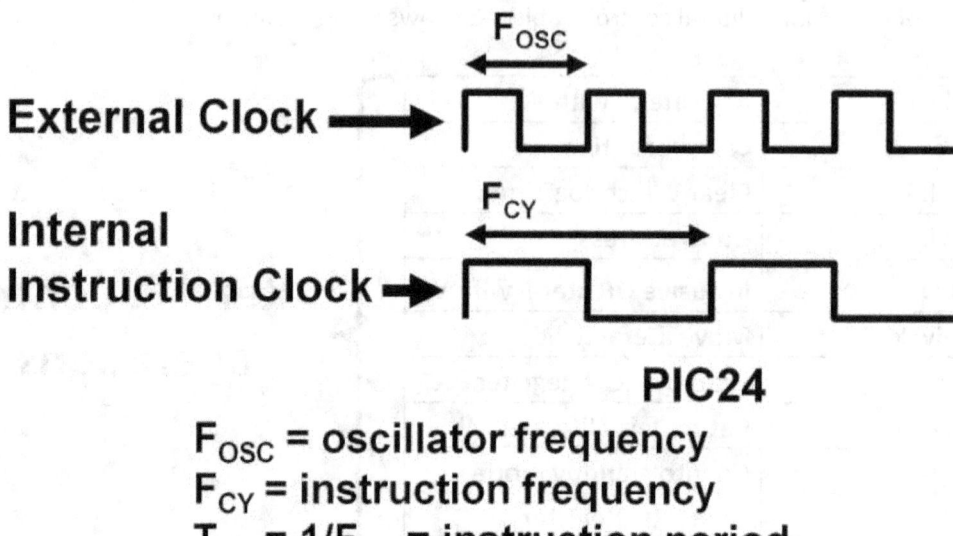

Figure 7.11: PIC24's oscillator (FOSC) and instruction (FCY) frequencies

The instruction clock is significant in embedded design. Instead of using frequency, MCUs use millions of instructions per second (MIPS) to define its performance. MIPS is defined by instruction clock, e.g., a 16 MIPS PIC® would need a 32 MHz external clock to operate according to figure 7.11. As for the number of cycles it takes per instruction, it depends on the instruction type. The majority of instructions take only one cycle; some take two. Mathematical instructions like division could take as many as twenty instruction cycles.

Internal Oscillator

MCUs come with internal oscillators; some even have a PLL frequency synthesizer on-chip. Designers can select, via software, an internal oscillator to supply the system clock, the clock for the CPU, and other peripherals. The flexibility of selecting the clock sources gives designers the flexibility to tailor their applications for optimal performance and power consumption. Figure 7.12 is a simplified PIC® MCU clock source block diagram. The LP, XT, HS, RC, and EC designations are as follows: EC (quartz crystal resonators), LP, XT, HS modes (ceramic resonators), and RC mode (resistor-capacitor circuits). Among ceramic resonators, LP: low-power crystal, 0 to 200 kHz, XT: crystal or ceramic resonator, 0 to 4 MHz, HS: high gain setting for the crystal.

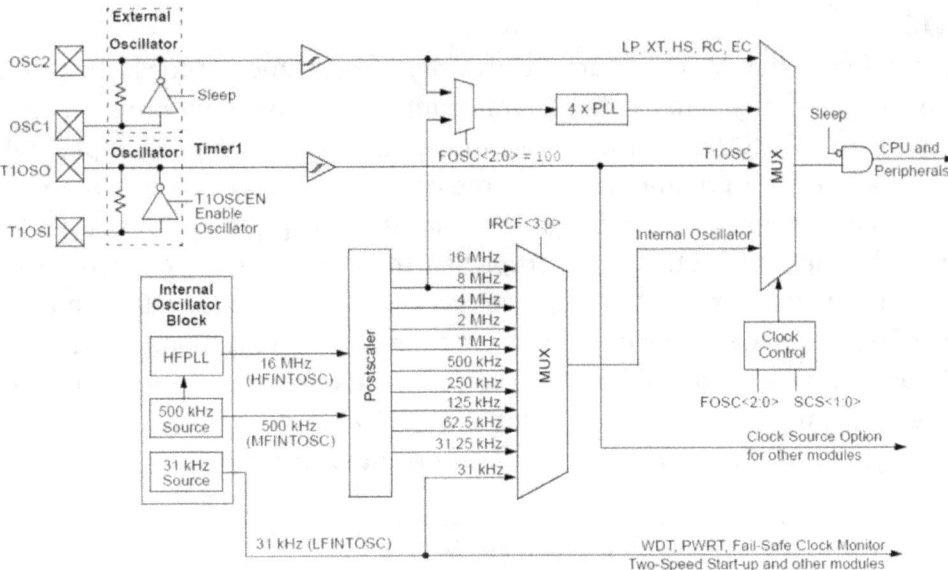

Figure 7.12: Simplified PIC® MCU clock source block diagram

The timer is an important peripheral in MCUs. The timer could run in the background without interfering with the main program. It can be used to time an event on an input and generate an output. During timing, the timer increments on every instruction clock with delay time determined by the timer register. You can think of the timer as a step counter. The step size depends on the internal or external clock source. Figure 7.13 shows a timing function. The start and stop time are fully configurable. The delay time is stored in the timer registers for retrieval.

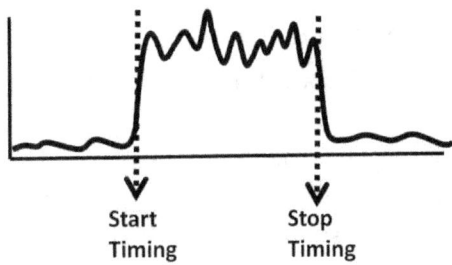

Figure 7.13: Timer's function

Other than timing, the timer can be used as a counter to count an event. When counting, the timer increments on every input's rising or falling edge independent of the internal clock. In this case, the timer is effectively counting signal transitions. Timers can record an event's arrival time; generate periodic interrupt; and measure pulse width, period, frequency, or duty cycle. Moreover, we can use a timer to generate a waveform. A simple application example using a timer could be toggling the LED connected to a GPIO pin many times per second. This requires timer configuration to overflow the number of times per second generating an interrupt. The ISR would include the code to blink the LEDs accordingly. Some programming examples will be shown later in the chapter.

Interrupt

Interrupt is another major MCU feature. It provides a real-time response to the MCU from either external or internal events. When an interrupt occurs, the main program stops and the interrupt flag goes high. MCU is instructed to jump to and then run the interrupt service routine (ISR). ISR is a user-defined program and can implement any MCU features the programmers would like. Upon ISR completion, the program resumes running the main program from where it was stopped by the interrupt. The interrupt flag then needs to be cleared by software. If interrupt is not enabled or if the interrupt never occurs, ISR will not be called and the interrupt flag will never be set. You can think of ISR as a program waiting to run under special conditions. These conditions are fully configured by the programs. There are many interrupt sources; table 7-4 lists some of them. The external interrupt is a dedicated external pin used exclusively for interrupt. See figure 7.14 for a practical external interrupt example.

Interrupt source
Oscillator Fail
Address Error
Stack Error
Math Error
ADC Convert Completion
Timer
SPI Error
External Interrupt
Comparator output change
USART receive completion

Table 7-4: External interrupt example

In this example, an RA2 pin is an external interrupt pin. If this pin is set up as the falling-edge triggered, then when the switch closes, RA2 gets pulled down. The interrupt flag then sets. The program stops and will jump to ISR. In the ISR program, it's written that the LED turns on by pulling up pin RC7 then turns off only if RA2 is high (switch opens). A second example could be that the switch is replaced by a push button. The ISR can be written such that once the interrupt flag is set, the LED turns on for a pre-determined period of time using a timer, then turns off.

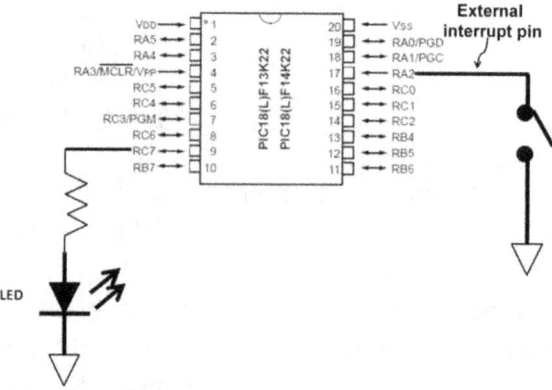

Figure 7.14: Practical example of external interrupt

Special Features

MCUs have many built-in features making it highly field programmable. Some MCUs have separate power supply pins to provide supply voltages to CPUs and peripherals separately. Isolating power supplies is essential in reducing noise from one area to another. Some MCUs have internal regulators allowing one external supply voltage (VDD) and power up multiple blocks within the MCU. This saves on MCU pin and board space while keeping the number of external supply components minimal. Figure 7.15 elaborates on this concept. A watchdog timer (WDT) is an independent timer that could recover from a software malfunction, (e.g., an infinite running loop). If the WDT is enabled followed by an overflow before it is reset via software, it assumes software control has been lost. It then automatically resets the PIC®. It can be used to wake a device from sleep (low power mode) without reset provided that the WDT is reset periodically before it times out. Power-On-Reset (POR) places the MCU in reset state when power is first applied. It then releases the device from reset after a period of time. This time is controlled by the internal power-up timer. The POR's job is to give enough time for the VDD to rise up to the minimum level. Both POR and WDT reset are software configurable. Brown-Out-Reset (BOR) resets the device if the VDD for the CPU core falls below a certain threshold. A PIC24 MCU VDD core minimum is around 2 V. During reset, if the VDD core rises up again, there is a delay time of about 20 us the MCU needs to wait before it gets released from reset. The idea of BOR is to prevent erratic behavior due to excessive system noise that may change VDD levels unexpectedly. The delay timer ensures that the voltage is stable before releasing the device from reset. The Oscillator Start-up Timer (OST) is an external crystal or resonator. The OST will automatically count 1,024 oscillator cycles before releasing the clock to the MCU making sure the oscillator has stabilized. Sleep mode is used to save power by minimizing clocks. The core and most peripherals can be stopped as a result of sleep. Some PIC® MCUs have XLP (extra-low power) technology to save power with current running as low as 20 nA in deep sleep mode. The device goes into sleep mode by executing a SLEEP instruction. Unlike an external Master Clear reset, there are several events that can wake up the device from SLEEP without resetting the device. The watchdog timer timeout, I/O pin, and peripheral output changes can wake a device up while not resetting the MCU and then resume program execution. Sleep mode is extremely useful in battery-powered applications to extend battery life. Examples of applications are portable medical devices such as blood pressure meters and digital thermometers; smart energy meters such as water, gas, heat, and electric meters; LCD drivers for graphic display; integrated USB applications; smart cards for authenticated systems; embedded wi-fi (wireless fidelity) modules; and power radio modules such as ZigBee® or RFID (radio frequency ID).

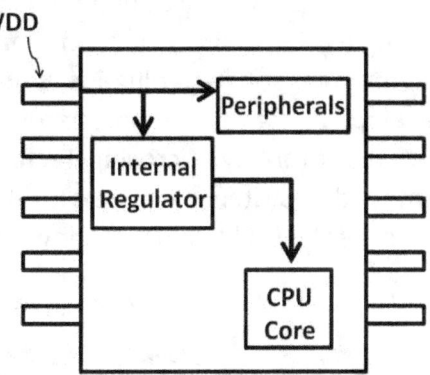

Figure 7.15: Internal regulator

Development Tools

Most MCU vendors offer free stand-alone software development kits (software platform) to embedded system designers and programmers. The development kits are a collection of software components all combined into one software package. The components include a program (code) editor, project manager, programming language support, source-level debuggers, and software plug-ins. Embedded system engineers can use these software features to develop, debug applications, hardware all in one software environment. MPLABX is the latest development tool by Microchip Technology. It's an integrated development environment (IDE) for embedded system engineers to develop code (software) in a single, cohesive platform. A snapshot of the MPLABX IDE is shown below (see figure 7.16).

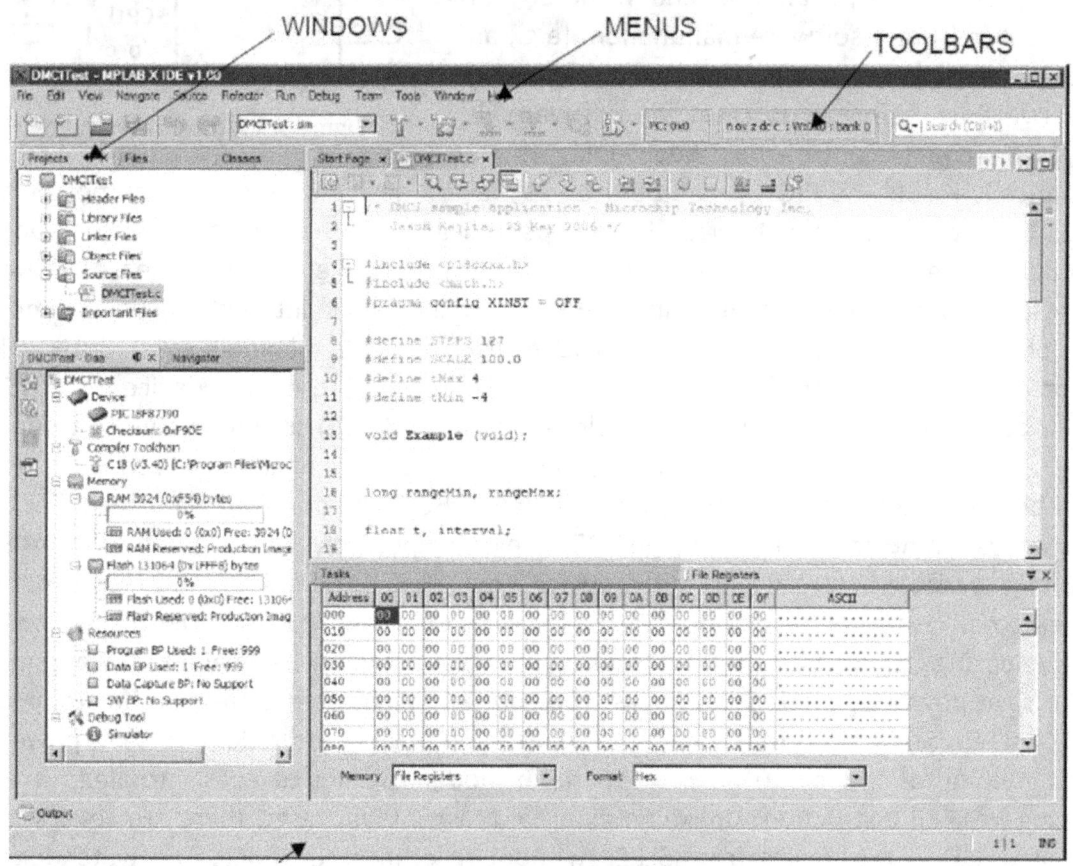

Figure 7.16: MPLABX IDE

The integration methodology goes beyond code development. Other than the standard file, project creation, management, and code editor windows, IDE comes with a compiler, an assembler, simulation, and hardware debugger functionality. A compiler is a software program that works with high-level languages such as C, C++, or Java. It transfers written code into machine language so that the MCU (hardware) can understand and run the applications accordingly. Assembler works on assembly language. Its function is similar to a compiler bridging the assembly and machine codes. Embedded system designers can choose any language they prefer. The difference between a compiler and an assembler is that an assembler works with low-level languages while compilers work with high-level ones. The major difference between high- and low-level languages is that a low-level language's syntax is closer to the machine code (0s and 1s). Assembly code is relatively difficult to understand at first glance whereas high-level code is easier to comprehend. Using the assembly code shown earlier,

addwf GPIO, F

It may not be obvious the above line of code add the contents of working and file registers. Comparing it to this C program below:

```
if ( X == 1)
{
printf ("X is equal to 1");
}
```

You can easily see that it's evaluating whether X is equal to 1. If it is, then the program prints out "X is equal to 1." Recall MCU families earlier in this chapter. Each family requires a compiler or assembler within the IDE to convert the programs and outputs to machines codes before the MCU can run its operations. Most compilers and assemblers are free. Many are both company-specific and device-specific, e.g., an 8-bit MCU design needs an 8-bit compiler, and the same compiler would not work on 16- or 32-bit code nor would it work in another company's IDE.

Debugger

Debugging the programs within an IDE is called simulation and debug. Simulation checks for the program's syntax errors using its programming algorithm built within the IDE software. The ultimate goal of any embedded design is to verify that the code works on the actual hardware using the debug function. To do so, the debugger is supported in the IDE to act as a bug-reporting tool between the IDE and target system (hardware). The debugger sends the code out to the hardware that runs the applications based on what the code is written for. It then reports back to the IDE through the debugger if there are any issues. It keeps track of data and program memory contents along with program status. If the program did not work the way the application was intended to, the programmers could then use the reported information to modify the program accordingly. This error reporting and program modification process repeats until the desired operations occur at the target hardware. Microchip Technology offers several choices of debuggers. PICKIT3 is a low-cost debugger with many useful features embedded system designers need. Many microcontroller companies offer variety of evaluation and development boards. Arduino is a palm-sized, single-board microcontroller popular among academia and hobbyists. Microchip Technology also supplies a variety of evaluation and development boards. Figure 7.17 on the next page shows PICKIT3 and an evaluation board that has a LDC screen on it.

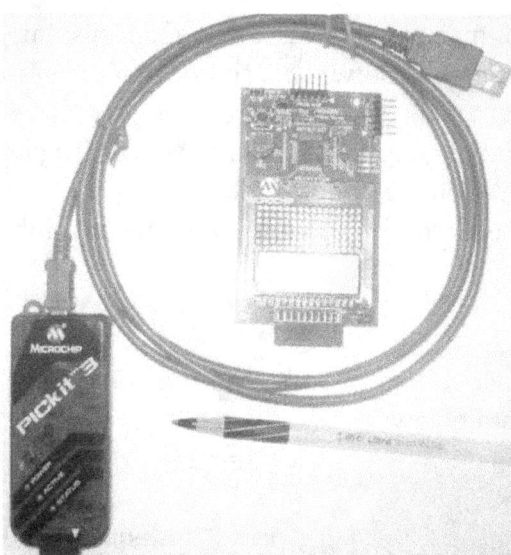

Figure 7.17: PICKIT3 and evaluation board with LCD screen

In-Circuit Debugger 3 (ICD3) is a CD-sized, mid-end debugger that offers a breakpoint feature. Figure 7.18 describes the debug methodology using ICD3 as debugger (see figure 7.18).

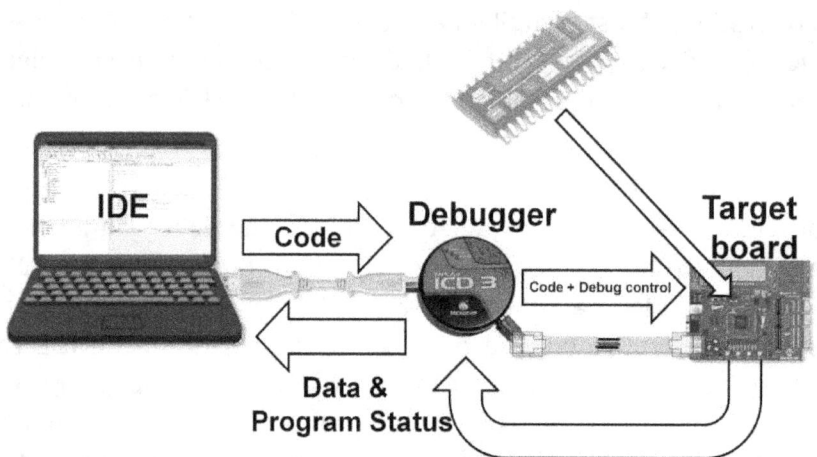

Figure 7.18: Debug methodology using ICD3

It supports full, high-speed USB connection, wide-input voltage ranges (2 V to 5 V), and multiple breakpoints so that engineers can pause the program during code executions. From figure 7.18, you can see that ICD3 acts as the communication bridge between the IDE and the target board. ICD3 allows users to add breakpoints in the program. The debugger pauses at where the breakpoint is located during debug. The program continues to be halted unless programmers click "continue" in the IDE. Breakpoint is a powerful debugging tool for pinpointing errors in a program effectively. Figure 7.18a shows an occurrence highlighted in a red box. When the debug starts, the program executes instructions one line at a time. At line 13, where the breakpoint is added, the program pauses during debugging. Programmers can now monitor and view program, data, and memory contents as they wish. Clicking the "continue" button in the IDE resumes program execution and move onto the next line of code (line 14).

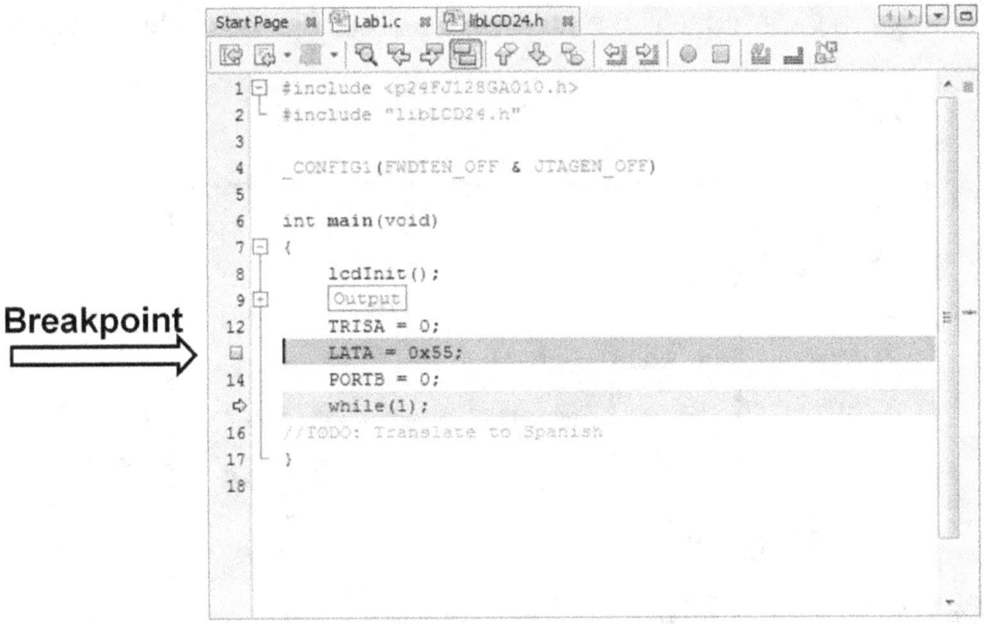

Figure 7.18a: Breakpoint on line 13

Other useful features in MPLABX are view, monitor variables or register contents. MPLABX offers several tools to show memory values used by the program at any time called Watches and Variables windows. With few mouse clicks, Watches and Variables windows can be displayed within the IDE. Programmers can choose any registers they wish to view. An IDE is a major component in designing embedded systems. Engineers and programmers need to get fully familiar with its features and functions in order to reduce product time to market.

Design Example: Comparator

To close this chapter, let's take a look at some embedded design examples. We'll start with a comparator. Just like the comparator mentioned in chapter 4, Analog Electronics, it compares two input voltages. Comparators are standard MCU peripherals often combined with capture and PWM modules. In this example, the objective is to detect the portable device's battery voltage. When the battery voltage falls below a threshold, the MCU comparator output pin will pull up. This output pin can be used to turn on an LED, notifying the user when the battery is low. The comparator output connects to an internal timer as a pulse counter. In the application block (see figure 7.19 on the next page), R1 and R2 set the battery voltage threshold. Assume battery voltage is 5 V when fully charged. 3 V is the threshold voltage you want to light up the LED. Using the voltage divider rule, resistor values will be:

$$R1 = 10 \text{ k}\Omega, R2 = 16.67 \text{ k}\Omega$$

The comparator output goes to an XOR gate. The same output goes to the timer, which acts as a counter (more on timer later). The second XOR input is a comparator control bit. It controls the polarity of the external XOR output. If the inverter bit is set to logic "1" (true), the XOR output is inverted. If the inverter bit is set to logic '0' (false), the XOR output is not inverted.

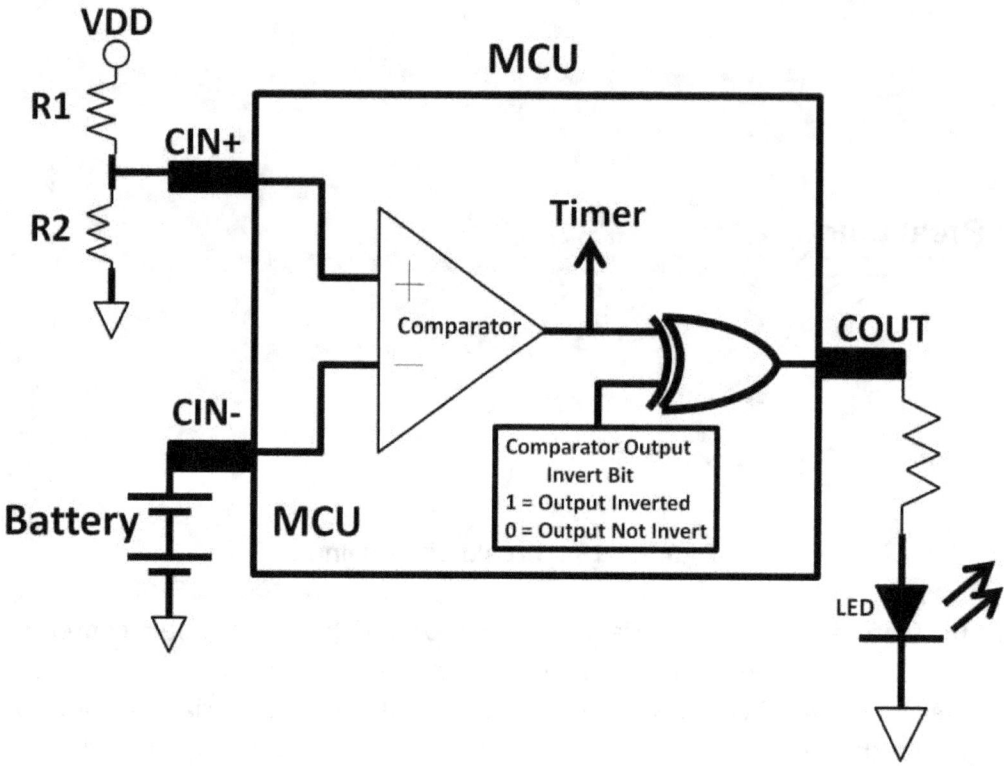

Figure 7.19: Comparator application

To program an MCU, engineers need to know what registers and control bits to configure. Figure 7.20 on the next page shows the control bit names and descriptions of the comparator control register (CMxCON). Once again, the control register name is different among MCU parts and vendors. The name below is just an example. Many MCUs have several comparators. The "x" variable used in the control bit name designates the comparator numbers. For example, the first comparator's register name would be CM1Con. The comparator polarity selection bit is CxPOL as shown below.

CxON	CxOUT	CxOE	CxPOL	—	CxSP	CxHYS	CxSYNC
bit 7							bit 0

bit 7　　　　　**CxON:** Comparator Enable bit
　　　　　　　1 = Comparator is enabled and consumes no active power
　　　　　　　0 = Comparator is disabled

bit 6　　　　　**CxOUT:** Comparator Output bit
　　　　　　　If CxPOL = 1 (inverted polarity):
　　　　　　　1 = CxVP < CxVN
　　　　　　　0 = CxVP > CxVN
　　　　　　　If CxPOL = 0 (non-inverted polarity):
　　　　　　　1 = CxVP > CxVN
　　　　　　　0 = CxVP < CxVN

bit 5　　　　　**CxOE:** Comparator Output Enable bit
　　　　　　　1 = CxOUT is present on the CxOUT pin. Requires that the associated TRIS bit be cleared to actually drive the pin. Not affected by CxON.
　　　　　　　0 = CxOUT is internal only

bit 4　　　　　**CxPOL:** Comparator Output Polarity Select bit
　　　　　　　1 = Comparator output is inverted
　　　　　　　0 = Comparator output is not inverted

bit 3　　　　　**Unimplemented:** Read as '0'

bit 2　　　　　**CxSP:** Comparator Speed/Power Select bit
　　　　　　　1 = Comparator operates in normal power, higher speed mode
　　　　　　　0 = Comparator operates in low-power, low-speed mode

bit 1　　　　　**CxHYS:** Comparator Hysteresis Enable bit
　　　　　　　1 = Comparator hysteresis enabled
　　　　　　　0 = Comparator hysteresis disabled

bit 0　　　　　**CxSYNC:** Comparator Output Synchronous Mode bit
　　　　　　　1 = Comparator output to Timer1 and I/O pin is synchronous to changes on Timer1 clock source. Output updated on the falling edge of Timer1 clock source.
　　　　　　　0 = Comparator output to Timer1 and I/O pin is asynchronous.

Figure 7.20: CMxCON register control bit names and descriptions

MCU programming is a subject that could take an entire book to cover. Readers who are interested in learning more about MCU programming need to first choose a programming language and study it. Below is a series of #define statements that configure the comparator. In C programming, #define statements are preprocessor statements. They are used as a substitution for text within code. For example, in line 1 of the example below, whenever COMP_INT_EN appears in the main program, it will be replaced by the bit values in binary: 0b1000000 ("0b" designates the number is a binary number).

```
#define COMP_INT_EN            0b10000000          //Enable Comparator
#define     COMP_INT_DIS       0b00000000          //Disable Comparator
#define COMP_INT_MASK          (~COMP_INT_EN)      //Mask Enable/Disable Comparator bit
//************ Comparator Output Enable/Disable **************
#define COMP_OP_EN             0b01000000          //Comparator Output Enabled on CxOUT
#define COMP_OP_DIS            0b00000000          //Comparator Output Disabled on CxOUT
#define COMP_OP_MASK           (~COMP_OP_EN)       //Mask Comparator Output Enable/Disable

//***************** Comparator Output Inversion Selection *******
#define     COMP_OP_INV        0b00100000          //Comparator with OP invert
#define     COMP_OP_NINV       0b00000000          //Comparator with OP non invert
#define     COMP_OP_INV_MASK   (~COMP_OP_INV)      //Mask Comparator Output Inversion Selection bit

//**************** Comparator Interrupt generation settings ******
#define     COMP_INT_ALL_EDGE  0b00011000          //Interrupt generation on any change of the output
#define     COMP_INT_FALL_EDGE 0b00010000          //Interrupt generation only on high-to-low transition of the output
```

Design Example: Timer

The timer is a popular peripheral in MCU. Many applications rely on timing functions. Time is an important parameter in embedded systems. It is represented by the count of a timer. Timer bit numbers depend on the MCU family. An 8-bit MCU comes standard with an 8-bit timer. An MCU timer can run on its own without interfering with the rest of the system. This feature makes MCU flexible and versatile in many applications. In terms of applications, a timer can record an event arrival time, generate an interrupt, and measure the pulse width, period, frequency, or even duty cycle of a signal. Most MCUs come with more than one timer. The block diagram below in figure 7.21 shows TIMER0 in a PIC16 part.

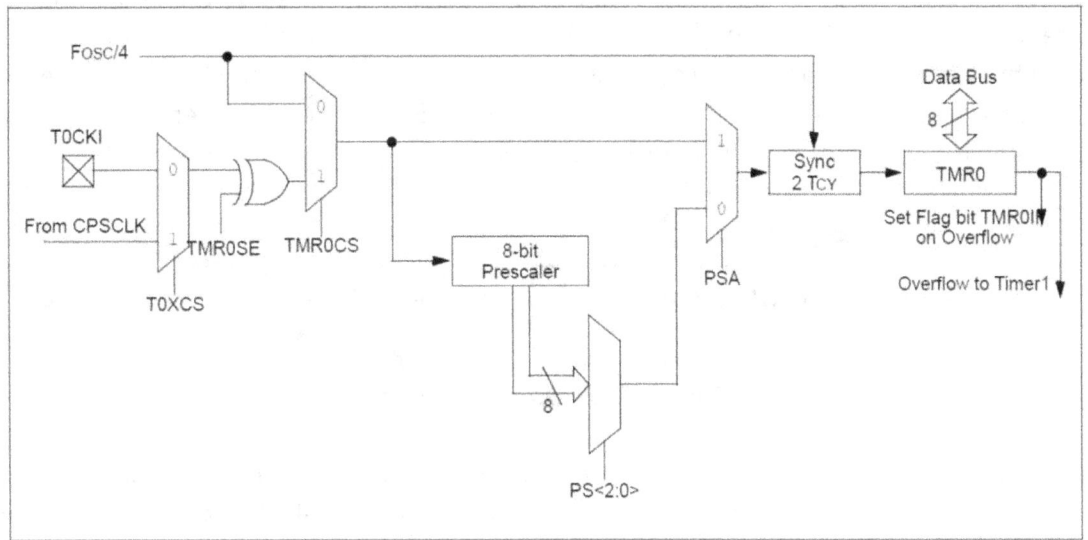

Figure 7.21: Timer0 block diagram

A timer requires a reference clock to run. In PIC16, TIMER0 can be driven by either FOSC / 4 (instruction clock) or external clock source (T0CK1). A T0XCS (clock select) bit decides whether FOSC / 4 or external clock is used. TMR0SE is the source edge select bit. If set to "1," TIMER0 increments on high to low transition. If it's set to zero (clear), TIMER0 increments on low to high transition. There is a prescaler function that slows the clock down. PS<2:0> means there could be 8 prescaler options **(2^3 = 8)**. The timer rate can be divided from 1:2 all the way to 1:256. For example, if the instruction clock runs at 1 MHz, PS bits are set to have a decimal value of 4, and the internal clock is now running at 250 kHz (1 MHz / 4 = 250 kHz). The next timer bit is a PSA bit. It determines whether or not you want to use a prescaler. If not used, no clock rate reduction occurs. For an 8-bit timer, when the timer rolls over to 255, an interrupt automatically occurs (256 incrementing started from 0 and ends at 255, **2^8 = 256**). We can then use the interrupt signal to control other functions. The OPTION register is the timer control register. Let's use TIMER0 to create a time delay of 2 ms. At the end of the 2 ms, an interrupt signal is generated. In this example, we are using a 16 MHz crystal oscillator. First we need to figure out the instruction clock cycle from the crystal. If you recall, the timer is fed by the internal clock. This clock comes from a crystal oscillator. We'd need to divide the crystal by four to get instruction clock cycle. That turns out to be 250 ns.

$$FOSC = 16\text{ MHz}$$

$$\frac{FOSC}{4} = \frac{16\text{ MHz}}{4} = 4\text{ MHz}$$

$$\frac{1}{4}\text{MHz} = 250\text{ ns}$$

To slow down the clock, we use the prescaler value of 32. This gives us the actual instruction clock frequency at 125 kHz **(4 MHz / 32 = 125 kHz) or 8 us clock cycle (1 / 125 kHz = 8 us)**. Consequently, each step the timer counts is now 8 us. To achieve 2 ms delay, we need to set TIMER0 register to start from 5 so that it will increment 250 steps rolling over at 255 **(5 to 255 = 250)**. Loading the option register in figure 7.22 would do exactly what we want.

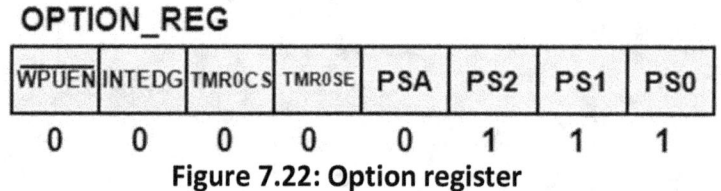

Figure 7.22: Option register

Below are the C program's implementations of this timer example. In this example, we name the C function, delay. The delay function will accept a parameter called x as a character type.

void delay(char x)

The void keyword means this C function does not return any values back to the operating system. Within the function, we declare i, initialize x, and clear the TIMER0 register. Since "i" is an index, we can step through 5 times. Recall that we need to roll over not just once, but 5 times in order to get 2 ms.

int i, TMR0 = 0, x = 1;

We disable the TIMER0 interrupt by assigning "0" to the timer interrupt enable bit so that the interrupt flag is first clear initially.

INTCONbits.TMR0IE = 0;

The next step is to load the option register with the appropriate bit values.

OPTION_REG = 0b00000111;

With a simple for loop, the program presets the TIMER0 register to 5 so that it overflows on the 250th pulse **(250 X 8 us = 2 ms)**.

```
for ( i = 0; i < x; i ++) {
    INTCONbits.TMR0IF = 0;
    TMR0 = 5
```

Inside the for loop, we first clear the interrupt flag, just to make sure it wasn't set from other parts of the program. We set x to 1 so that this for loop only executes once. Indexing it once is all we need to roll over the timer register by setting TIMER0 register started from 5. The while statement monitors the interrupt flag. It will set if the timer register rolls over at 255, and then the program will exit out of the for loop. The result of this function is that we successfully create a 2 ms time delay.

```
    while (!(INTCONbits.TMR0IF)); }
```

Summary

Microcontrollers' market and definitions were first described in this chapter, followed by MCU types, parameters, architecture, instruction cycle, and instruction set definitions. MCUs equip with many peripherals that are highly configurable to allow embedded system designers to tailor specific applications. Practical MCU applications using comparators, timers, debuggers, IDEs, and programming techniques were covered. Embedded system engineers need to master both hardware and software skills. A good understanding of MCU parameters and the datasheet will lead to increased design effectiveness and efficiency, reducing product design time to market.

Quiz

1) Name five popular MCU applications.

2) Name three differences between MCUs and computing CPUs.

3) What are the two types of memory found in MCUs? What are the differences between them?

4) What are the most popular MCU product families in number of bits? What does this number mean?

5) List five popular MCU peripheral modules.

6) If an MCU uses an external oscillator running at 32 MHz, what is the instruction cycle frequency?

7) List three MCU special features and briefly describe their functions.

8) Write an assembly code that subtracts the value of "5", which is stored in the working register, from the GPIO register. The result of the subtraction will be stored in the file register (F).

9) If the instruction clock cycle in the timer design example (see page 269) is divided down by a 1:8 prescaler instead of 32, what is the final clock frequency?

10) Write a C program to create a 1 ms time base using TIMER1 module.

Chapter 8: Programmable Logic Controllers

Programmable logic controllers (PLCs) are digital computers used in industrial and commercial applications. Their main functions are to control machines and automate complex processes and motions. The PLC market is fragmented with many PLC system manufacturers. Some leading PLC suppliers are Bosch and Siemens, Allen-Bradley, General Electric, Panasonic, and Mitsubishi Electric. Figure 8.1 shows an advanced PLC from General Electric called the Programmable Automation Controller (PAC). Its dimensions are approximately 4 feet long by 1 foot wide.

Figure 8.1: GE Programmable Automation Controller (PAC)

Due to the needs of operating in industrial environments, PLCs are built to be rugged while maintaining many components found in personal computers. For example, PLCs, like computers, have memory, CPUs, input and output terminals connecting to input and output devices, and built-in power supplies. A PLC lacks, however, a hard drive, keyboard, and monitor. On the other hand, PLCs come with control programming software allowing PLC designers to create custom PLC programs on computers to tailor their design needs. The programming language used in PLCs is ladder logic, a graphical programming language as opposed to text-based programming languages such as C, C++, or JAVA. In this chapter, we will cover PLC history, overview, operations, functions, applications, ladder logic program creation, techniques, and several PLC project examples.

History

The development of PLCs first started with General Motor (GM) in the late 1960s within its Hydra-Matic division. The original objective of the PLC was to replace bulky, costly relays, eliminating cables in the manufacturing systems. The benefits are that they reduce cost and increase the range of functions, versatility, and flexibility while achieving higher reliability. A relay acts like a switch applying electromagnetic theory to operate. Its main function is to control mechanical movements (see figure 8.2). By closing the switch (above the relay control AC source), an AC control signal is applied to the electromechanical relay. It energizes the coil via the electromagnet magnetizing the armature, causing it to deflate downward. As a result, the end points at the right-hand side of the armature making contact achieving a normally-closed (NC) condition. If there is no relay signal applied to the magnet, the armature is de-energized. It then tilts upward anchored by the spring. It now makes contact to the normally-open contact (NO) point. Relays are used in heavy load applications such as motors. The benefit of using relays is that the control signal is electrically isolated by the magnet, providing isolation protection between loads and users. This is essentially a safety measure as many industrial applications involve high-power input and output devices.

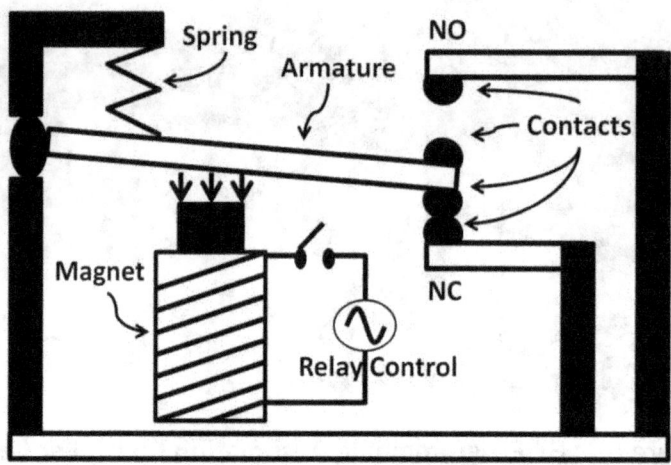

Figure 8.2: Relay diagram

The control of electrical systems within GM's Hydra-Matic division in the 60s relied heavily on conventional hard-wired relays. Imagine a cabinet full of these relays clogged with cables. Figure 8.3 shows a cabinet with relays and cables.

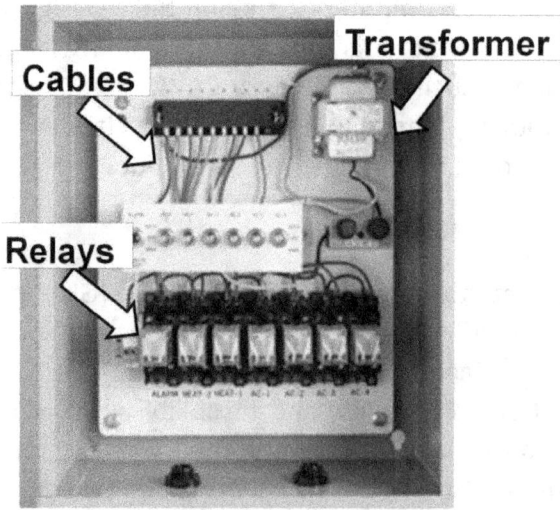

Figure 8.3: Relay control panel

These external components added not only material and labor costs but also complexity to the system as well as difficulty in troubleshooting, design, change, and repair. Lowering costs, increasing flexibility, and ease of maintenance became major initiatives within GM. GM engineers then developed the concept of PLCs utilizing computer technologies in a system that would automate complex industrial application processes, and at the same time the systems would operate optimally in harsh industrial environments filled with dirt, dust, moisture, and in some cases, chemicals, shocks, and vibrations (e.g., automotive manufacturing). Since its inception of PLCs, Allen Bradley, as a company, took the concept further developing the PLC terminology. Allen Bradley (now a Rockwell Automation company) PLCs have since became the industry standard.

PLC Benefits

PLC functions include timing, counting, calculating, and digital and analog signal processing. These basic functions form the foundations of PLC structure. The major advantage of PLCs is that the majority of the functions are contained within the PLC hardware. This largely reduces the size of the overall systems and the likelihood of making any wiring mistakes. The designers of the PLC programs have full capability to create ladder logic programs through software displayed on a computer monitor. Any program modifications are easily made by software. This is much better than physical wiring changes, and lowers labor and material costs. Many modern PLCs are equipped with communication capabilities allowing PLCs to communicate with each other through wired or wireless connections. With wireless capabilities, users can remotely log into the system for troubleshooting purposes. Each PLC must include at least one CPU (microprocessor). This gives PLCs the ability to process data and information in a short period of time compared to bulky relays. Many PLCs work in conjunction with sensors in manufacturing environments. Often times, manufacturing facilities have fast-moving conveyer belt systems (see figure 8.4). The input device is the sensor, whereas the turn motor acts as the output device in response to the input sensor.

Figure 8.4: PLC application: Conveyor belt system

PLC Components

Figure 8.5 shows a conceptual PLC block diagram. It includes six basic components/modules: Input and output modules, power supply, CPU, memory (program, data), and a programming device. The input and output modules can be combined into one module (I/O module).

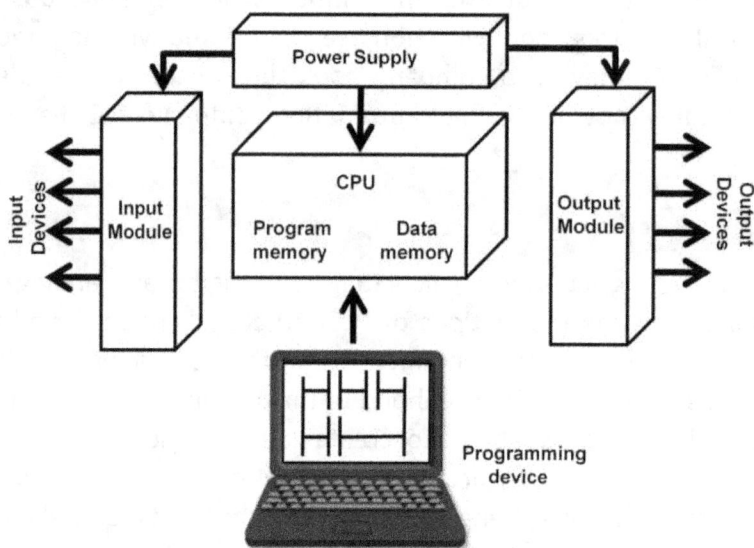

Figure 8.5: PLC block diagram

The input module (slot) serves as a gateway between external devices and the internal circuitries of the PLC. Input devices are hardwired to the input terminals of the input module. Input modules receive electrical signals from the input devices and then transmit them internally to the PLC for processing. Many I/O modules are scalable where an input/output slot can have multiple terminals, from four to sixteen or more. Common input devices are relays, toggle switches, push buttons, and sensors of all types. These types of input devices produce signals that are either logic high or low. The I/O modules that accept and produce discrete signals are called discrete (clearly defined level) input modules. I/O modules that accept and output analog signals such as temperature, pressure, or humidity values are called analog I/O modules. Figure 8.6 shows the symbols and actual examples of PLC input devices. In some PLCs, both input and output modules are integrated into one module. These printed circuit board (PCB) modules can be plugged into and out of the PLC. This removable feature makes the PLC system scalable and very easy to repair without replacing the entire PLC. Figure 8.1, shown previously, contains multiple input/output modules. PLC input and output devices potentially carry high current and voltage. Power protection circuits such as opto-isolators are needed to isolate high-power field devices from lower-voltage PLC electronics. Opto-isolators rely on light-sensitive transistors to turn on or off internal PLC circuits. Because the transistor is triggered by an LED, which is physically isolated within the transistor, it provides electrical isolation of the external power from the internal low-voltage PLC circuits.

Figure 8.6: Switch, push buttons, and symbols

The output modules connect to the output devices through physical wires. Just like input slots, outputs slots can have many terminals. Examples of output devices are motor starters, solenoid valves, and indicator lights. The output devices and symbols (motor starter; solenoid; and red, green, yellow, and blue lights) are shown in figure 8.7.

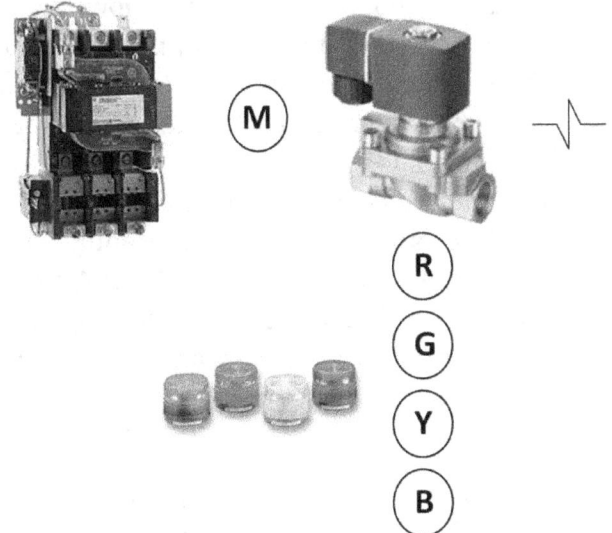

Figure 8.7: Motor starter, solenoid valves, and output device symbols

The power supply provides the electrical power to all modules converting from AC to DC. The DC voltage supplies power to all internal PLC circuits. The PLC power supply typically does not support external input or output devices, only internal ones. The AC ratings are different from one country to another. Common ratings are from 120 V to 240 V AC. The way the CPU works is very similar to a conventional computing CPU in terms of performing logic operations and interfacing with data and program memory as well as fetching and executing commands from the PLC programs. The major difference of PLC CPUs and computing ones are that PLC CPUs' performance is generally lower than that of computing CPUs in terms of clock speed.

PLC Programming and Ladder Logic

PLC programs are written in ladder logic. After a PLC program is written and inputted by PLC designers using ladder logic software installed on computers (programming devices), PLC designers can test the programs right on the computer screen without connecting to the actual PLC. This is called software simulation. If the actual PLCs are available, designers can upload the PLC programs to the PLCs via a standard computer interface such as Ethernet or RS-232. End users can then run, control, and execute the programs using the programming device. During program execution, while the PLC connects to the programming device, it reports the program status back to the programming device and displays it on the computer screen for troubleshooting and debugging purposes. When the PLC programs execute, they operate in repetitive loops. First, the CPU reads the status of all input devices. Then it executes the PLC program. Finally, the PLC programs update and control the output devices. This process continues until the program is paused or stopped by the programmer or end user. The programming device serves as a platform for PLC designers to enter ladder logic programs in program mode. In the field, hand-held devices are used in place of programming devices providing portability benefits. A ladder logic program could include many elements such as normally-open (NO) and normally-closed (NC) contacts, output symbols, and logical functions. An NC contact has a forward slash symbol in it. Figure 8.8 shows an example of a ladder logic program, which is entered by the designer. The program is stored in the program memory of the PLC. In this example, each horizontal branch is called a rung which comprises contact and output symbols. S1 to S4 are on rung 1. S5 to S7 are located on rung 2. A contact can be either normally-open or normally-closed contacts (further explanations will follow shortly). It is possible to add a parallel branch on a rung. S1 and S4 form a branch instruction. The circles on the right end of each rung are the output symbols representing the output devices. These output symbols ultimately control the output device via the PLC's output terminals. The CPU processes a ladder logic program stored in the memory, one rung at a time, starting from the top of the program and reading the inputs of contacts from left to the right.

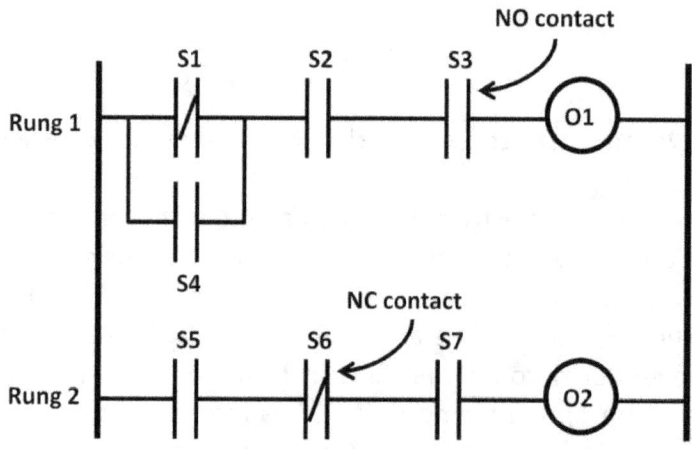

Figure 8.8: PLC Ladder logic program

In order for the programming elements to be entered correctly, the PLC needs to be set to program mode. The designers need to first understand the proper operations of the NO and NC contacts as well as the output symbols. As mentioned previously, NO and NC contacts correspond to input devices. And the NO contact needs to be correctly addressed to an input terminal which connects externally to an input device. Let's assume the input device is a push button. When the button is pushed, the input status of the NO is logically high causing the contact output status to be in a logic high state as well. When the contact is not energized, (i.e., the push button is not pushed), then the contact input status is low leading to logic low output status (see figure 8.9).

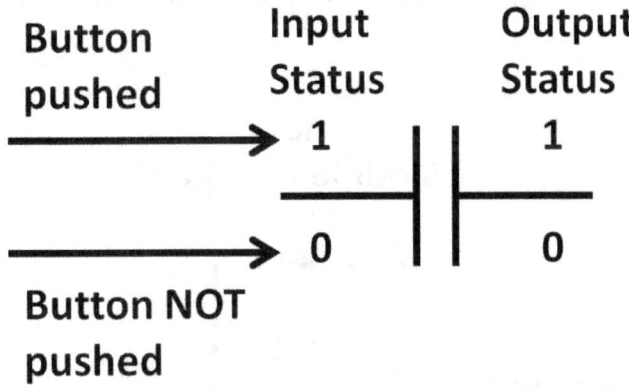

Figure 8.9: Normally-open contact status

If a push button is addressed to a NC contact, when the button is pushed, the input status of the NO is logically high. Because it's a normally-closed contact (the forward slash with the contact symbol), by energizing the contact, logic low contact output is achieved. When the contact isn't energized, (i.e., the push button isn't pushed), the contact input status is low. In this case, the contact output status is high. You can imagine that a NO operation works just like an inverter (see figure 8.10).

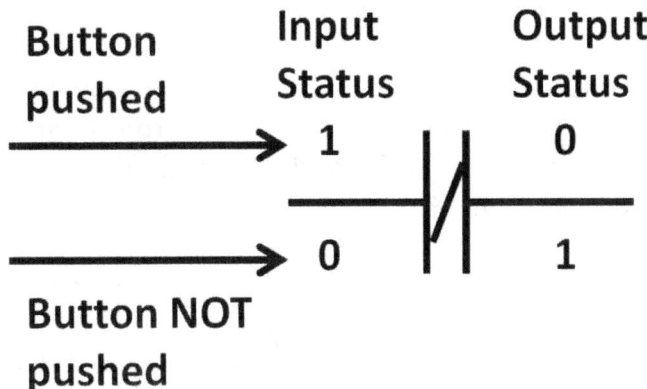

Figure 8.10: Normally-closed contact status

All contacts, outputs, and logic function elements in ladder logic programs need to be addressed accordingly in terms of address format. For NO or NC input contacts, the following format can be realized:

$$I: 0/2$$

This address means that it is an input device denoted by the initial "I" letter. Followed by ":", the number "0" corresponds to the module number. Recall that a PLC could contain multiple input or output modules; this zero represents the very first module. The right digit "2" corresponds to the terminal number which represents the third terminal of module 0 (see figure 8.11).

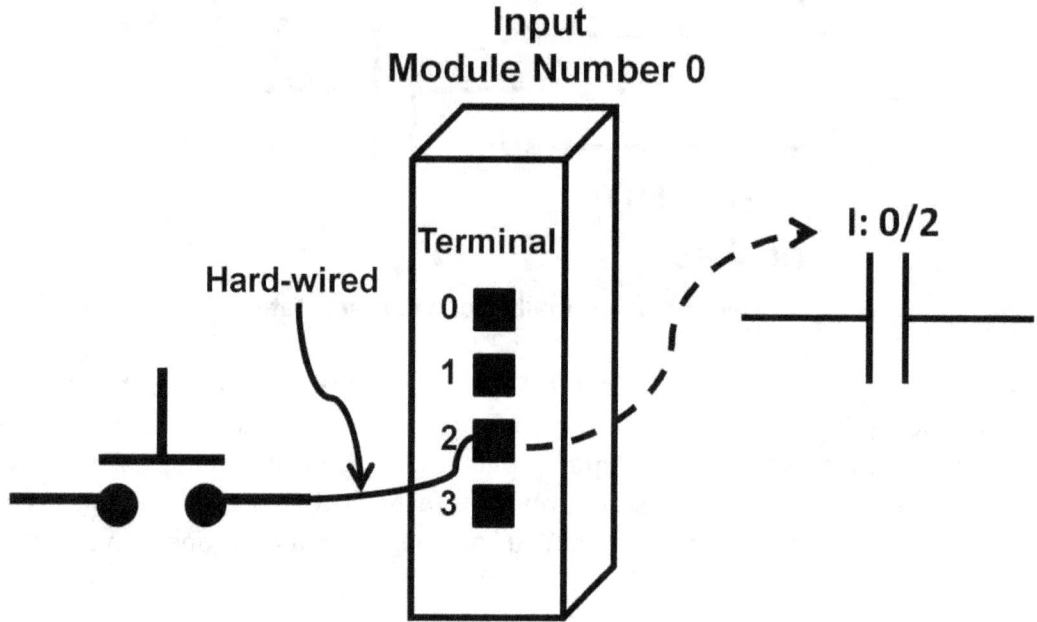

Figure 8.11: Input contact address

PLC programmers need to make sure input contact is addressed correctly so the intended input device is used according to the PLC programs. Many PLC errors stem from incorrect contact addressing. The above format applies to Allen Bradley's brand of PLC only. Keep in mind there are many other address formats from other PLC manufacturers.

The output symbol shares a programming technique similar to that of the input contact. The output symbol would require the correct address (see figure 8.12). In this case, the output symbol O: 0/2 within the PLC programs maps to the first output module (module 0) and the third terminal (terminal number 2) which connects physically to an output device (motor).

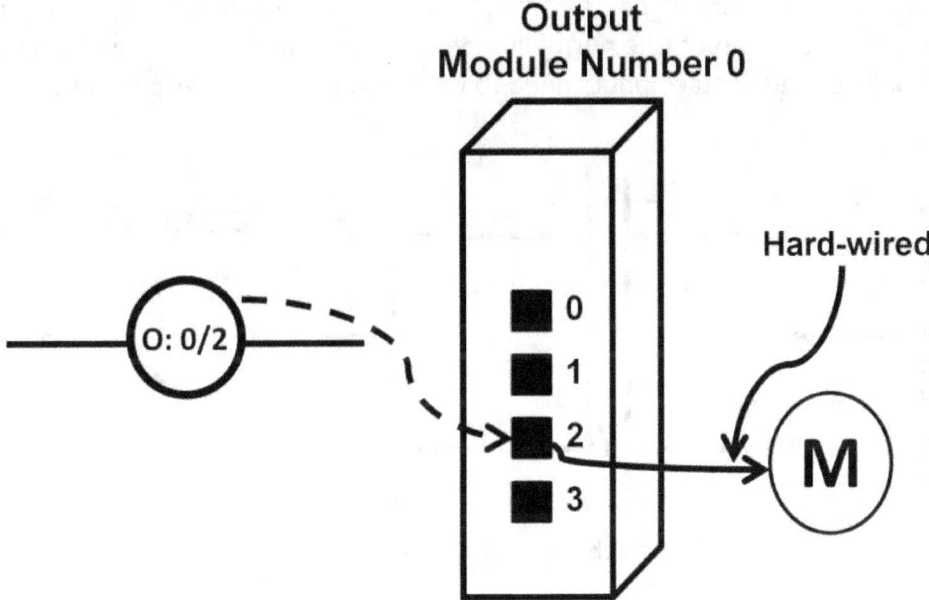

Figure 8.12: Output contact address

As mentioned earlier, contacts can be connected in parallel forming a branch. In figure 8.13, contacts 1 and 2 form a branch. The output symbol logic status is controlled by input contact's outputs. The output symbol will be in a logic high state if either outputs of contact 1 or 2 is high, (i.e., it functions as an OR gate). The output symbol status-control mechanism is best described using continuity. It will be further explained in the next section.

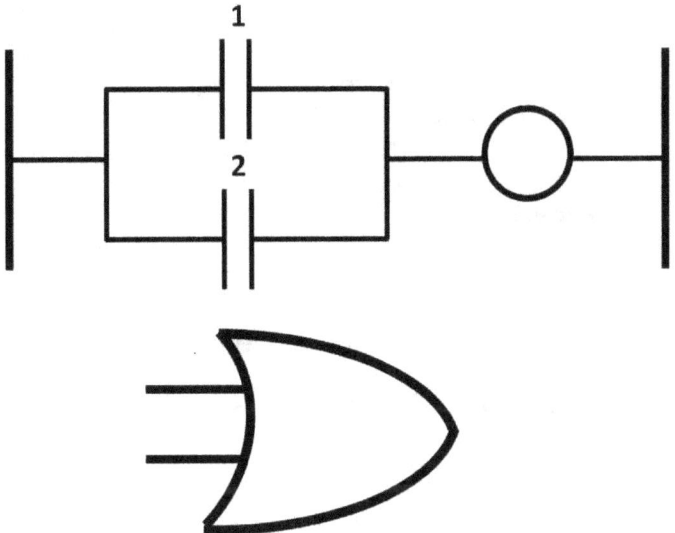

Figure 8.13: Branch = OR gate

Both NO and NC contacts can be connected as a branch (parallel). Figure 8.14 shows an example. To enable the output, the normally-open contact input needs to be high (output high) and the normally-closed contact inputs need to be low (output high) respectively.

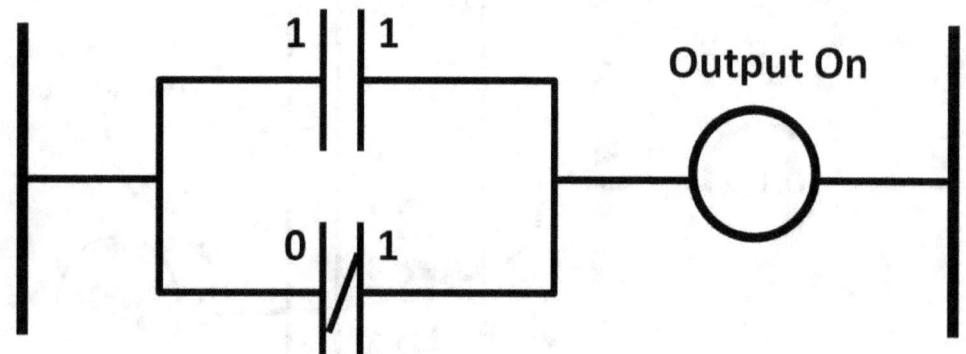

Figure 8.14: Branch instruction using NC and NO contacts

If contacts are connected in series, the output is only energized when all 1, 2, and 3 contact output statuses are high, (i.e., an AND gate operation) (see figure 8.15).

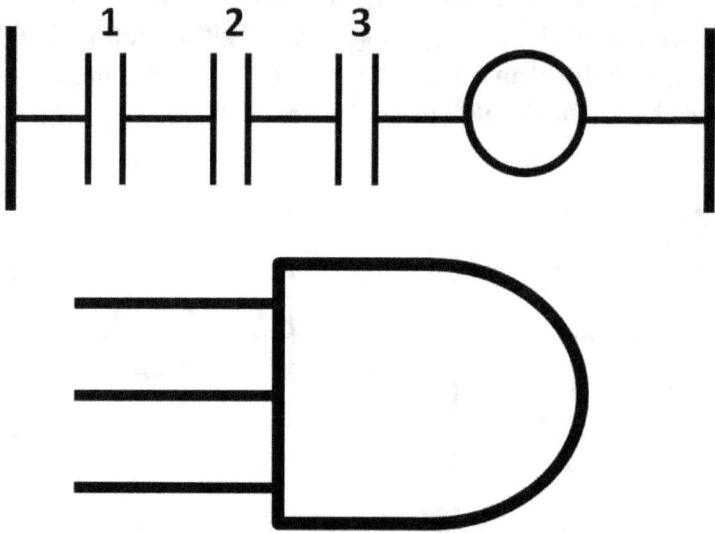

Figure 8.15: Series contacts = AND gate

Just like branch instructions, a combination of NO and NC contacts can be connected in series. To energize output in figure 8.16, the NO contact input needs to be high while the NC contact inputs need to be low.

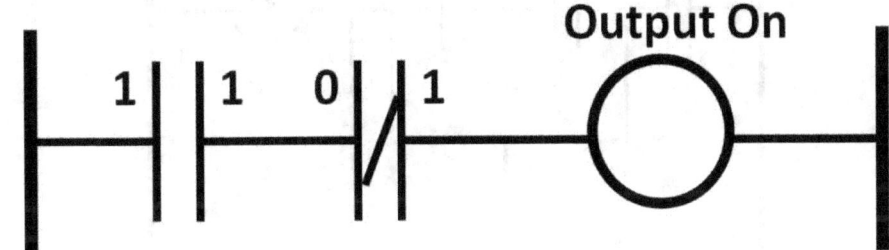

Figure 8.16: Series contact with NO and NC contacts

PLC Programming Example

Let's use a practical example to further study how ladder logic programs work. This application senses the temperature and humidity of a warehouse. If the temperature and humidity levels go above a predetermined value, the air-conditioning (A/C) system and fan would turn on. Additionally, there is a manual overdrive button allowing warehouse workers to turn on the A/C system and the fan regardless of temperature or humidity levels. For safety reasons, an emergency stop button is in place to disable all operations. Before starting to program ladders logic, designers need to first identify what the input and output devices are. In this design, there are four input devices: a manual overdrive button, an emergency stop button, temperature sensors, and humidity sensors. For outputs, there are two output devices: an A/C system and a fan. See table 8-1 for field devices names, types, and address assignments.

Field devices names	Type	Address
Manual overdrive button	Input	I: 0/0
Emergency stop button	Input	I: 0/1
Temperature sensor	Input	I: 1/0
Humidity sensor	Input	I: 1/1
Air-conditioning system	Output	O: 0/1
Fan	Output	O: 0/2

Table 8-1: PLC application input, output devices, and addresses

Figure 8.17 shows the ladder logic program for the previous applications.

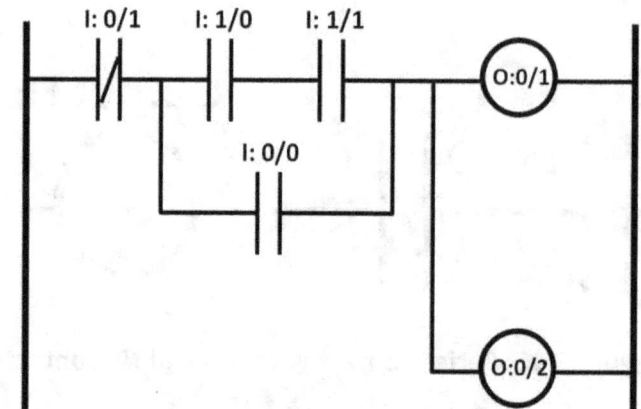

Figure 8.17: Air-conditioning and fan ladder logic program

When this program executes in run mode, it goes through one rung at a time reading the input contact status moving from the left to the right on each rung. This program only contains one rung even though the branch instruction is formed among input contacts and output symbols. In order to enable both parallel connected outputs, continuity needs to be established, meaning logic outputs on each contact need to be high starting from the left of the rung and continuing all the way to the right. Figure 8.18 demonstrates one way to establish continuity, denoted by the dotted line. The output status of I: 0/1, I: 1/0, and I: 1/1 all need to be high in order to turn on O: 0/1 and O: 0/2. Although these contacts are connected in series, each contact is independent and does not affect the contact next to it, (e.g., a high output at I: 0/1 does not cause the I: 1/0 input to go high). The status of each contact solely depends on the field device associated with the contact addressed to it. In this scenario, when the emergency button (I: 0/1) is not pushed, if both temperature (I: 1/0) and humidity (I: 1/1) sensors are tripped, A/C (O: 0/1) and fan (O: 0/2) turn on.

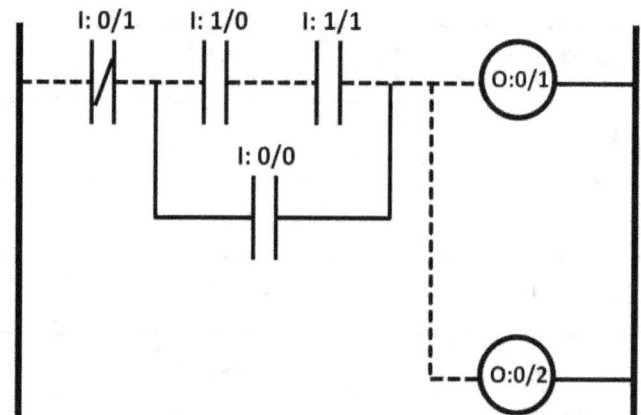

Figure 8.18: Continuity scenario one

There is a second scenario in which both outputs would be on. It's shown in figure 8.19, denoted by the dotted line. In this scenario, when I: 0/1 and I: 0/0 output statuses are high, O: 0/1 and O: 0/2 are on. This means when the emergency button (normally-closed) is NOT pushed while at the same time the manual overdrive button is pressed, the A/C and fan turn on. Lastly, when the emergency stop button is pressed the output of I: 0/1 goes low, and regardless of the input status of the rest of the contacts, the outputs stay off. In this particular scenario, I: 0/0, I: 1/0, and I: 1/1 form a parallel (branch) instruction.

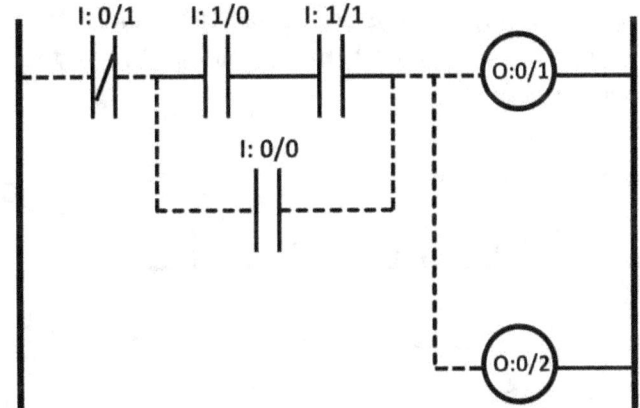

Figure 8.19: Continuity scenario two

Combinational logic circuits (see figure 8.20) can be used to describe and model PLC programs. Some PLC designers first use combinational logic as design tool before inputting the actual PLC programs. The equivalent logic circuit of the previous example consists of two AND gates and one OR gate. The inverter models the normally-closed contact. If the emergency stop is not pushed, inverter input is low and output is high. If both temperature and humidity sensors are tripped, the AND gate output is high energizing the A/C system and fan. If the emergency button is pressed, the AND gate input is low leading to low OR gate output. If the emergency button is not pressed and the manual overdrive button is, the AND gate in the bottom results in logic high status turning on O: 0/1, O: 0/2.

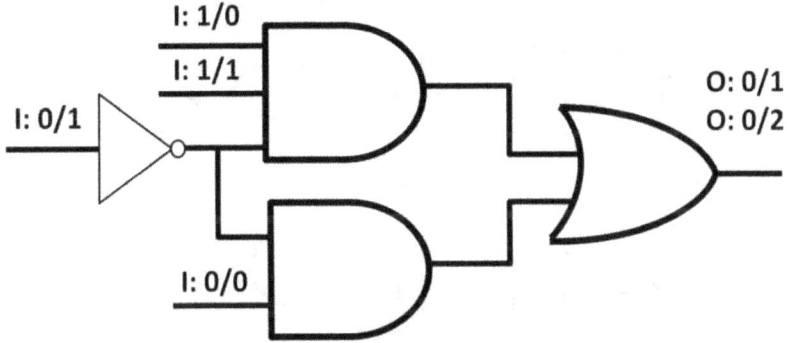

Figure 8.20: PLC design combinational logic circuit

There are maximum limits on contact numbers on a rung. If an application requires more than the maximum contact numbers to be on at the same time to enable an output, an internal output symbol can be used (see figure 8.21). Suppose five is the maximum number of contacts allowed on a rung. This design requires all eight NO contacts to be energized to turn on O: 0/2. On rung 1, five NO contacts are used to control the internal output (B3: 1/1). The internal output address is then used to address an NO contact on rung 2 (far left) along with the remaining three contacts. These four contacts now control the output symbol (O: 0/2). When the first five contacts are closed, (i.e., the outputs of 1 to 5 are all high), the output statuses of these five contacts energize B3: 1/1, which is addressed to the first input contact on rung 2. This makes the input status of this contact high. If the remaining three contacts' inputs (6, 7, and 8) are high as well, O: 0/2 will turn on.

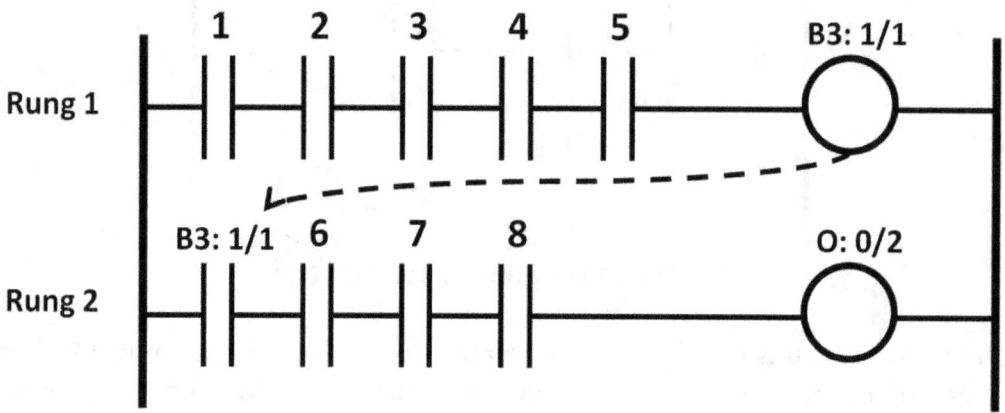

Figure 8.21: PLC design using internal output

PLC Programming Syntax

Similar to text-based programming, there are syntax rules in PLC programming that designers need to be aware of. The following diagram shows common ladder logic syntax. First, the output symbol needs to be on the far right-hand side of a rung (see figure 8.22).

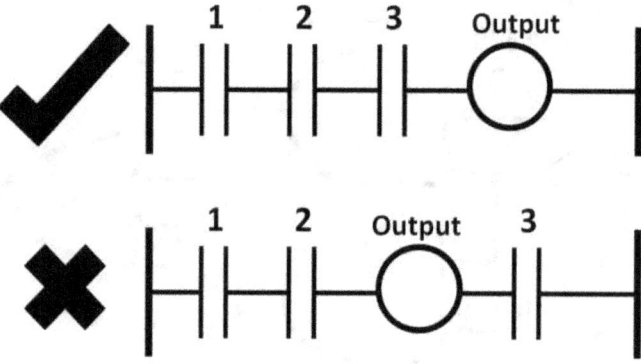

Figure 8.22: Output symbol syntax

It's valid to have one output on a rung by itself. It's invalid, however, to have only input contacts (either NO or NC) on a rung (see figure 8.23).

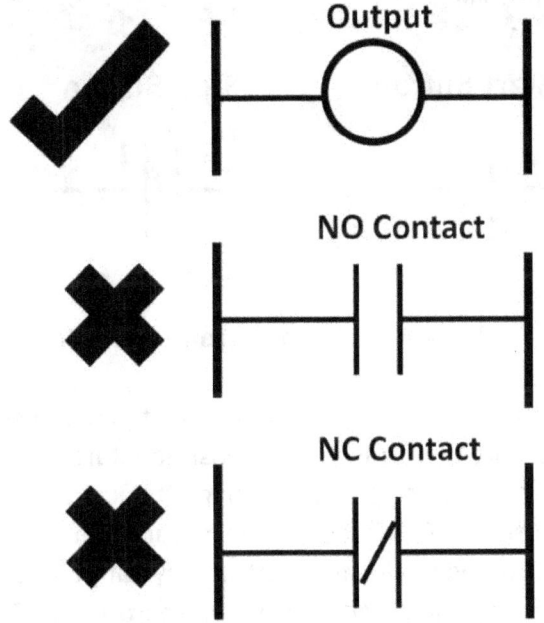

Figure 8.23: Input, Output symbol by itself syntax

To control multiple outputs at once, a parallel output symbol can be used. It's invalid, however, to have multiple outputs in series on a single rung (see figure 8.24).

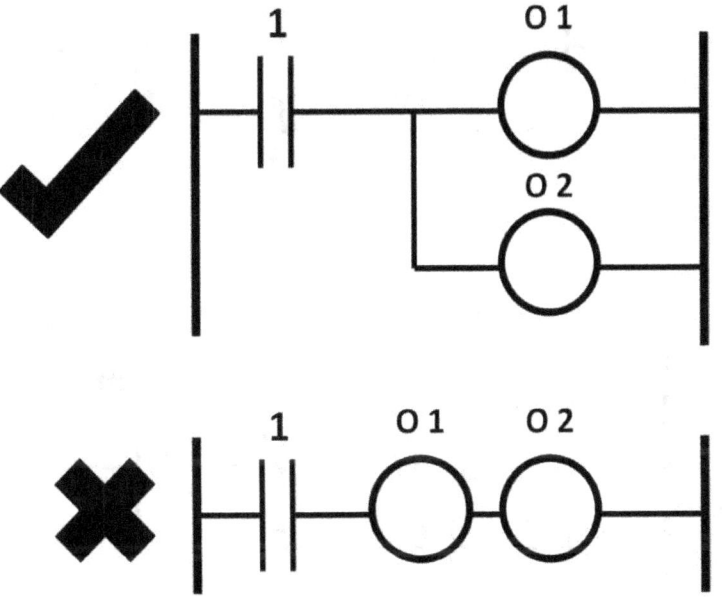

Figure 8.24: Multiple outputs

A branch instruction can be useful when push buttons are used. In most cases, push buttons require users to hold the button down or else the button is off. In figure 8.25, if the start button turns on O: 0/3, the user needs to hold the button down the entire time. This is inconvenient and does not offer much flexibility.

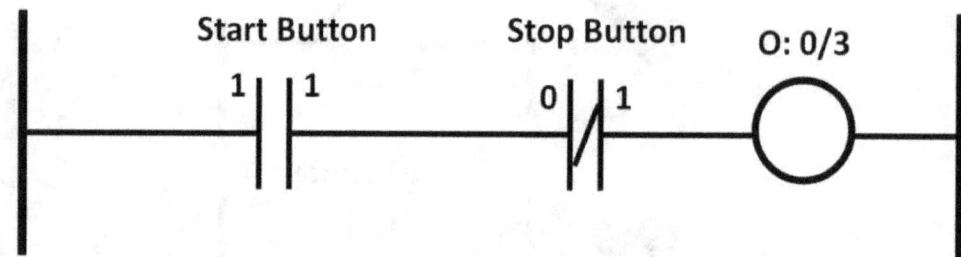

Figure 8.25: Push button application

A special type of branch instruction called a seal-in circuit can be used to resolve this issue (see figure 8.26). In step 1, the start button is not yet pushed (start button output is low). Although the stop button (an NC contact) is not pressed making its output to logic high state, there is no continuity path to turn on the O: 0/3 because the output status of the start button is low. In step 2, the start button is pressed while stop button stays off, and output turns on. In this case, the bottom branch contact has the same address as the output (O: 0/3). Enabling the output causes the branch contact output status to go high (dotted line).

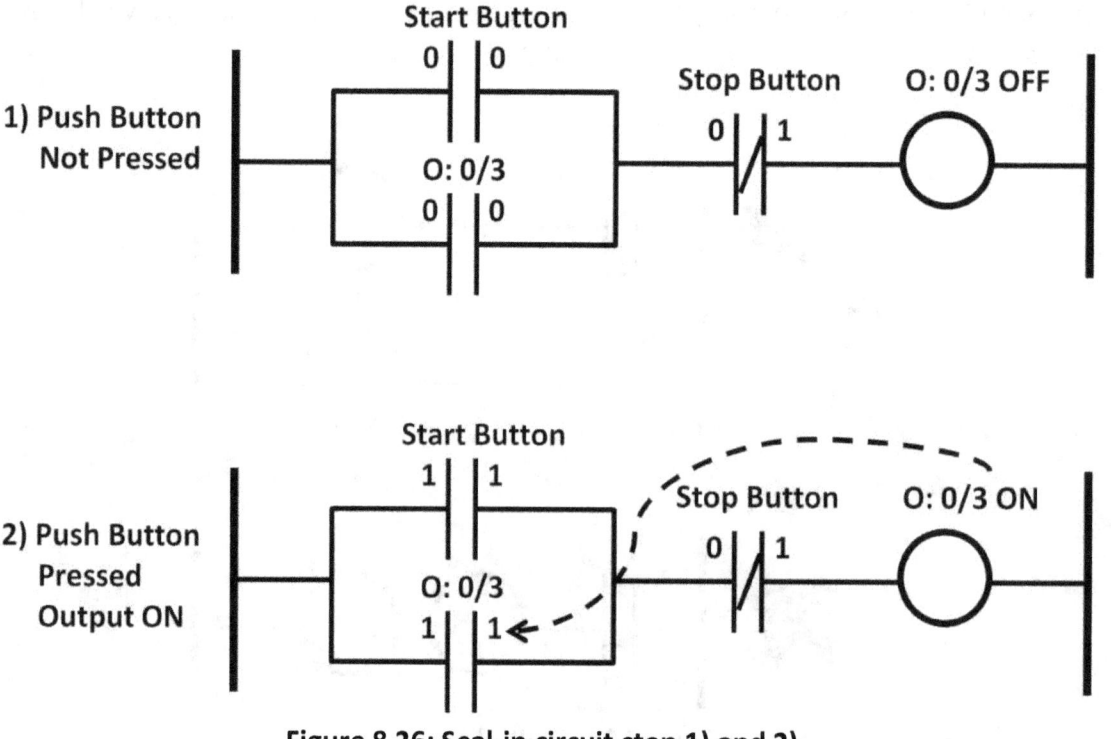

Figure 8.26: Seal-in circuit step 1) and 2)

In step 3 (see figure 8.27), the start button is released, and the output remains on due to the continuity path (dotted line) by the branch contact while the stop button remains off. In step 4, the stop button is pressed causing its output status to go low breaking the continuity path. It de-energizes O: 0/3 causing the branch contact input to de-energize.

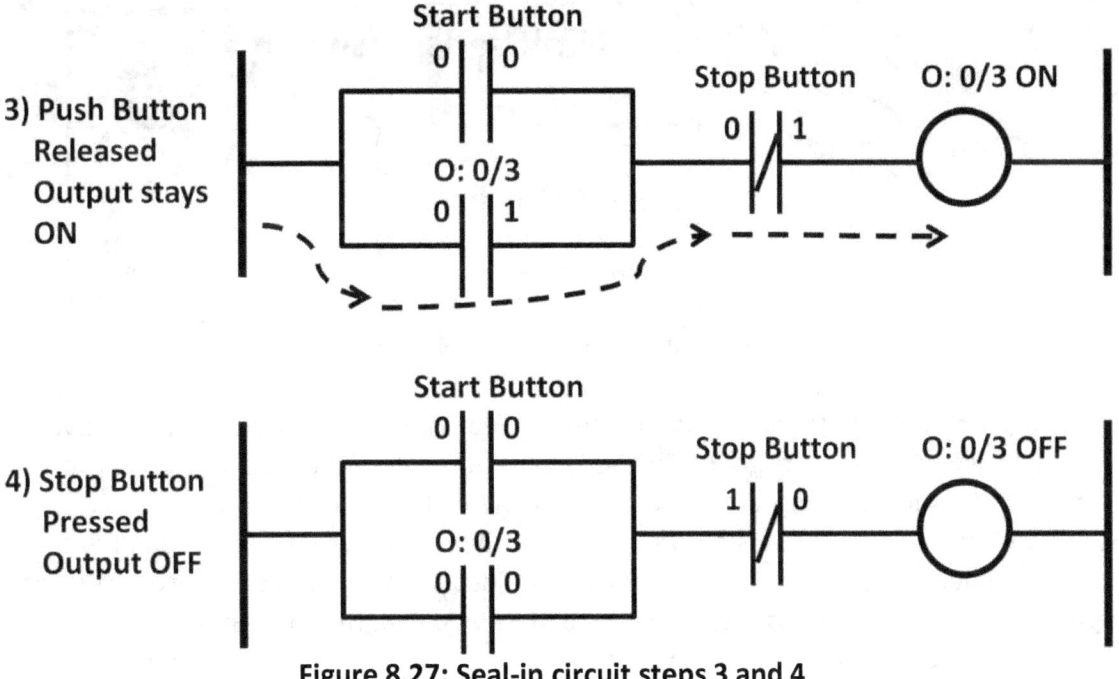

Figure 8.27: Seal-in circuit steps 3 and 4

After inputting the PLC programs, designers can debug the programs within the software platform without physically connecting to the field devices. The on/off state of each device can be easily viewed in the software. This software debug feature allows designers to focus on ladder logic program development, isolating other design variables from field devices. When designing applications that use PLCs to control and automate processes and motions, be sure to capture all system-level electrical and physical specifications from all input and output devices. All control process flows first must be understood and documented before programming. Combinational logic can be applied as pointed out before. The following design example demonstrates the steps to designing and implementing a successful PLC system (see figure 8.28 on the next page).

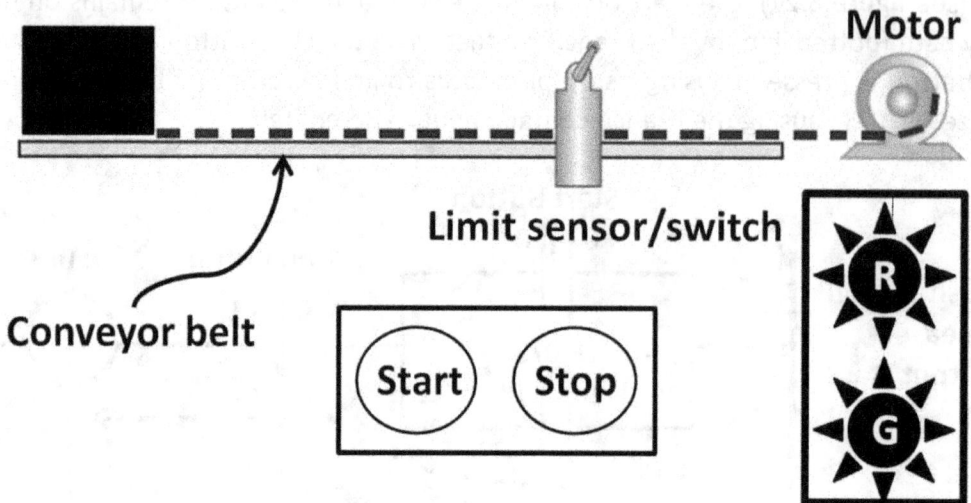

Figure 8.28: PLC Conveyor system

This design is commonly found in manufacturing and assembly facilities where conveyor belt systems are used to transport goods. In this application, when the start button is pressed, the motor starts to turn, moving the conveyor belt. After the box of goods moves to the limit sensor position, the motor stops automatically. While the conveyor belt is running, the green light is on. When it stops, the red light turns on. These process steps are used to design PLC programs. Aside from process steps, PLC designers and programmers must understand each and every field device's electrical specifications making sure the proper input and output modules are capable of receiving and driving the output devices. Questions may arise regarding start and stop button timing requirements, the red and green lights' current, voltage limitations, motor loading, power specifications, limit switch sensitivity level, etc. A list of the field devices with corresponding addresses is shown in table 8-2.

Field device names	Type	Address
Start button	Input	I: 0/1
Stop button	Input	I: 0/2
Limit sensor	Input	I: 0/3
Green light	Output	O: 0/1
Red light	Output	O: 0/2
Motor	Output	O: 0/3

Table 8-2: PLC conveyor application input and output devices and addresses

Figure 8.29 is the ladder logic program of the conveyor belt system. The internal output, B3: 0/1 on rung 1, is controlled by the start/stop buttons and the limit switch. If the start button is pressed while the stop button and limit switch are de-energized, then the internal output, B3: 0/1, is on. On rung 2, the same B3: 0/1 address is mapped to an NC contact (dotted line), which is now logic high, making the second rung input contact's output low. This logic low output keeps the red light off. On rung 3, both the green light and motor are on as a result of B3: 0/1 being on. As the box reaches the limit switch position, input of I: 0/3 is high causing its output to be low (rung 1). This cuts off continuity on rung 1 turning B3: 0/1 off. On rung 2, the red light turns on due to the low input and high output of the NC contact (B3: 0/1). On rung 3, the green light and motor turn off. To start the conveyor again, the user needs to press the start button. The box needs to be taken out of the switch position. This PLC program design is just one way to perform the design tasks. Remember that two completely different PLC programs could perform the same functions. A right or wrong design is not really the question. Rather, a design that has the shortest scan time, has higher efficiency, and is easier to troubleshoot and maintain will be the best design.

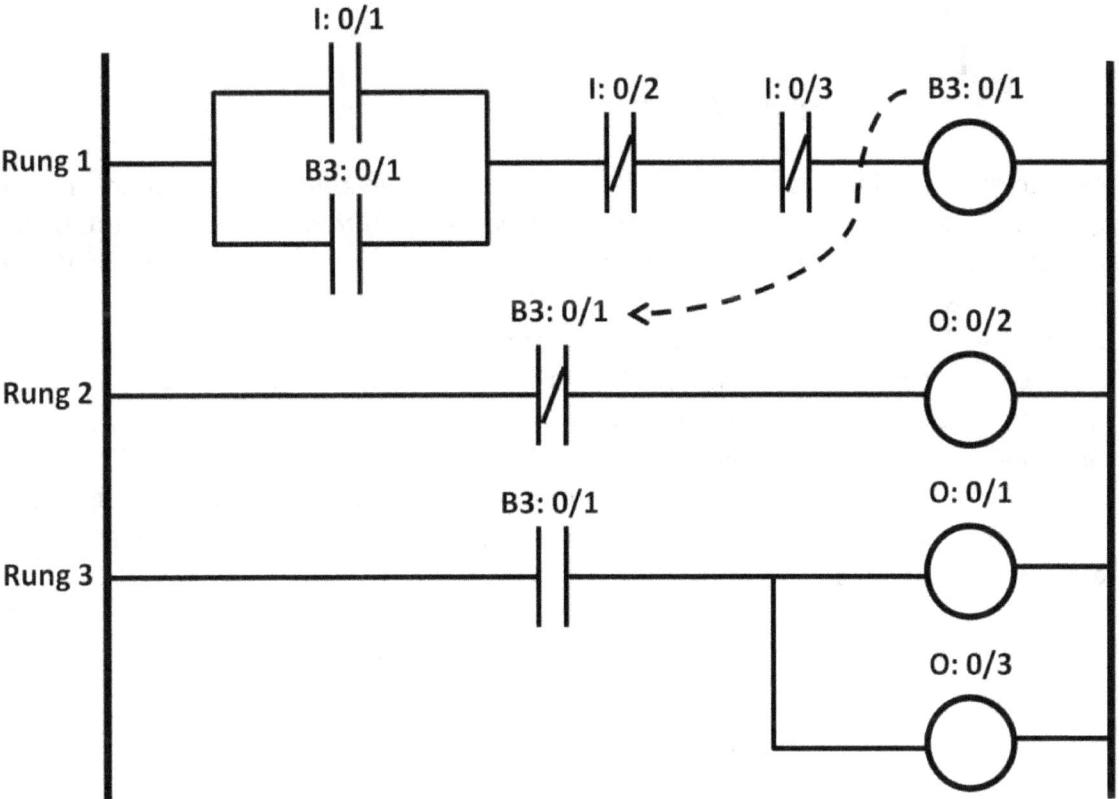

Figure 8.29: Conveyor belt system ladder logic program

Timers

Let's now go over timers in ladder logic programs. Time is a critical parameter in PLC systems. The timer is a step counter and the step size is determined by time base, which is easily changed in PLC software. On- and off-timers are available to perform timing functions. The timer ladder logic symbol of an On-timer controlled by an NO contact is shown in figure 8.30.

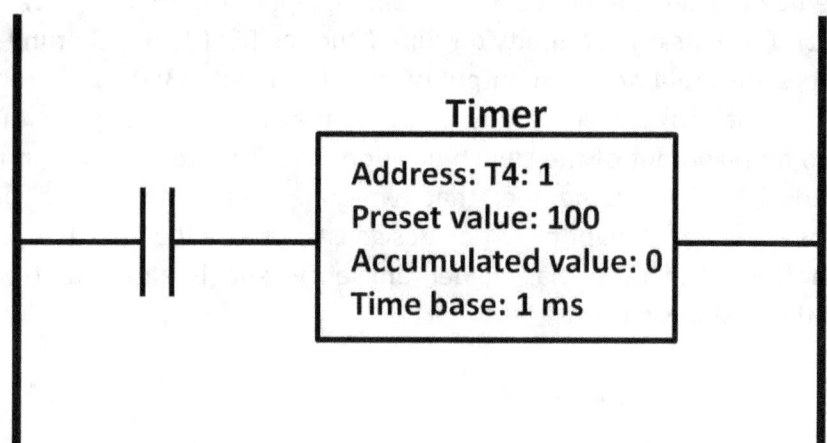

Figure 8.30: Timer symbol and parameters

A timer requires a unique address to differentiate itself from others. In order to program a timer, a preset value, time base, and accumulative values need to be set up. In figure 8.30, the timer address is T4: 1. The delay timer is calculated by multiplying the preset value by the timer base, (i.e., **100 X 1 ms = 100 ms**). If a delay time of 1 s is desired, the preset value can be programmed to 1,000, (i.e., **1,000 X 1 ms = 1 s).** The time base can be changed to other values by software conveniently. Accumulated value represents the delay time measured in real time. It usually starts from zero seconds. A programmer can read the accumulated value during program mode in real time. In addition to timer parameters, a timer comes with output bits that can control other contacts. These bits are enable (EN), timing (TT), and done (DN) bits. The specific address of each bit is associated with the timer's address. Using T4: 1 as the timer address, timer bit names and addresses are shown in table 8-3.

Timer bit names	Address
Enable	T4: 1/15
Timing	T4: 1/14
Done	T4: 1/13

Table 8-3: Timer bit names

On-Timer

Using figure 8.30 as a reference, we can construct a timing diagram to further understand how a timer works (see figure 8.31). When the NO contact's output is logic high, the EN bit follows while the timer starts to time (the TT bit goes high while the timer is timing). At the end of 100 ms, the timer has reached the time delay value set by the preset value. As a result, the done bit (DN) goes high, and TT now goes low because the timer is no longer timing. If the NO contact output remains high, the DN bit stays high. If not, the DN bit goes low, as does the EN bit. Essentially, EN follows the output status of the NO contact.

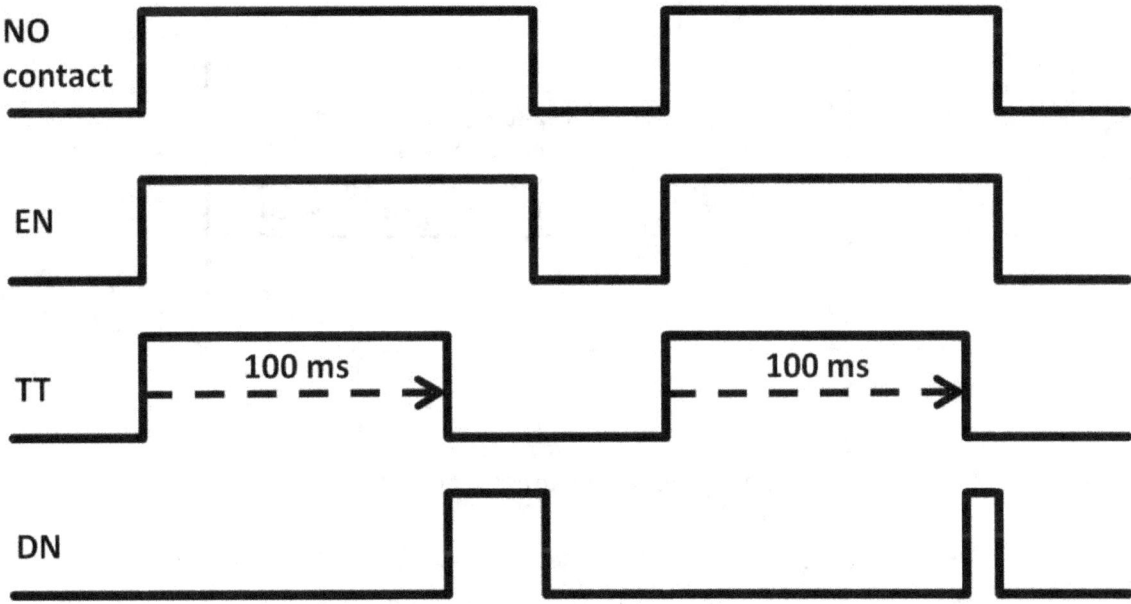

Figure 8.31: On-timer timing diagram

On-Timer Application

Let's use an application example to further examine the on-timer (see figure 8.32). This design equips with start and stop buttons and a motor. 10 s after the start button is pressed, the motor turns on. B3: 0/1 on rung 1 forms a seal-in circuit. The stop button (normally-closed) is used as an emergency stop button. B3: 0/1 on rung 2 controls T4: 1. The DN bit controls the motor.

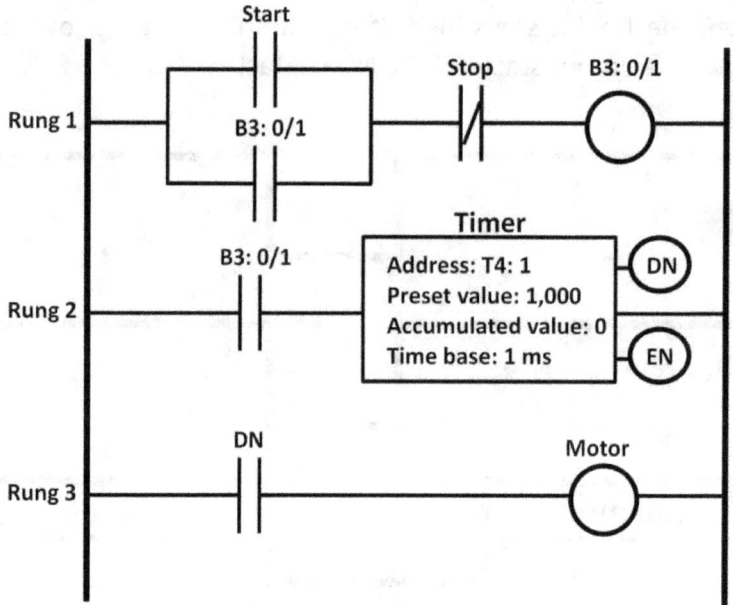

Figure 8.32: On-timer example

To generate a 10 s delay time, a preset value is set to 1,000. The DN bit turns the motor on 10 s after start button is pressed (see figure 8.33). The motor remains on even after 10 s. The motor only turns off when the stop button is pressed, resetting the EN and DN bits of the timer. This turns off the motor.

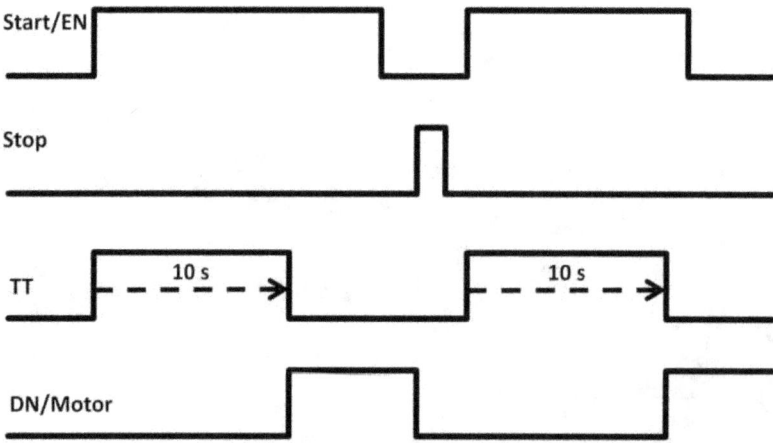

Figure 8.33: On-timer example timing diagram

Off-Timer

The second timer type is the off-timer. According to figure 8.34, the theory of off-timer operation is that the DN bit is energized as soon as the off-timer input is high, (i.e., the EN bit also goes high). When the timer input is false, the off-timer starts to time (TT goes high). After a period of time set by the preset value, the timing bit (TT) and DN bit go low. Figure 8.34 shows the off-timer symbol. In other words, the off-timer takes time to turn the output off.

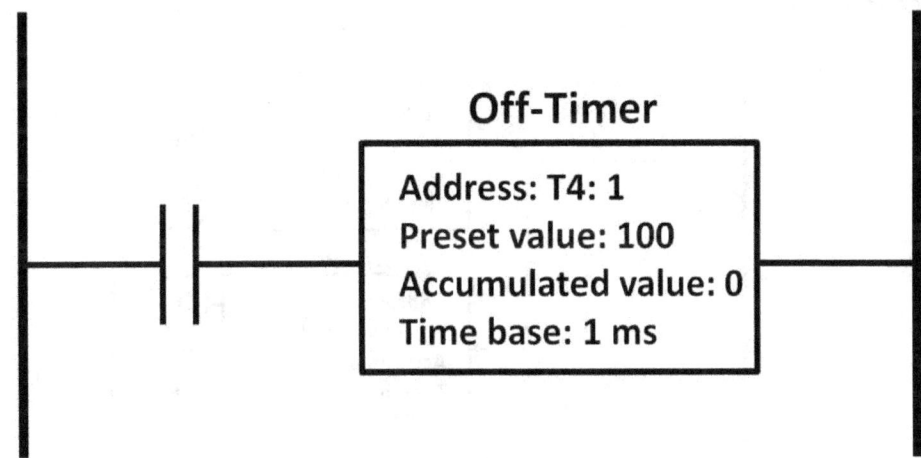

Figure 8.34: Off-timer symbol

Figure 8.35 is the off-timer timing diagram.

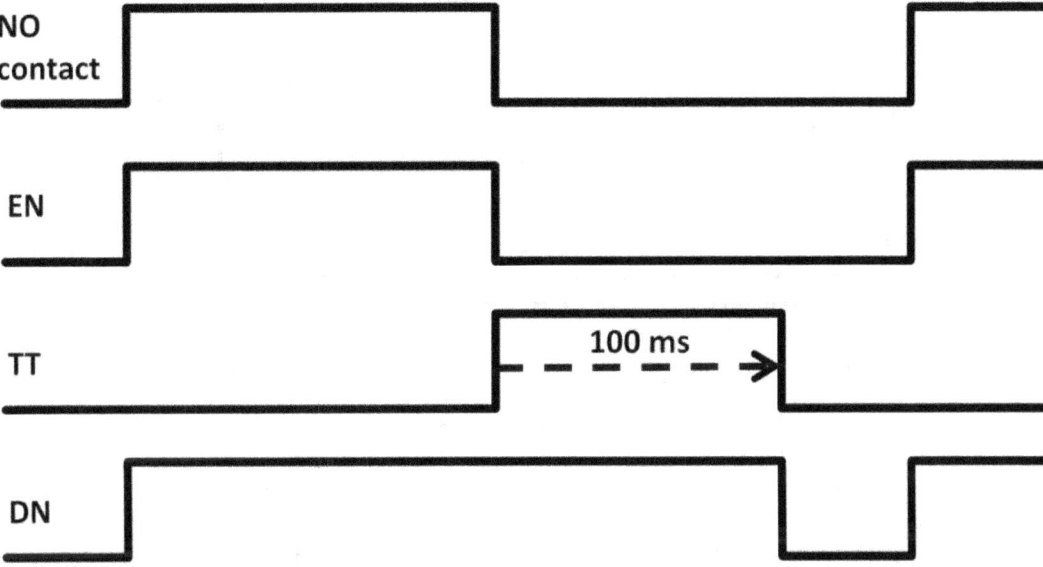

Figure 8.35: Off-timer timing diagram

Off-Timer Application

Let's now use a design example to better understand the off-timer. In this application, when the switch is pressed, unlike a push button, the switch stays pushed (closed); both lights turn on immediately. After the switch is pressed again, it opens (off). At that moment, the off-timer starts to time. The first timer T4: 1 turns Light-1 off 10 s **(1,000 X 10 ms = 10 s)** after the switch is pressed the second time. Light-2 turns off after 20 s **(2,000 X 10 ms = 20 s)**. Figure 8.36 shows the PLC ladder logic diagram of this off-timer application. Figure 8.37 shows the timing diagram of this application.

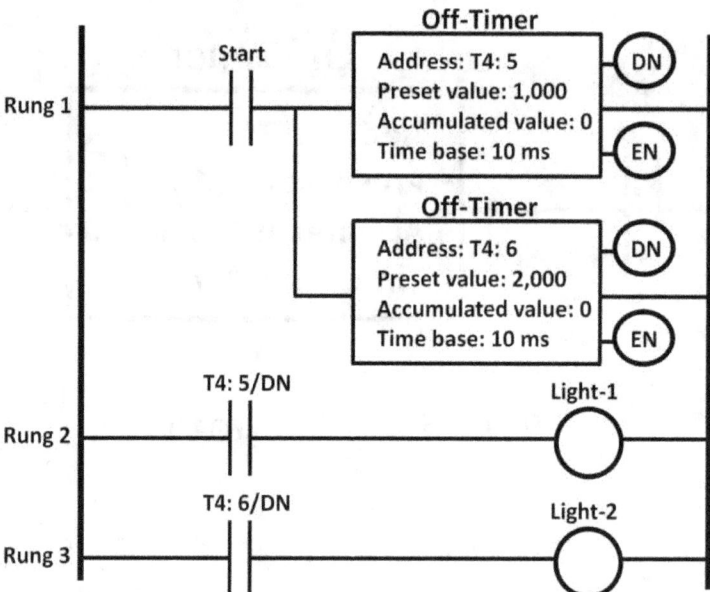

Figure 8.36: Off-timer application example

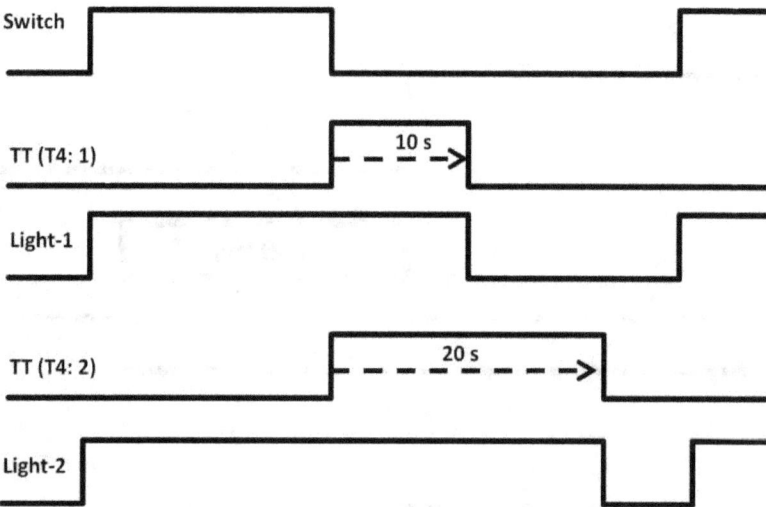

Figure 8.37: Off-timer application timing diagram

Counter

In addition to the timer, the counter's instructions are available in PLCs. The counter in a PLC can count items up or down. It's essential to use counters in many applications such as counting the total number of parts produced in a factory using proximity sensors, or to keep track of cars coming in and out of a parking lot. Proximity sensors are input devices that produce a discrete signal when an object passes by the sensor. It can be used in conjunction with counters. There are two types of PLC counters—up and down-counters. The ladder logic counter symbols are shown in figure 8.38.

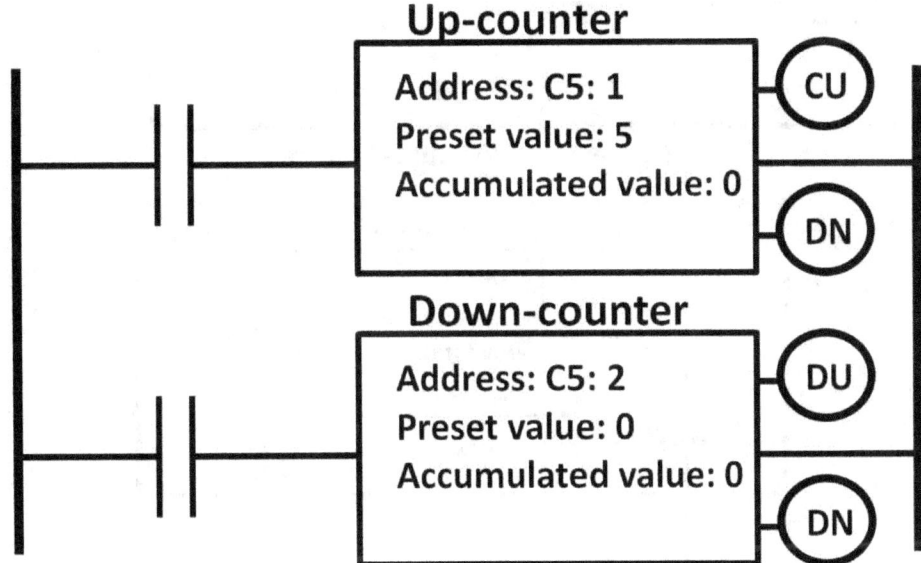

Figure 8.38: Up- and down-counter symbols

In the up-counter, the preset value is configurable. It's now set to 5 in figure 8.38. The CU (counter-up bit) bit goes high when the counter input is true. When the NO contact is active, the CU bit is high and the accumulated value goes up by one. The counter DN bit eventually goes high when the accumulated value reaches the preset value 5. A unique characteristic of the counter is that the accumulative counter value continues to go up even after it reaches the preset value. For example, if the NO contact is active again, the accumulative value would go to 6. The following counter diagram shows how an up-counter operates (see figure 8.39 on the next page). In order to reset the counter's DN and accumulative value, a separate reset instruction is required to independently reset the counter's DN bit. Figure 8.40 shows an example of using a counter reset instruction. The counter DN is reset to zero when switch 2 is closed. The down-counter works very much like the up-counter except that the accumulated value decreases by one every time the down-counter input is active. The DU (down-counter bit) is true every time when the down-counter input is high.

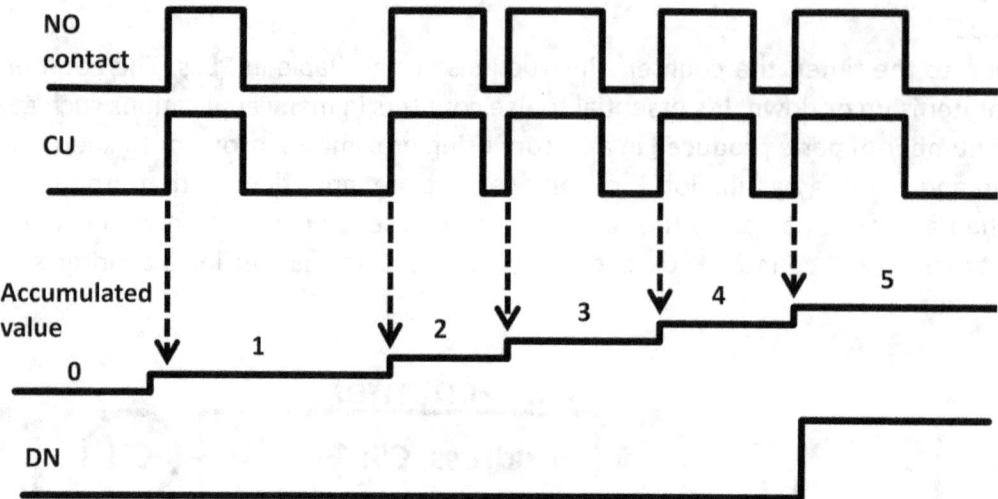

Figure 8.39: Up-counter diagram

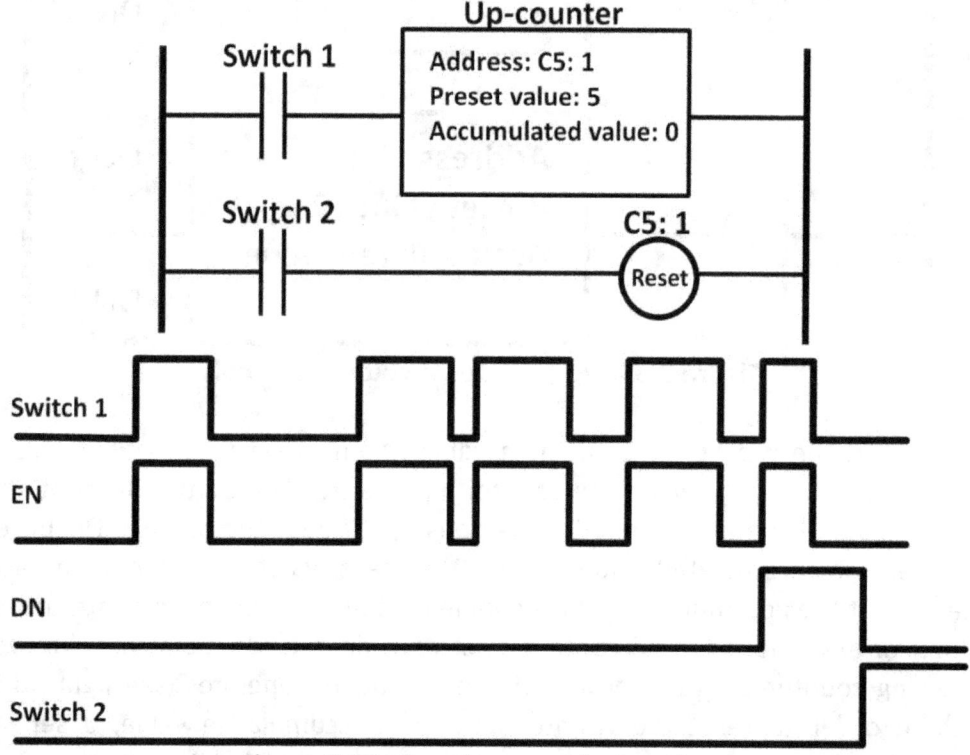

Figure 8.40: Counter reset instruction

Counter Application

Let's use a counter and timer to design a bottle-counting application, shown in figure 8.41. This design involves using a conveyor belt system and a proximity sensor counting the number of bottles passing through the sensor for a fixed period of time (one hour). It's an important part of the manufacturing process to evaluate factory throughput.

Figure 8.41: Bottle conveyor system

The ladder logic program is shown in figure 8.42.

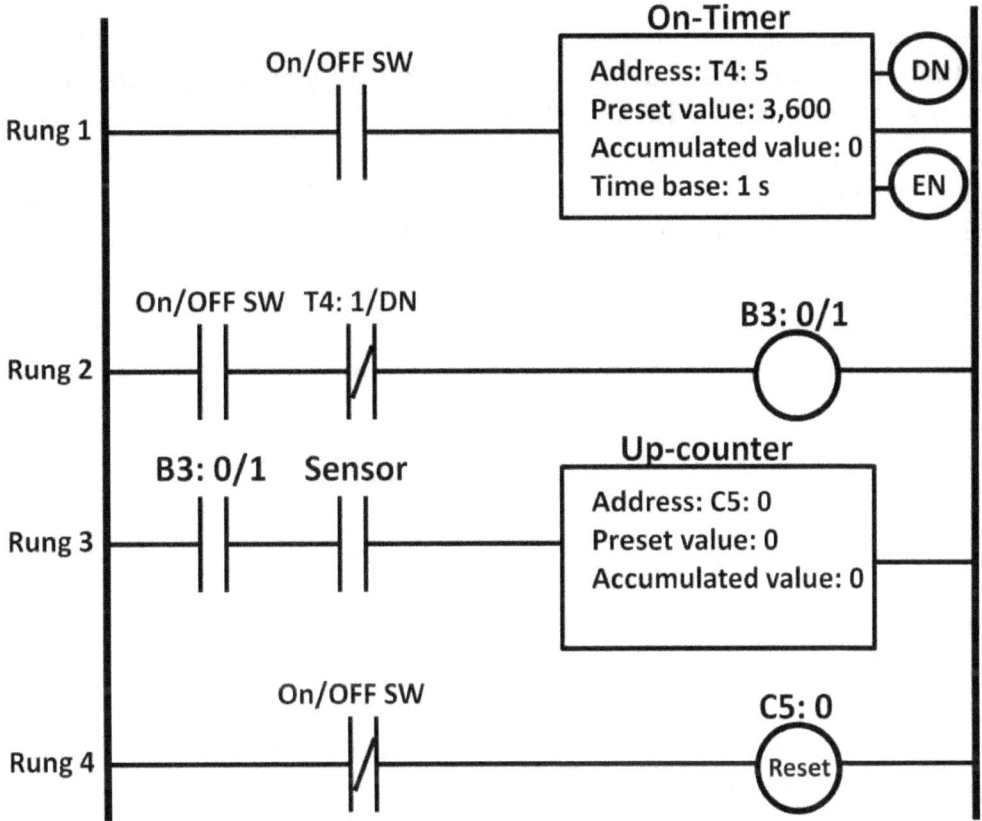

Figure 8.42: Conveyor counting application

When the on/off switch is pressed, it's active in the ON position. When the on/off switch is pushed again, it's in the OFF position. As soon as the switch is pressed, the timer on rung 1 starts timing for 3,600 s (one hour). During the hour, B3: 0/1 stays on making the B3: 0/1 contact on rung 3 active. The sensor contact on rung 3 detects a bottle passing through it. It increments up-counter's accumulative value on every bottle passing through the sensor. The accumulative value of the counter increases by one when a bottle passes through the proximity sensor. Right after the timer accumulated value reaches 3,600, the timer's DN bit goes high, breaking rung 2's continuity due to the NC contact addressed to timer's DN bit. The count value can now be read from the accumulative value in the counter. To reset the counter value and DN bit, push the switch again to trigger rung 4's reset instruction. To start counting for one hour again, press the switch to start the timer all over.

Program Control Instructions

There are many control instructions in ladder logic to control program flows. These instructions allow designers to enable or disable a block of programs or call a specific section of the program rungs. It gives flexibility in designing, implementing, and debugging only certain part of the program, reducing development time.

Jump to Label Instructions

Jump (JMP) is an example of a program control instruction. It works with label symbol (LBL). Once the JMP instruction is enabled, it jumps to the LBL symbol anywhere in the program defined by the PLC program. Figure 8.43 shows an example. As the program progresses from rung 1 to rung 2 and so on, if push button 2 is pressed, the JMP instruction takes the PLC program to rung 4 skipping rung 3, then continues on to rung 5. The status of the push button 3 and O: 0/3 will not be examined and processed. The input and output status of rung 3 remains the same.

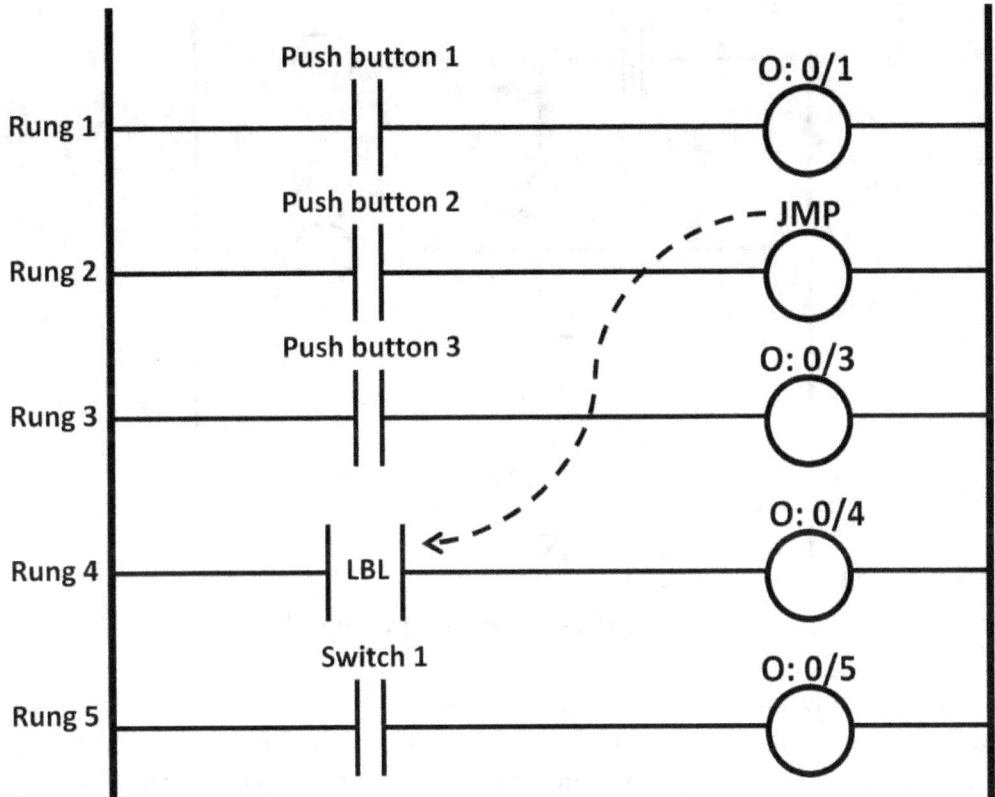

Figure 8.43: Jump-to-label instruction example

The purpose of JMP, LBL instructions may be that rung 3 does not affect the outcome of rungs 4 and 5. By skipping rung 3, the JMP instruction isolates parts of the program making it easier to troubleshoot and saves scan time during program executions.

Jump to Subroutine Instructions

The next control instructions are the jump to subroutine, subroutine, and return (JSR, SBR, and RET) instructions. When the JSR is called upon, it jumps to the SBR instruction within the program. The SBR could contain one or more rungs. At the end of the user-defined subroutine, the RET instruction is needed to return back to the rung right below the SBR. Figure 8.44 shows the concept of using the JSR, SRB, and RET instructions. Suppose your PLC program requires task 1 to be performed multiple times. Multiple tasks will need to be included in the program, taking up program memory space and reduce scanning time. This is not an efficient way to design PLC programs. If the applications require fast timing response, the extra scan time may ultimately fail system specifications.

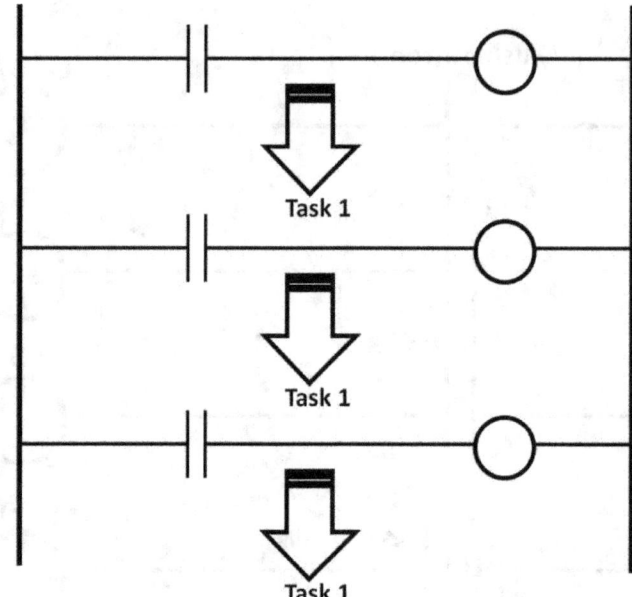

Figure 8.44: Same task performed multiple times

Using subroutines, only one task is needed in the entire program. If the task needs to be performed, the jump to subroutine instruction calls the task as a subroutine. Once the task has been completed, the return (RET) instruction takes the program back to the rung right below the JSR instruction. The program then continues to execute (see figure 8.45).

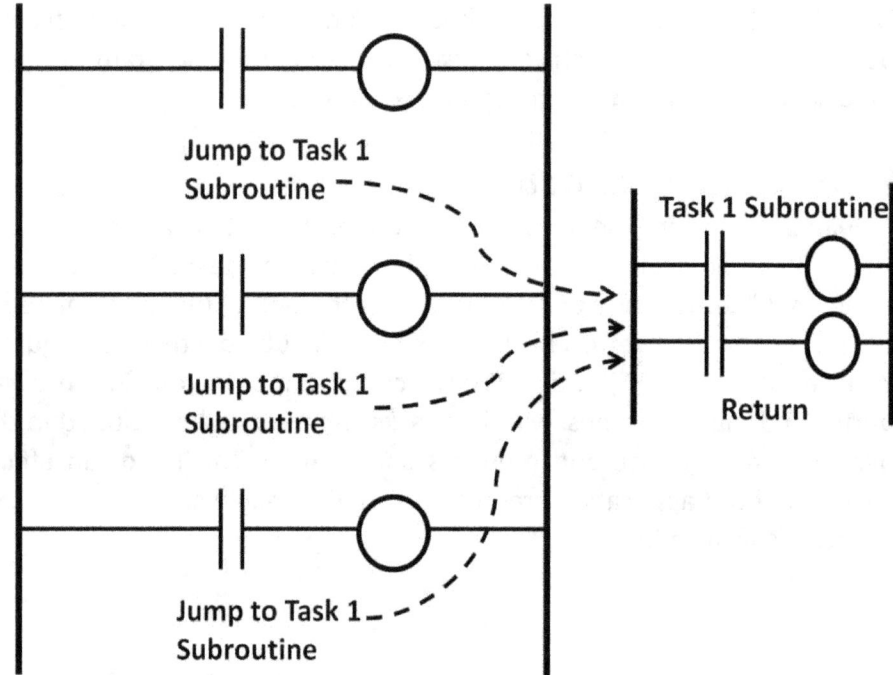

Figure 8.45: Jump to subroutine concept

Nested Subroutines

The jump to subroutine reduces program sizes and scan time, and it eases troubleshooting efforts. It's possible to implement a subroutine within a subroutine. Figure 8.46 demonstrates an example called a nested subroutine.

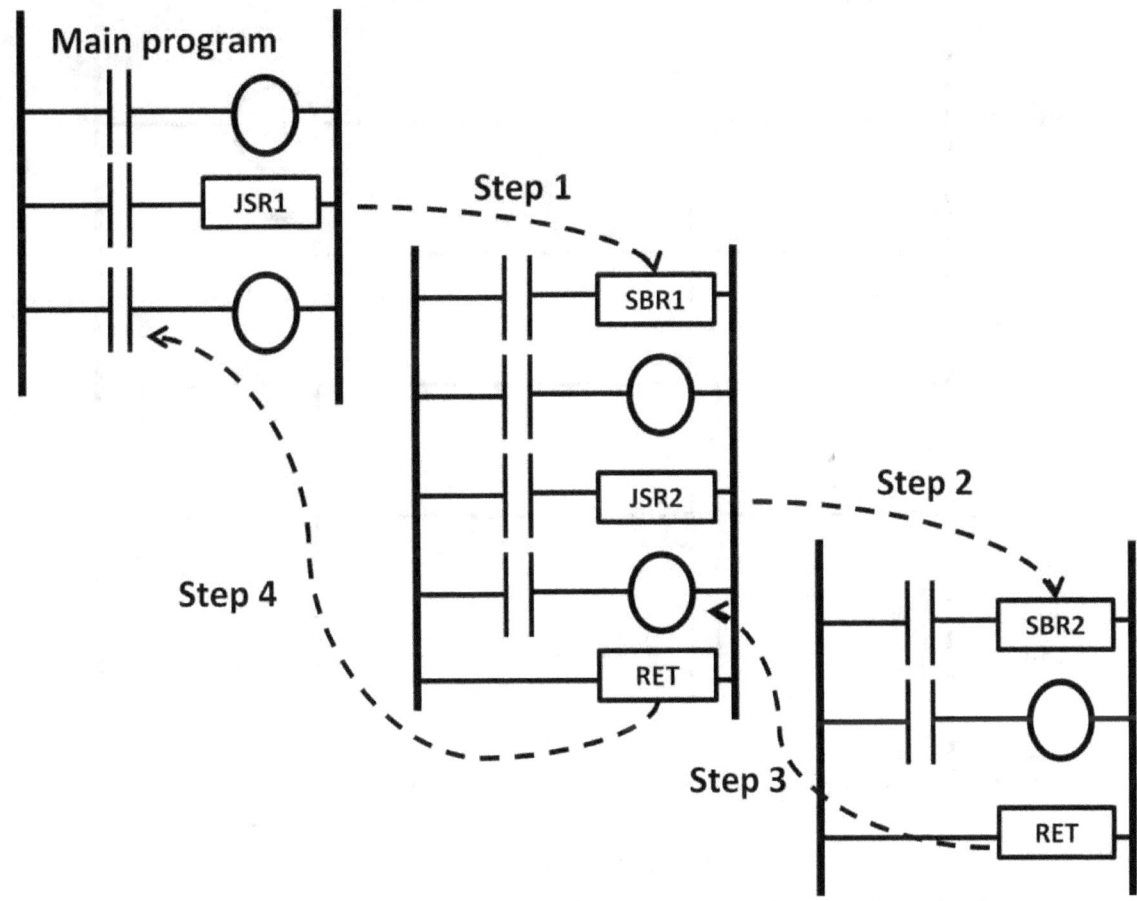

Figure 8.46: Nested subroutine (subroutine within a subroutine)

In this example, the first subroutine is called by JSR1 (Step 1). Within subroutine 1 (SBR1), JSR2 calls the SBR2 subroutine (Step 2). The RET instruction returns the subroutine back to the rung below JSR2 (Step 3). Step 4 returns SBR1 back to the rung below JSR1 in the main program. Note that RET returns only to the subroutine it was called from, not the one prior to that. In figure 8.45, RET in SBR2 only returns to the rung below JSR2, not JSR1. Careful planning is required so that the correct subroutines are called.

Temporary End

Temporary end (TND) is yet another useful ladder logic debug feature. Temporary ends serve as breakpoints throughout the program allowing designers to run the program to pause and continue one section at a time. A TND can be controlled with or without input contacts. Figure 8.47 shows a TND concept.

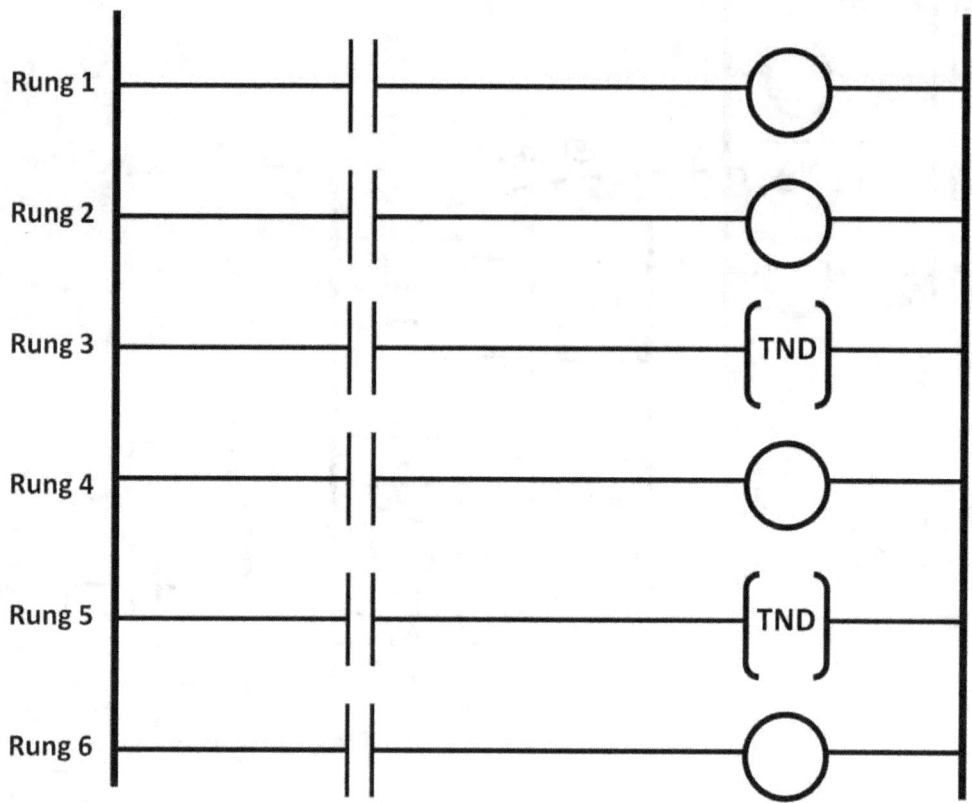

Figure 8.47: TND concept

As the program progresses to rung 3, if the rung 3's normally-open (NO) contact is closed, the TND will take effect. At this point, the program pauses. When the rung 3's contact goes low, it disables the TND, then the program continues to scan and moves to rung 4. If rung 5's contact is not active, it continues onto rung 6. Designers have full control over when to halt the program during debug.

Data Manipulation Instructions

PLC contains program and data memory similar to microcontrollers. The ladder logic programs are stored in the program memory. Data, constants, and numbers are stored in the data memory. There is a need for data to be able to move around so that ladder logic can be used more effectively. Copy and move instructions are available for this purpose.

PLC Data Structure

We first need to understand data structure and how it is stored within the PLC memory. Then, we will use some practical examples to understand how data manipulation instructions increase programming effectiveness. One data word consists of multiple bits. The 8-bit word (one-byte word) is the most common type, although a 16-bit word is found in PLC memory. A block of an 8-bit word is shown in figure 8.48.

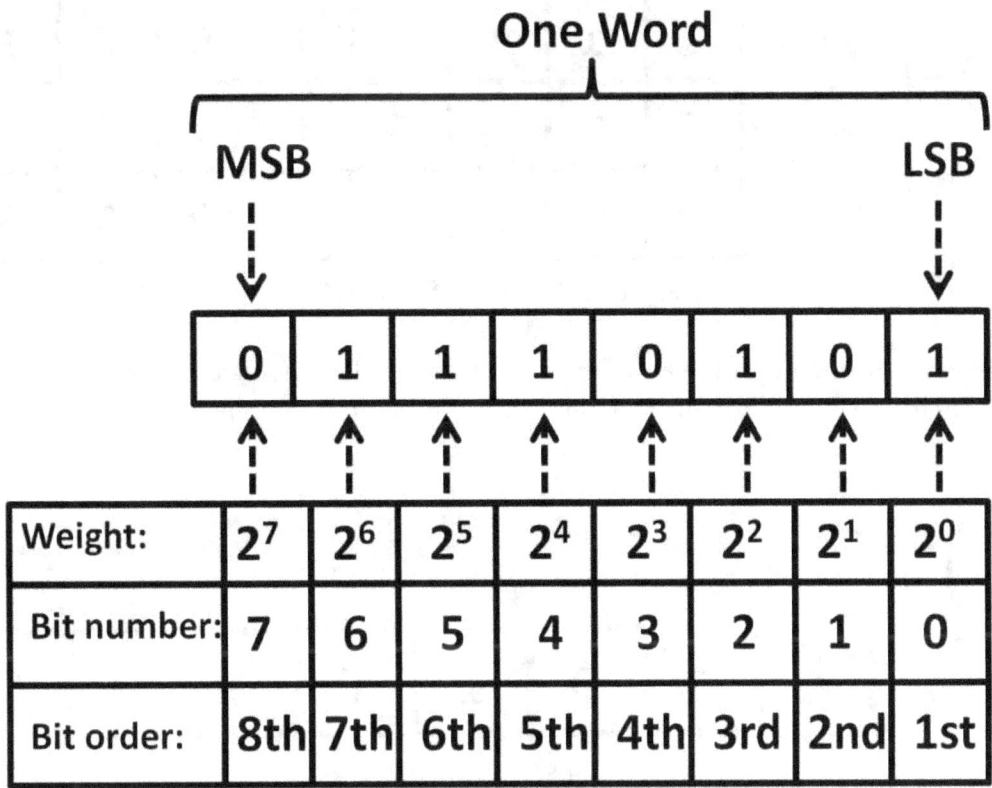

Figure 8.48: Bit word, weight, and number

In the same way as digital electronics, the first bit (bit number 0) on the right is the least significant bit (LSB). The last bit (bit 7) on the left is the most significant bit (MSB). This word has a decimal value as follows that depends on the weight of each bit:

$$2^7 \times 0 + 2^6 \times 1 + 2^5 \times 1 + 2^4 \times 1 + 2^3 \times 0 + 2^2 \times 1 + 2^1 \times 0 + 2^0 \times 1 = 117$$

If the PLC data memory size is 4 KB (4,096 bytes), it equates to 512 words, i.e., **512 X 8 = 4,096 bytes.** Multiple words make up a file. The number of words in a file is user defined. Figure 8.49 on the next page shows a 4-word file named File 0.

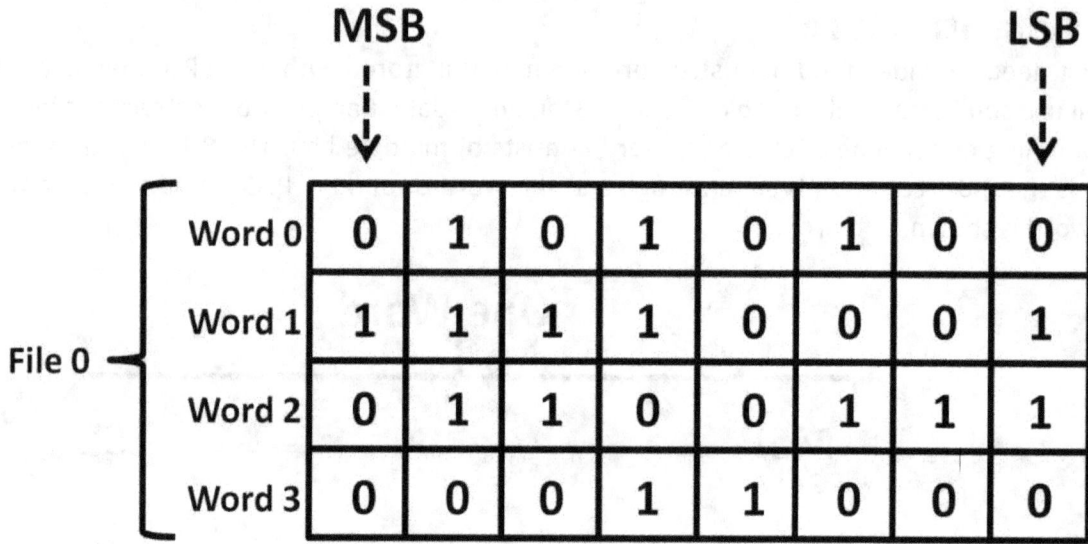

Figure 8.49: Word file in PLC

Each word in data memory has an address so that the ladder logic program knows exactly where to fetch it from. Address formats are different among PLC vendors. Figure 8.50 shows a word address example.

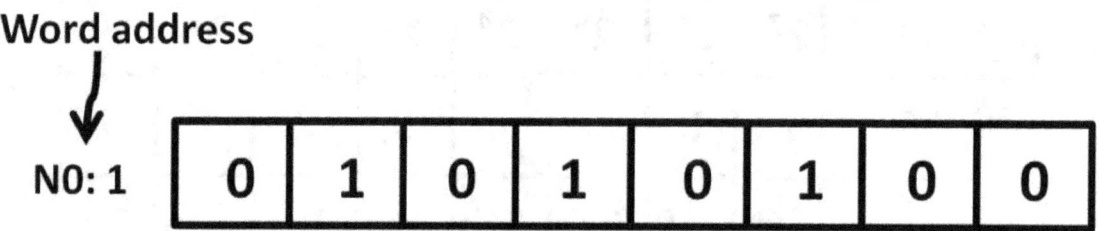

Figure 8.50: Word address

MOV Instruction

To manipulate and move data around, we use a MOV instruction, shown in figure 8.51. Data 01010100 is first loaded into N0: 0 by the ladder logic software. When the switch is closed, the MOV instruction copies data from N0: 0 to N0: 1 replacing 11110001 in N0: 1 with 01010100. Note that MOV instruction is a copy instruction. The original data in N0: 0 remains as 01010100.

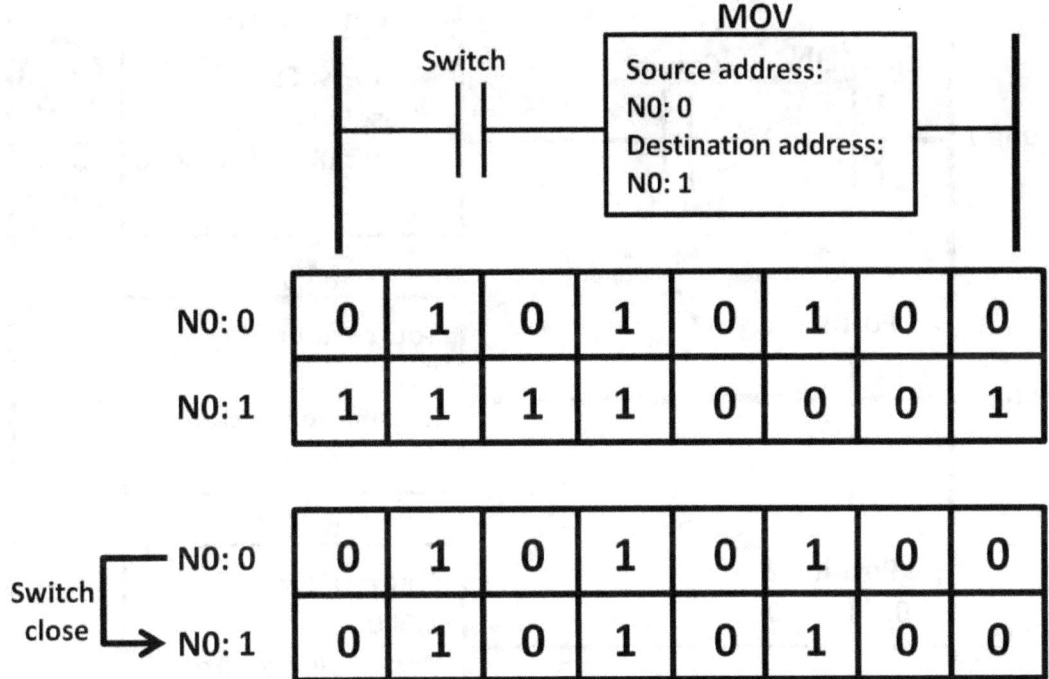

Figure 8.51: MOV instruction

MOV Instruction Application

A counting application that implements MOV instructions is shown in figure 8.52 on the next page. A count button and turn switch are used as input devices. The two position switches (positions 1 and 2) set the count values. By turning to position 1, the counter preset value is set to 500. Position 2 sets the count to 1,000. Using a MOV instruction easily transfers the count values to the counter without using a second counter. When the program first starts, on rung 1, the DN bit is low because the counter preset is less than the accumulative value. This makes the NC contact output high on rung 1. If the count button is pushed, the up-counter accumulative value goes up by one every time the count button is pressed.

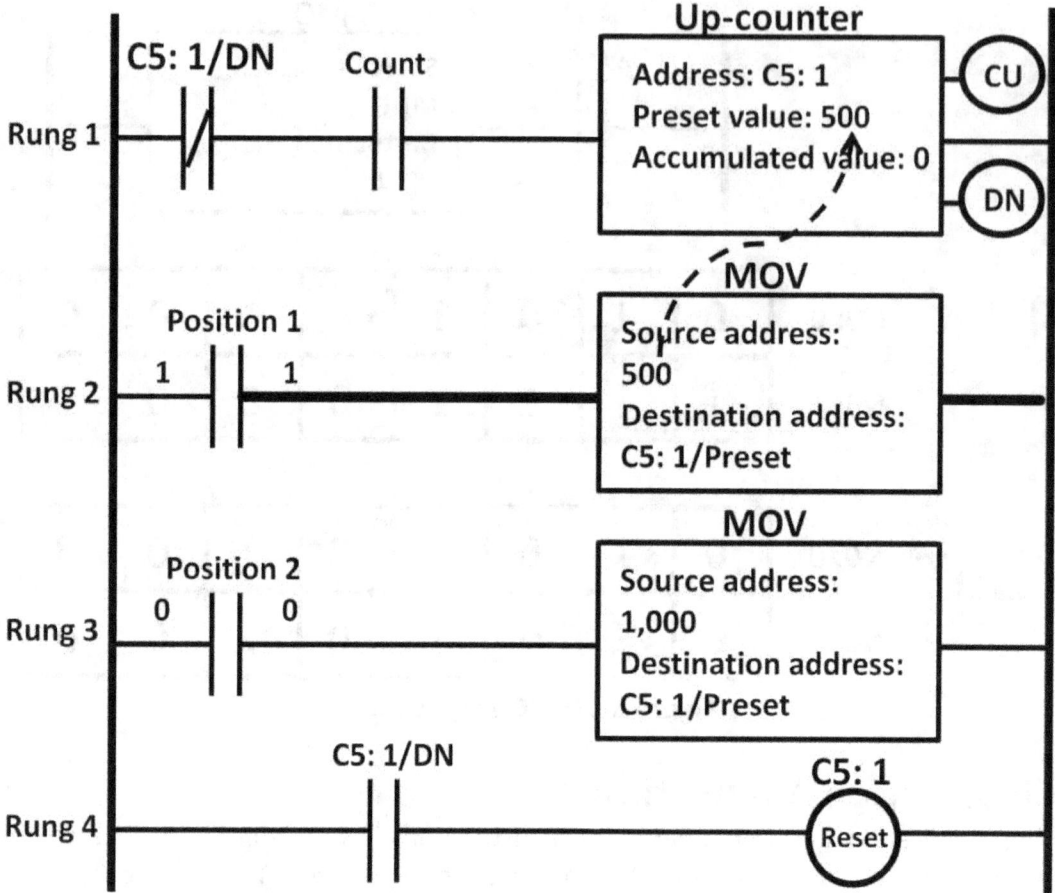

Figure 8.52: Count application and MOV instruction

Rung 2 sets the counter preset value to 500 if the position 1 switch is closed (dotted line). If the position 2 switch is closed, the preset value is set to 1,000 instead because both MOV instructions and destination addresses map to the C5: 1 counter. Once the accumulative value reaches either preset value depending on whether position 1 or 2 is pressed, the DN bit goes high. This stops rung 1's continuity. Accumulative value and DN get reset. A new count can now start over again.

Data Compare Instructions

PLCs come with logic instructions to perform compare functions. Compare instructions are input instructions. If the comparison result is true, the compare instruction output goes high. A list of compare instructions that compare numerical values is shown on the next page: equal to (EQ); not equal to (NEQ); less than (LES); greater than (GRT); less than or equal to (LEQ); greater than or equal to (GEQ). Let's first examine the EQU instruction. In figure 8.53, the EQU instruction turns on a red light when source A (10) is equal to source B (timer accumulative value).

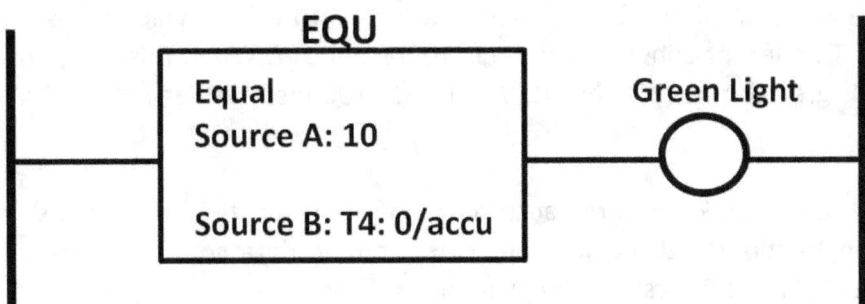

Figure 8.53: EQU instruction example

A not equal to (NEQ) instruction compares two source values. If they are unequal, the instruction output is true. In figure 8.54, source A contains a value of 0.5 in the word N2: 1 data memory. Source B is assigned to I: 0/3 that connects to a thermocouple. The transfer function of the thermocouple gives 0.5 V at room temperature 27°C. If the temperature is not 27°C, the red light turns on.

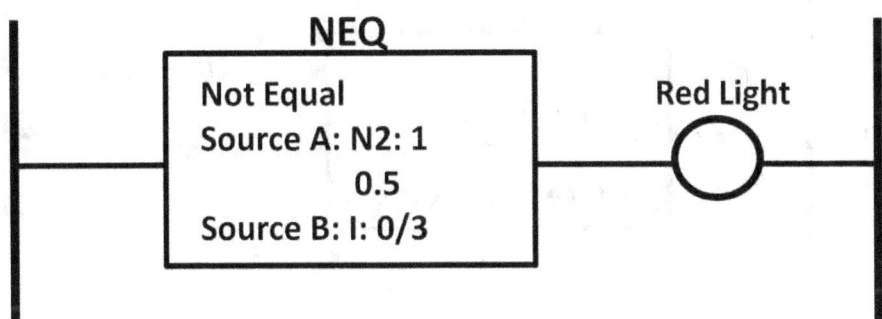

Figure 8.54: NEQ instruction example

The great than (GRT) instruction compares sources A and B. If source A is greater than B, the output is logically true. Figure 8.55 demonstrates an example.

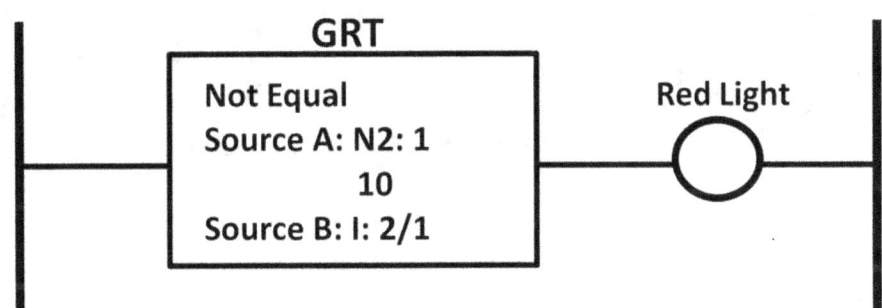

Figure 8.55: GRT instruction example

Source A has a value of 10 in N2: 1. Source B connects to I: 2/1, which takes its value from a weight sensor. This sensor conversion ratio is 10 lbs per 1 V. When this program runs, as soon as the weight is greater than 100 lbs (10 V at I: 2/1, 100 lbs / (10 lbs / V) = 10 V), the red light turns on.

LES, LEQ, and GEQ work similarly according to their function definitions. As with data manipulation instructions, PLC applications can combine data compare instructions with any other instructions. Figure 8.56 shows an example.

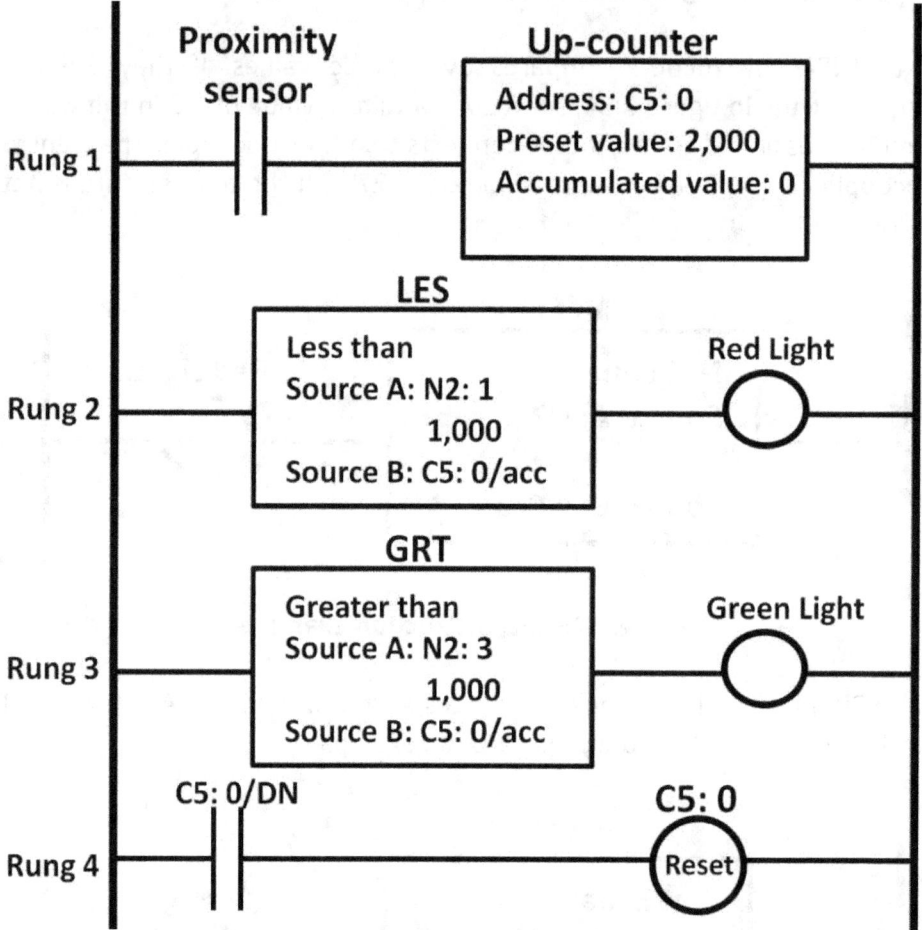

Figure 8.56: Up-counter, LES, GRT application

This is again a counting application with additional functions. On rung 1, the up-counter preset value is set to 2,000. The proximity switch triggers when an object passes through it increasing the accumulative value by one. During the first 1,000 counts (less than 1,000), rung 2 is true turning the red light on. Once the count value goes above 1,000, the red light turns off and the green light turns on due to rung 3's GRT instruction being true. By 2,000 counts, the counter-done bit goes high. This causes rung 4 to reset the counter, C5: 0.

Math Instructions

Arithmetic functions can be performed using math instructions. PLC math instructions are output instructions that include addition (ADD), subtraction (SUB), multiplication (MUL), and division (DIV) instructions. Figure 8.57 shows a math instruction example.

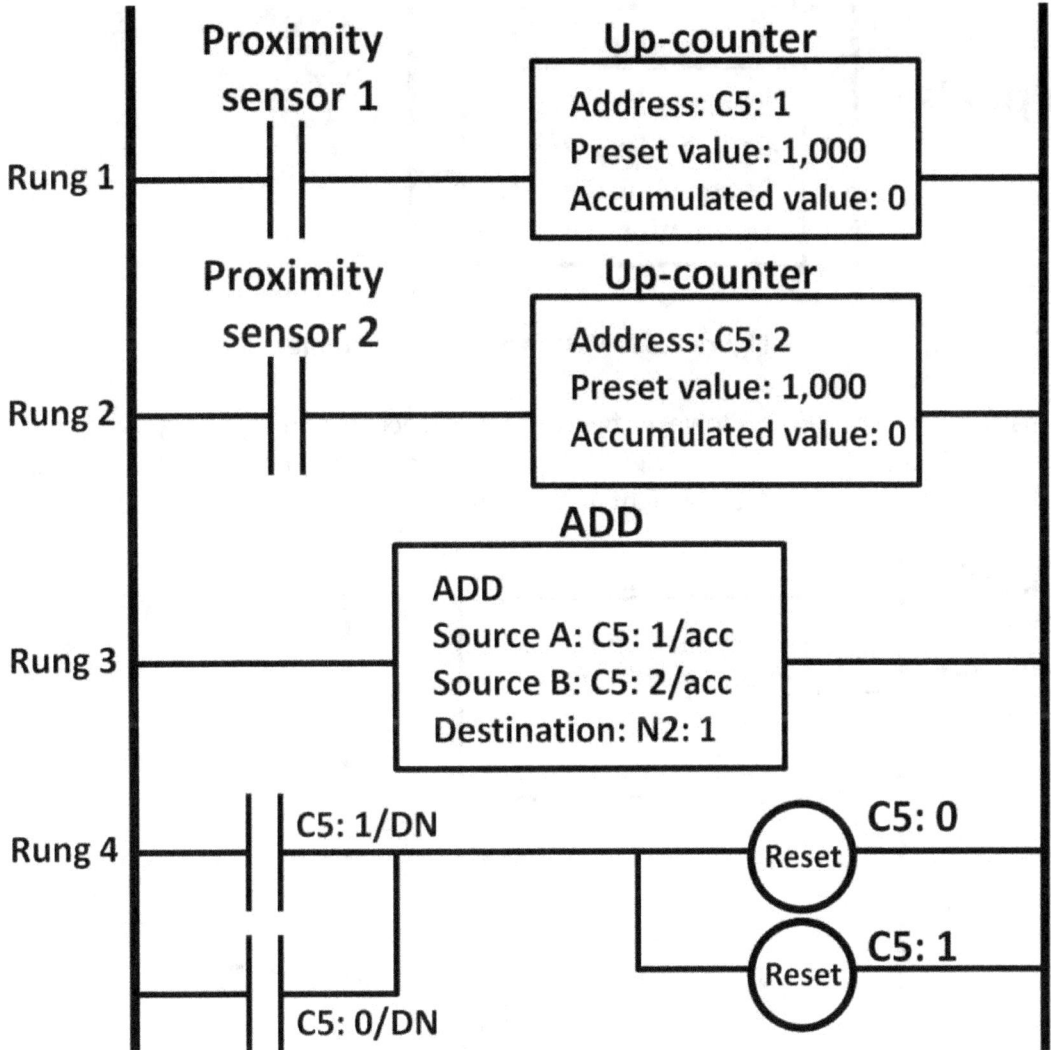

Figure 8.57: Math instruction example

Rungs 1 and 2 control up-counters that are individually triggered by two proximity sensors (1 and 2). The total counts from both counters are calculated by the add instruction on rung 3. The result is stored in the destination N2: 1. The ADD instruction on rung 3 does not have any output symbol connected to it. This is perfectly valid because it's an output instruction. After both counters reach preset values of 1,000, both counters get reset (rung 4).

The next math instruction example is shown in figure 8.58. It's a Vrms converter application using a MUL instruction. I: 3/7 takes an average peak voltage as an input. The PLC converts it to an average rms voltage using a MUL instruction (multiply by 0.707) and then displays it on a 7-segment display output device (O: 0/1).

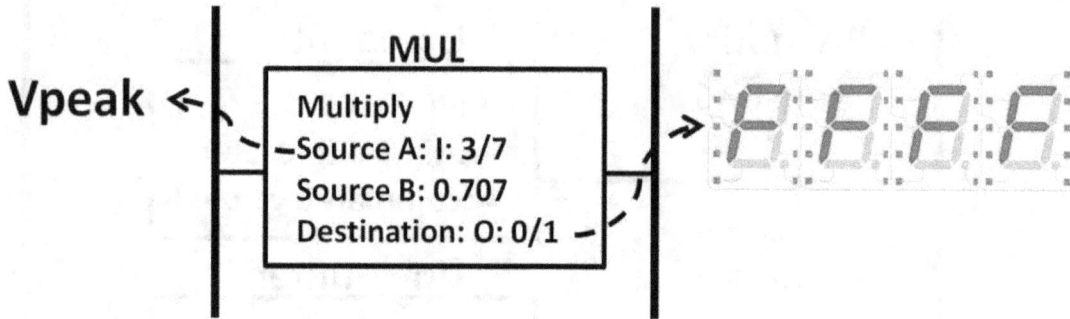

Figure 8.58: MUL instruction example

A DIV instruction example is shown in figure 8.59. It receives an input signal from a weight transducer (I: 2/1) that produces a weight value in kilogram (kg). The DIV instruction converts it to lbs. If the button is pressed, the weight in lbs is displayed on an LCD display (O: 1/1).

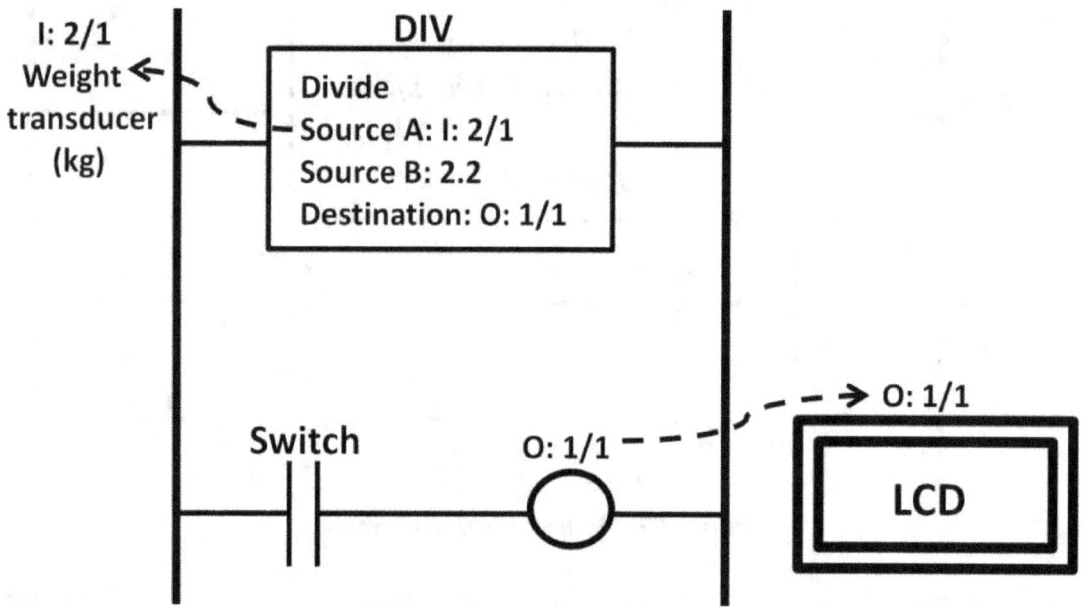

Figure 8.59: DIV instruction example

In some cases, you may want to invert a value from positive to negative or vice versa. A negate (NEG) instruction can perform such a function. Figure 8.60 shows its operations.

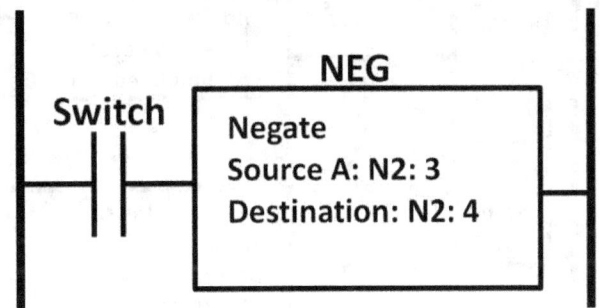

Figure 8.60: Negate instruction example

After the switch is pressed, the contents of N2: 3 are inverted from 1 to 0 and the result gets stored in N2: 4. For example, if N2: 3 data is 00000000, it will be inverted to 11111111. The result is stored in N2: 4. Figure 8.61 is an application combining math and data compare and manipulation instructions. This is a car wash application. In this design, there are two car wash types for customers to choose from: standard and supreme. Standard service takes five minutes. Supreme service takes ten minutes. This application automates the car wash process controlling the on-time for the water, foam-dispensing pumps, and the wind-drying motor, depending on whether the standard or supreme button is pressed by the operator. This PLC program keeps track of the total number of cars washed. In addition, if the total number of washed cars reaches 490 (500 − 10), the maintenance light turns on. The field device names, types, and addresses are shown in table 8-4 below. In this example, all the water, foam-pumps, and blowers are presumably on at the same time for simplicity reasons. In reality, there will be separate timers to control the on-time for each of the three output devices individually.

Field devices names	Type	Address
Start button	Input	I: 0/0
Standard button	Input	I: 0/1
Supreme button	Input	I: 1/0
Water pump	Output	O: 0/1
Foam pump	Output	O: 0/2
Blower	Output	O: 0/3
Maintenance	Output	O: 0/4

Table 8-4: Car wash device names, types, and addresses

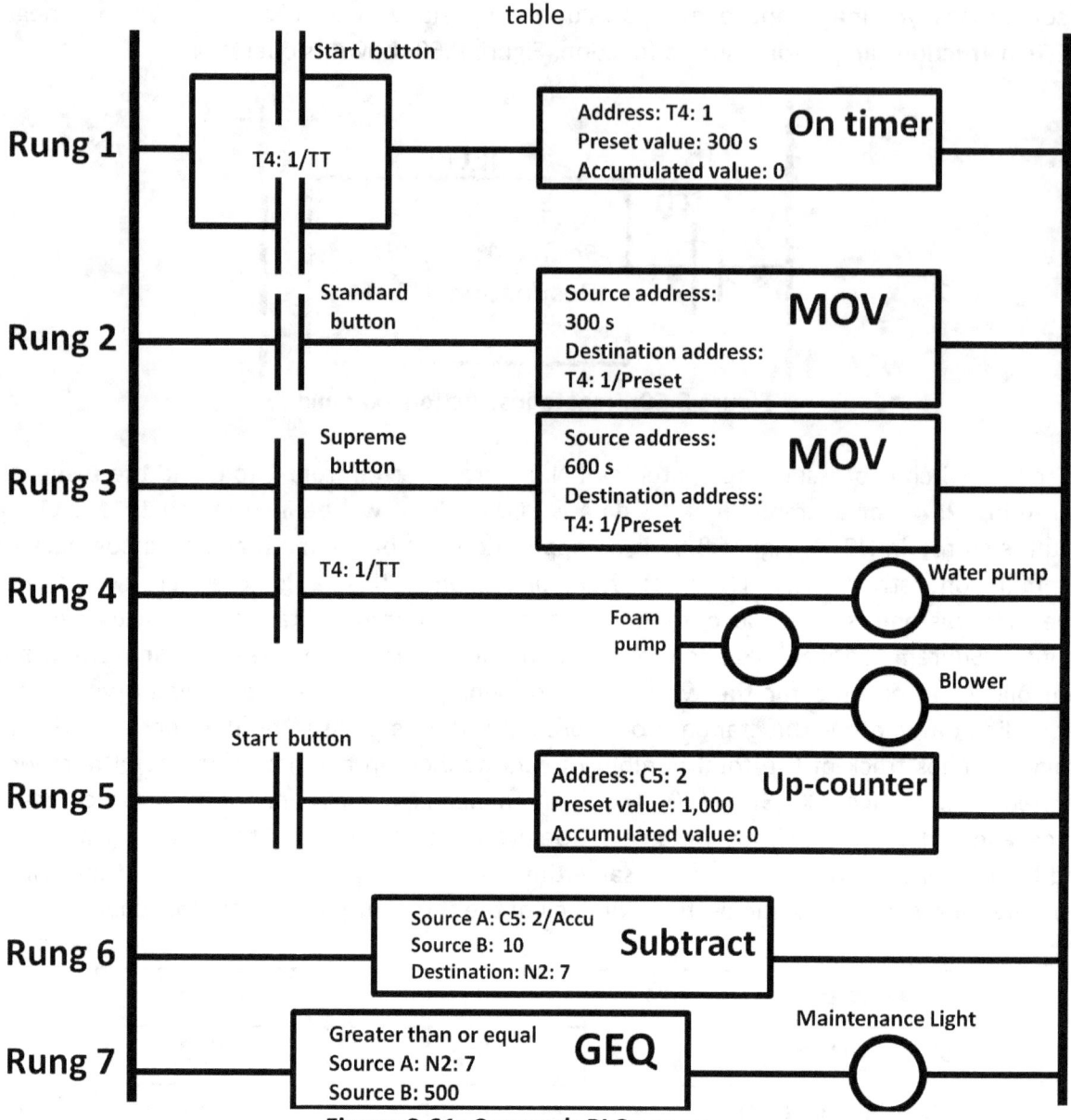

Figure 8.61: Car wash PLC program

On rung 1, pressing the start button starts the timer. The TT branch forms a seal-in to keep the timer running. Depending on whether the standard button (rung 2) or supreme button (rung 3) is pressed, either 300 s or 600 s is copied to the timer's preset value. During the timer's on-time, rung 4 turns the water, foam-pump, and blower on. Rung 5 keeps track of the total cars washed using an up-counter. Rungs 6 and 7 determine if the total car number has reached 490 by using a subtract instruction. If so, the maintenance light turns on (rung 7's GEQ instruction).

Sequencer Instructions

Plenty of industrial and commercial applications execute instructions continuously in a loop. Industrial washing machines, large-scale warehouse conveyor systems, merchandise-processing systems, and traffic light systems are few examples. Sequencer instructions reduce the number of rungs needed and simplify sequential operations in PLC applications. A sequencer output instruction (SQO) symbol is shown in figure 8.62.

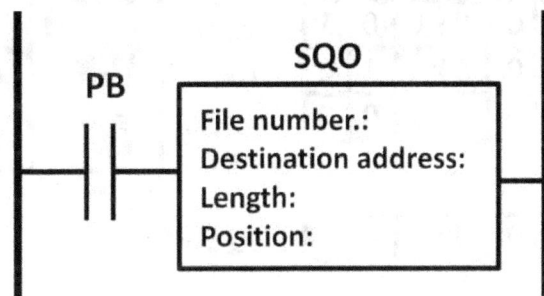

Figure 8.62: SQO instruction

To use an SQO instruction, PLC programmers need to assign values to the items within the SQO instruction. These items are file number, destination address, length, and position fields. File number corresponds to the starting address of the sequencer file. This file contains words that PLCs execute upon. For example, figure 8.63 shows an SQO file (#B3: 1) comprising five words. The "#" sign designates it's a file instead of a word. The first word (word 0) will be transferred to the destination address (e.g., an output device) if the SQO input is logic high (push button PB is pushed). The second word (word 1) is transferred to the output device if PB is pushed again. The 6th time PB is pushed, SQO loops back to word 0 and the sequence repeats again. The number of words and contents of each word are user defined. How often the data transfer occurs depends on the ladder logic program. The length item defines the total number of words (steps) that will be transferred. If the length is two in a 5-word file, only the first two words will be transferred even though there are five words in the sequencer file. Position determines the starting word location, which typically starts at position one.

Sequencer file number #B3: 1								
Word 0	1	1	0	1	0	1	0	1
Word 1	1	0	1	1	0	1	0	0
Word 2	0	1	0	0	0	1	1	1
Word 3	0	0	0	1	1	0	0	0
Word 4	1	1	0	1	1	0	0	1

Figure 8.63: Sequencer file

Let's apply an SQO instruction to a simplified traffic light application (see figure 8.64). This application turns on and off red, yellow, and green lights in a sequence using an SQO instruction controlled by push buttons 1, 2, and 3 (PB1, PB2, and PB3).

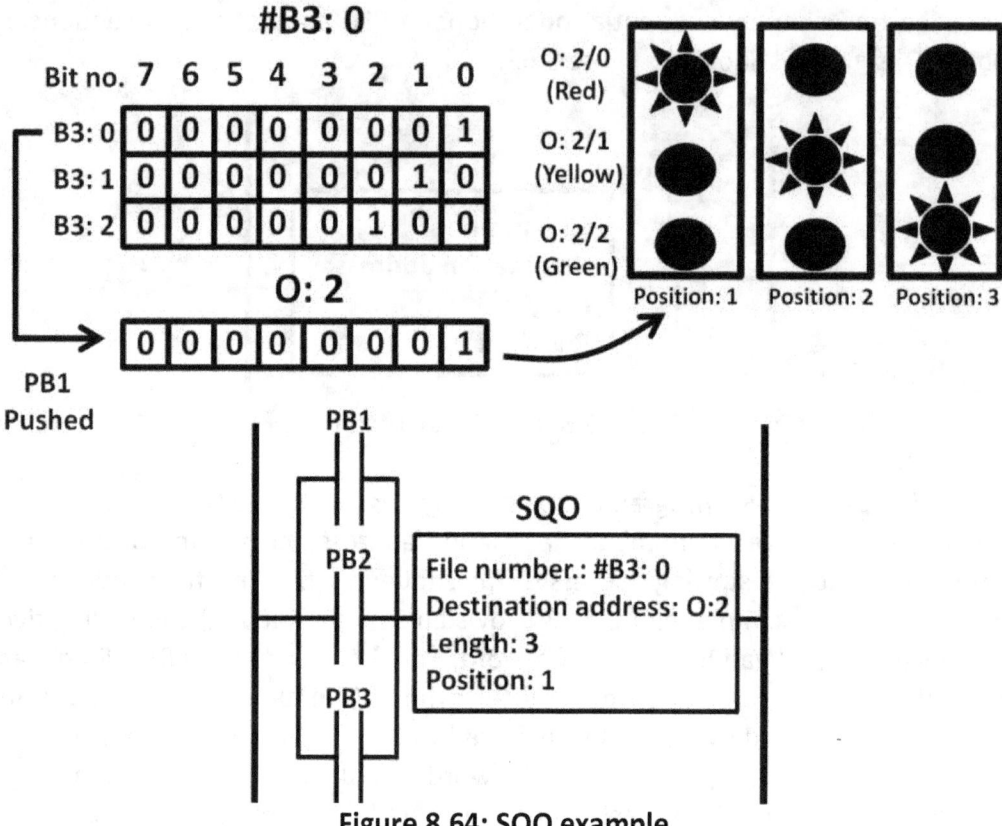

Figure 8.64: SQO example

In this example, #B3: 0 is the file number made up of three words starting with address B3: 0. The second and third word addresses are B3: 1 and B3: 2. PB1, 2, and 3 are controlled by three separate timers (not included in this example). These three push buttons form a parallel branch which controls the SQO. The timers control PB1, 2, and 3 one at a time. The destination address O: 2 connects to three traffic lights. The first three bits of O: 2 (O: 2/0, O: 2/1, and O: 2/2) connect to red, yellow, and green lights respectively. If PB1 is pushed, SQO loads data from B: 3.0 (001) to the destination address O: 2. This lights up the red light and turns off the yellow and green lights. PB2 then goes high triggered by another timer. B3: 1 (010) now loads its content into O: 2 turning on the yellow light and shutting off the red and green lights. Lastly, PB3 is pushed loading B3: 2 (100) into O: 2, turning on the green light while turning off the red and yellow lights. This process repeats itself. The on-time duration of the lights is easily controlled by the timer's preset values. This example demonstrates that using SQO instructions, only one rung is needed to perform repetitive operations without the use of multiple rungs. This reduces program complexity and eases troubleshooting efforts.

Trends

PLC technology development continues to evolve. Sophisticated large-scale industrial control systems such as Supervisory Control and Data Acquisition (SCADA) have gained popularity in recent years. SCADA is capable of controlling large and multiple sites such as semiconductor fabs (factory) with wireless communication capabilities. In addition to process and motion controls, SCADA systems offer real-time process information, database creation, data analytics tools, and maintenance information for trending and throughput analysis. Increasing CPU power allows parallel PLC processing without sacrificing process speed and accuracy. Some modern, complex PLC systems utilize human machine interface (HMI), which is an apparatus to show human operators real time process data and pictures of the actual system components (input and output devices) while the system is running. Figure 8.65 shows an HMI example of a filling system. This system transports tanks on a conveyor belt while they are filled up by the materials stored in the funnels. Buttons 1 and 2 control the opening and closing of the funnels. Level sensors 1 and 2 monitor the tanks' levels. Button 3 triggers the siren if the tank level passes the level set by the sensor 2. This graphical interface is displayed on the monitor in real time. Buttons can be pushed with a click of a mouse with a PLC controlling the conveyor belt, on/off switch for the funnel, level sensors on the tank, and siren.

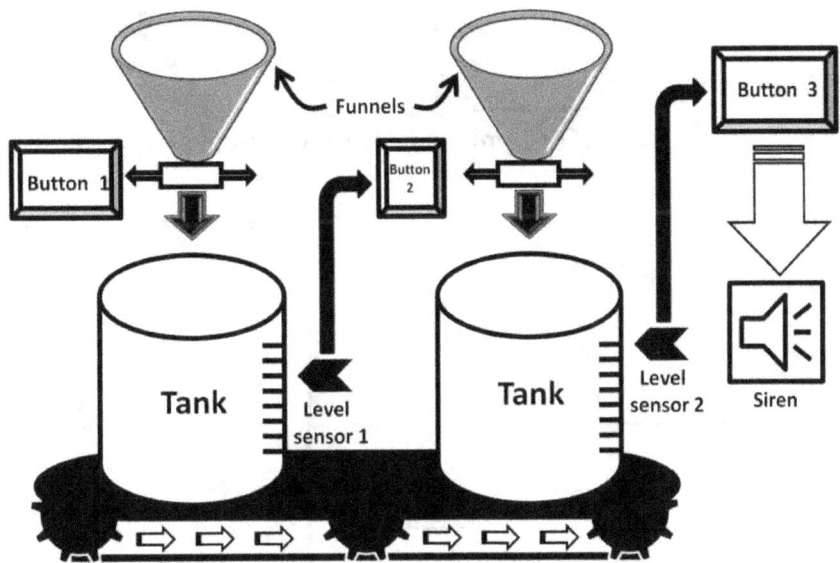

Figure 8.65: SCADA example

Summary

In this chapter, we covered PLC history, components, input, and output devices. Ladder logic syntax and programming techniques were introduced. Several PLC instruction types were discussed including timers, counters, math, data manipulation, comparisons, and sequencer instructions. PLC memory structure, practical PLC program examples, and industry trends were presented throughout the chapter.

Quiz

1) List three benefits of PLCs over traditional relay systems and five PLC components.

2) List three differences between computers and PLCs.

3) List five input and output device examples.

4) Design a PLC ladder logic program for a semiconductor fab conveyor system (see figure 8.66). A silicon wafer box (lot) waits for 30 minutes at process point 1. The level sensor detects whether the wafer lot has been filled up to 12 wafers. Once it's filled, the lot will be transported to the process checkpoint 2 where it stops and waits for 15 minutes for further processing. At the end of end 15 minutes, the green light turns on.

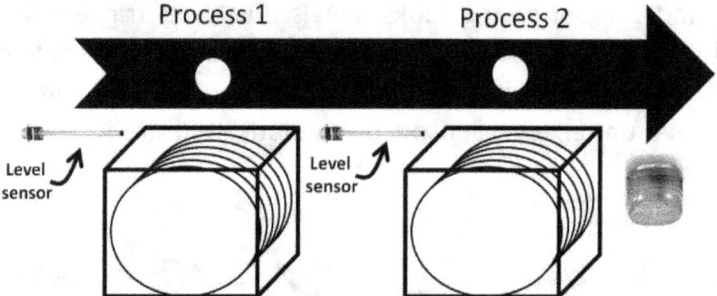

Figure 8.66: Semiconductor fab conveyor system

5) Figure 8.67 shows a periodic clock generator consisting of three on-timers. Complete the timing diagram on the right.

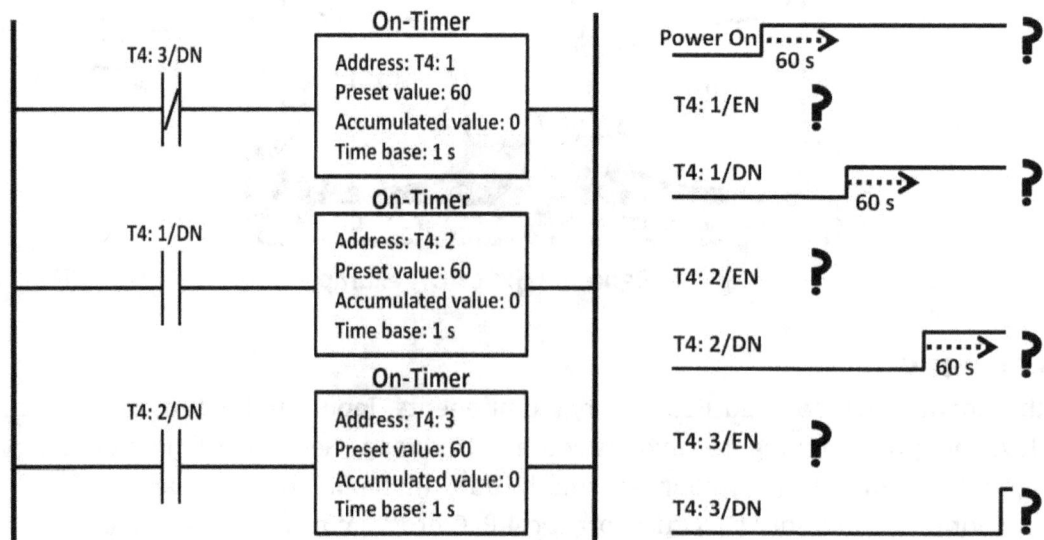

Figure 8.67: Periodic clock signal generator

Chapter 9: Mental Math

Electronics often use basic arithmetic to solve engineering problems, identify solutions, and perform technical analysis. Although many calculations regarding electronic engineering deal with large numbers, most electronic engineering solutions can be obtained quickly, efficiently, and accurately by using mental math, pen, and paper instead of using calculators. Despite the advanced features offered by calculators, most electronics calculation used in daily engineering tasks involves only short, simple-form calculations. Calculators should only be deemed necessary when working with multi-order math models. The misconception of using a calculator is undermined by the fact that numbers and math symbols could be entered incorrectly. Combine that with improper use of parentheses resulting in wrong answers, delaying progress, and slowing productivity. Becoming proficient with math techniques described in this chapter enhances your mathematic, analytic, and problem-solving skills while you demonstrate competency and increase productivity. In this chapter, basic arithmetic and numbering systems used in electronics calculations are first reviewed. Then, you will learn simple techniques to improve your mental math ability to calculate electronics arithmetic. Topics include large- and small-number multiples, submultiples, percentage-decimal conversion, divided-by-fractions, one-over reciprocals, multiply-divide power (exponent) rules, and dB-to-log conversion. Examples are provided throughout the chapter directly related to electronic engineering calculations.

Multiples and Submultiples of Units

Table 9-1 includes the names of the multiples and submultiples, their symbols, and the factors frequently used in calculating for electronics. The incentive of using multiples is the ability to express extremely large numbers in simplified forms. For example, the state-of-art CMOS transistor's leakage current is measured as low as femto (1×10^{-15}) amperes. It's much easier to interpret 1 fA (1×10^{-15} A) than 0.000000000000001 A. Here is a second example: an AC source's frequency is 2,000,000 Hz. It's simpler to write it as 2 MHz because **2,000,000 Hz = 2 X (1×10^6) Hz = 2 MHz**.

Name	Symbol	Factor
Femto	f	1×10^{-15}
Pico	p	1×10^{-12}
Nano	n	1×10^{-9}
Micro	u	1×10^{-6}
Milli	m	1×10^{-3}
Kilo	K	1×10^{3}
Mega	M	1×10^{6}
Giga	G	1×10^{9}
Tera	T	1×10^{12}
Peta	P	1×10^{15}

Femto through Milli are Submultiples; Kilo through Peta are Multiples.

Table 9-1: Multiples and submultiples of units

Decimal Numbers

Decimal numbers are any numbers written with a decimal point ".", such as 2.3, 5.78, or 0.005. The decimal point separates the ones place (left) from the tenths place (right) in decimal numbers (see figure 9.1). If a DMM's resolution is 0.0001 V, it can display down to one ten-thousandth of a volt on the DMM's display.

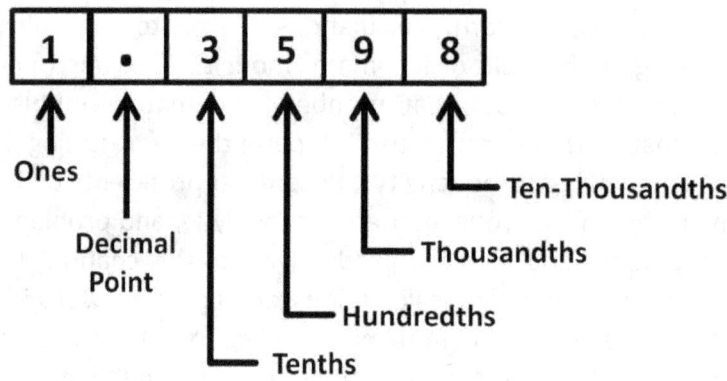

Figure 9.1: Decimal places

Whole Numbers

Whole numbers are non-negative integers that are made up of digits to the left of the decimal point. For example, the whole number of 1,288.00 is 1,288. To identify tens (10s), hundreds (100s), and thousands (1,000s) easily, a comma is used at every third place, starting at the decimal point and moving towards the left. For example, a resistor size of 17,452,223 Ω is more easily recognized than 17452223 Ω.

Multiples Number Conversion

Converting a low multiple to a high factor one makes it easier to read and understand. For example, a smartphone CPU's clock speed is 2,000,000,000 Hz. Using multiples, it can be written as 2,000 MHz. However, it would be even simpler to convert to fewer digits with a larger factor multiple (giga). The conversion process is shown in figure 9.2. First, place the decimal point to the right of 2,000. The difference in the power number (exponent) between "M" and "G" is 3 (9 − 6 = 3). The next step (step 2) is to move the decimal point 3 times (the result of the power number difference) to the left. The last step is to rewrite the number as 2.0 from 2,000 and replace "M" with "G."

1) Place decimal point → 2,000.0 MHz
2) Move decimal point 3 places to the left

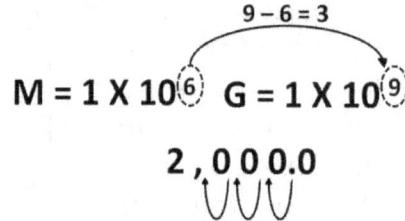

3) Rewrite 2,000 MHz to 2.0 GHz

Figure 9.2: Small to large multiple conversion

1) Place decimal point → 1.0 MHz
2) Move decimal point 3 places to the right

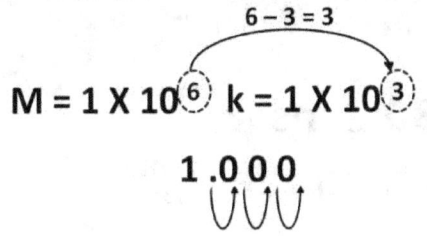

3) Rewrite 1 MHz to 1,000. kHz

Figure 9.3: Large to small multiple conversion

To convert a higher multiple factor to a smaller one, the process is reversed. For example, 1 MHz can be rewritten as 1,000 kHz. The conversion steps are shown in figure 9.3. M (mega) is 1×10^6. To convert it to a lower factor (power of 3), we first find the difference in power numbers (6 − 3 = 3). The next step (step 2) is to move the decimal point to the right (instead of left) according to the result of the power number subtraction. Lastly, rewrite the number as 1,000 and replace "M" with "k." From these two examples, you can see that converting a low multiple to higher one requires moving the decimal point to the left; converting a larger multiple to smaller one requires moving the decimal point to the right. The number of times to move the decimal depends on the result of the subtraction between the power numbers.

Submultiples Number Conversion

Similar techniques can be applied to convert submultiples. For example, 2,500 nA is easily converted to 2.5 uA. The conversion process is shown in figure 9.4. First, place a decimal point to the right of 2,500. Then move the decimal point 3 places to the left. It moves 3 times because the difference between the power numbers is 3 (9 − 6). The last step is to rewrite as 2.5, and replace "n" with "u."

1) Place decimal point → 2,500. nA
2) Move decimal point 3 places to the left

$$9 - 6 = 3$$

$$n = 1 \times 10^{-9} \quad u = 1 \times 10^{-6}$$

2,500.

3) Rewrite 2, 500 nA to 2.5 uA

Figure 9.4: Small to large submultiples conversion

To convert large submultiples to smaller ones, the process is reversed. For example, 30.2 ms is rewritten as 30,200 us. The process steps are shown in figure 9.5. First identify the decimal point. Second, move the decimal point 3 (6 – 3) places to the right (instead of left). Fill in the empty spaces with zeros. Finally, rewrite as 32,000 and replace "m" with "u."

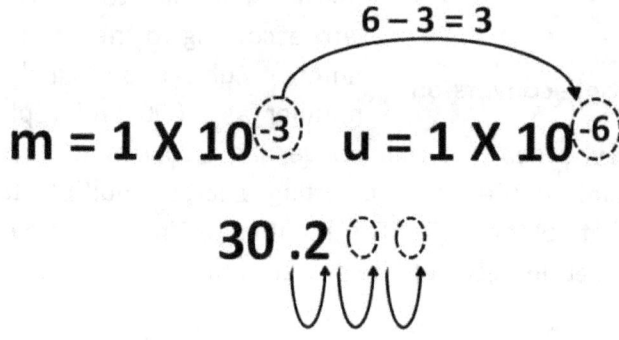

1) Identify decimal point → 30.2 ms
2) Move decimal point 3 places to the right

$$6 - 3 = 3$$

$$m = 1 \times 10^{-3} \quad u = 1 \times 10^{-6}$$

$$30.200$$

3) Fill spaces with zeros, rewrite to 30,200. us

Figure 9.5: Large to small submultiples conversion

Table 9-2 summarizes multiples to submultiples conversion methods.

Multiples		Submultiples	
High to low order	Low to high order	High to low order	Low to high order
Move decimal point to the right N times.	Move decimal point to the left N times	Move decimal point to the right N times	Move decimal point to the left N times
e.g. 1.0 M → 1,000 k	1.0 k → 0.001 M	1.0 m → 1,000u	1.0 u → 0.001 m

Table 9-2: Multiples and submultiples conversions

(N = Subtraction result between 2 power orders)

One-Over Reciprocal with Multiples and Submultiples

Once you get familiar with multiples and submultiples number conversions, we can apply them to practical electronic engineering calculations. Fractions are used often in calculating electronic arithmetic. Conversion of a fraction to a non-fraction produces quick and accurate results. For example (see figure 9.6), an LED flashlight requires two AA batteries connected in series **(1.5 V X (2) = 3 V)**. When an LED turns on, its forward voltage drop is 2 V. To limit current drawn at 10 mA for a certain brightness level, a current-limiting resistor is placed in series with the LED. The resistor size is calculated and shown in figure 9.6.

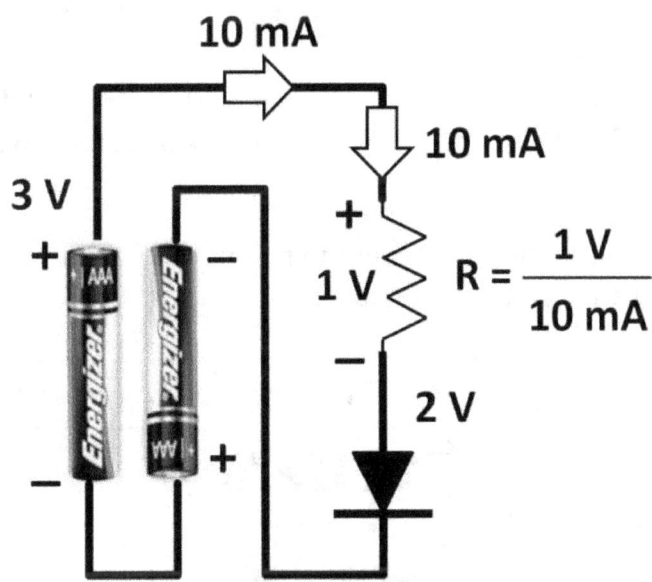

Figure 9.6: LED flashlight current limiting resistor size

$$R = \frac{1}{10 \times 1 \times 10^{-3}}$$

$$R = \frac{1}{10} \times \frac{1}{1 \times 10^{-3}} \quad \text{- 3 to + 3}$$

$$R = 0.1 \times 1 \times 10^3$$

$$R = 0.1 \times 10^3 = 0.1 \text{ k} = 100$$

Figure 9.7: Fraction to non-fraction, submultiples

To convert the resistance from fraction to non-fraction, the steps are shown in figure 9.7. In this fraction, we first convert milli to 10^{-3}. The denominator contains a submultiple number 10 mA (10 X 1 X 10^{-3} A). Since all numbers in the denominator are separated by multiplication signs, it can be broken down into two fractions: 1 / 10 and 1 / (1 X 10^{-3}).

1 / 10 = 0.1. For 1 / (1 X 10^{-3}), convert negative 3 power to positive 3 power then remove the fraction. The final result is 0.1 X 1 X 10^3, which is equal to 0.1 kΩ. Using the large multiple conversion rule, previously discussed, move the decimal point 3 times to the right turns 0.1 kΩ to 100 Ω.

To convert a fraction with multiples in the denominator to a non-fraction, the conversion process is reversed. For example, figure 9.8 calculates period from a 2 GHz clock. Separate 1 / 2 from G (1 X 10^9). **1 / 2 = 0.5**. The power (exponent) of positive 9 in the denominator now becomes negative 9. We then remove the fraction. The result is 0.5 X 10^{-9} = 0.5 ns. If you want to write 0.5 ns to lower submultiples (e.g., pico), use the rule described earlier. Move the decimal point 3 places to the right.

$$Period = \frac{1}{2\,G}$$

$$Period = \frac{1}{2} \times \frac{1}{1 \times 10^9} \quad 9 \text{ to } -9$$

$$Period = 0.5 \times 1 \times 10^{-9}$$

$$Period = 0.5 \times 10^{-9} = 0.5 \text{ ns}$$

0.5 ns = 500 ps

Figure 9.8: Fraction to non-fraction, multiples

This technique enables you to convert fractions to non-fractions easily, quickly, and accurately. Table 9-3 summarizes the fraction to non-fraction conversions.

Fraction	TO	Non-fraction
$\frac{1}{n}$	⟷	G
$\frac{1}{u}$	⟷	M
$\frac{1}{m}$	⟷	k
$\frac{1}{k}$	⟷	m
$\frac{1}{M}$	⟷	u
$\frac{1}{G}$	⟷	n

Table 9-3: Fraction to non-fraction conversions

Multiplication and Division with Multiples and Submultiples

Multiplication is used all the time in electronic engineering calculations. Multiplication with multiples and submultiples works opposite to multiplication of fractions. Instead of subtracting power numbers, multiplication involves adding power numbers (exponents). For example, if **frequency = 10 MHz**, and **Inductance = 25 uH**, calculate inductive reactance, **XL = 2 π f L**. Figure 9.9 shows the steps. 2 π is simply 6.28. The multiplication of the "k" multiple and "u" submultiple involves adding exponents to each other **(3 + (– 6) = – 3)**. The resulting exponent is – 3. The rest of the calculations are simple multiplication **(6.28 X 250 X 1 m = 1, 570 m = 1.57 k).** Division with multiples and submultiples in electronic engineering math involves subtracting exponents. For example, when a push button is pressed, there is the presence of on-resistance. The voltage across a push button when it's pressed is measured at 2 mV with 100 uA flowing through it. Figure 9.10 shows the steps to calculate the on-resistance of the push button. **2 / 100 = 0.05**. Negative 3 (milli) power less negative 6 (micro) power is positive 3, which becomes the final exponent.

XL = 2 π f L

XL = 2 (3.14) (10 k) (25 u)

$XL = 6.28\ (10 \text{ X } 1 \text{ X } 10^{3})\ (25 \text{ X } 1 \text{ X } 10^{-6})$

3 + (-6) = -3

$XL = 6.28 \text{ X } 250 \text{ X } 1 \text{ X } 10^{-3} = 1{,}570 \text{ m} = 1.57 \text{ k}$

Figure 9.9: Multiplying multiples

$$\text{On-resistance} = \frac{2 \text{ m}}{100 \text{ u}}$$

$$\text{On-resistance} = \frac{2 \text{ X } 1 \text{ X } 10^{3}}{100 \text{ X } 1 \text{ X } 10^{-6}}$$

– 3 – (– 6)
= – 3 + 6 = + 3

$$\text{On-resistance} = 0.05 \text{ X } 1 \text{ X } 10^{3} = 50$$

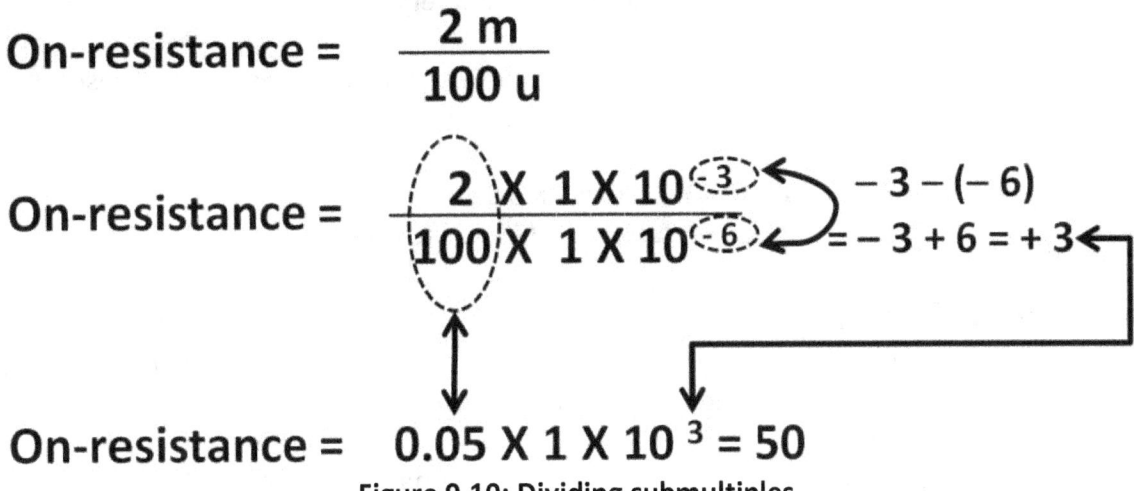

Figure 9.10: Dividing submultiples

Percentage to Decimals

In electronics, we often use percentages to calculate power efficiency, duty cycle, device tolerance, accuracy, error, resolution, gain change, voltage variation, current change, and power difference. Converting percentages to decimals quickly helps you analyze problems effectively. A number with a percentage sign means the original number gets multiplied by 100. To convert a percentage to a decimal number, first identify the decimal point. Then divide the number by 100 (move decimal point two places to the left). For example (see figure 9.10), the duty cycle of an AC signal is 75%. To convert it to a decimal number, first identify the decimal point location (to the right of 5). Then, move it 2 places to the left and remove percentage sign. To convert the number back to a percentage, reverse the process by moving the decimal point two places to the right and add percentage sign. The example in figure 9.11 converts the number 2 to a percentage. After moving 2 decimal places to the right, fill the empty spaces (dotted) with zeros, and add a percentage sign to complete the conversion.

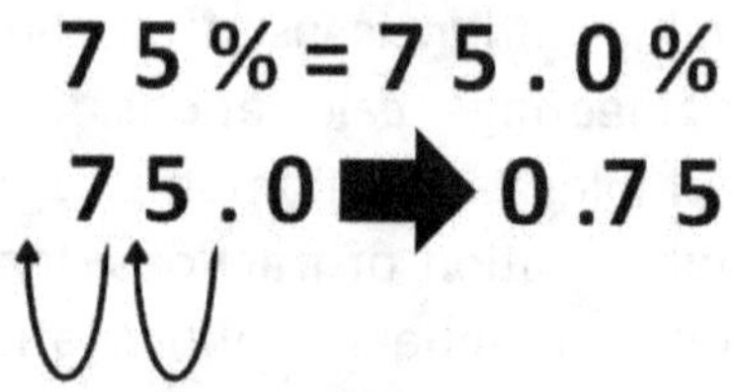

$$75\% = 75.0\%$$
$$75.0 \Rightarrow 0.75$$

Figure 9.10: Percentage to decimal number

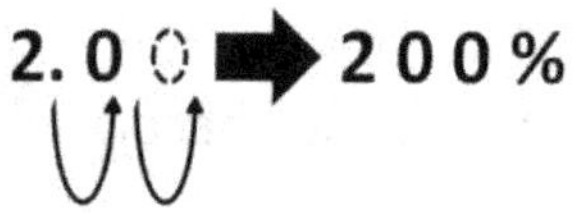

$$2.0\,0 \Rightarrow 200\%$$

Figure 9.11: Number to percentage

For example, a carbon resistor has a +/− 10% resistance tolerance. If the nominal resistance is 33.33 kΩ, what is the range of resistance values? 10% is quickly converted to 0.01:

$$33.33 \text{ k}\Omega \times 10\% = 33.33 \text{ k}\Omega \times 0.01 = 0.33 \text{ k}\Omega = 33 \text{ }\Omega$$
$$(33.33 \text{ k}\Omega - 33.33) < R < (33.33 \text{ k}\Omega + 33.33)$$

Log to Real Number

We use logarithms (log) in voltage, current, and power dB calculations. The log of a number is equal to the exponent (power) of the base number. For example, **log$_{10}$ 100 = 2** because 10 to the power of 2 is 100 (see figure 9.12). The base number can also be other numbers except for 10. **Log$_2$ 16 = 4** because 2 to the power of 4 is 16. If the base number is not shown, then by default, the base number is 10.

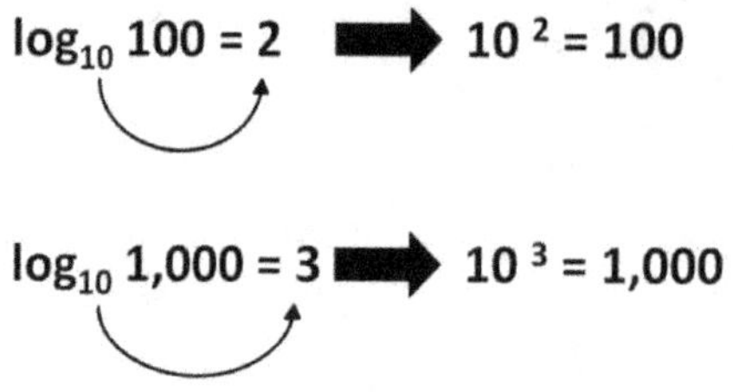

$$\log_{10} 100 = 2 \Rightarrow 10^2 = 100$$

$$\log_{10} 1{,}000 = 3 \Rightarrow 10^3 = 1{,}000$$

Figure 9.12: Log with a base of 10

Extending from this concept, a log table is shown in table 9-4. Log 0 is invalid because 10 to the power of any value will be larger than zero. From this table, you can easily estimate the range of log numbers. For example, if you try to estimate value of log 20, you can easily tell it's between 1 and 2.

Log 0	invalid
Log 1	0
Log 10	1
Log 100	2
Log 1,000	3
Log 10,000	4
Log 100,000	5

Table 9-4: Log table

If you recall dB calculations using log, a fraction is often used within log. For example, an amplifier with gain of 100 dB, Vout / Vin can be quickly evaluated:

$$100 \text{ dB} = 20 \log \left(\frac{\text{Vout}}{\text{Vin}}\right)$$

$$5 = \log \left(\frac{\text{Vout}}{\text{Vin}}\right)$$

$$\frac{\text{Vout}}{\text{Vin}} = 100,000$$

With power efficiency calculations, output power is always less than input power (Pout < Pin) due to electrical signal losses. Using an LED as an example, its power efficiency (less than 15%) is much higher than that of an incandescent lamp (less than 2%). A typical LED burns roughly 6 W to 8 W of power. Incandescent lamps' power ratings differ greatly depending on the type. The most common ones consume 60 W of power. We can use dB to express the input and output power as a ratio instead of as absolute value. For example, the output power measured is 10 times less than the input, i.e., Pout / Pin = 1 / 10 = 0.1. Power in dB is calculated as:

dB = 10 log (0.1) = – 10 dB

From this example, you can see that when the log number is less than 1, it equates to a negative number. A similar log table like table 9-4 is developed for the log numbers that are less than 1 (see table 9-5).

Log 0.1	−1
Log 0.01	−2
Log 0.001	−3
Log 0.0001	−4
Log 0.00001	−5

Table 9-5: Log number less than one

Summary

In this chapter, we first covered multiples, submultiples, decimal numbers, and percentages. We then used practical examples using common electronic engineering tasks to convert between higher and lower multiples to and from submultiples. We then applied these multiples and submultiples conversion techniques to multiplication and division that are frequently used in electronic engineering calculations. This chapter closed with using logarithmic numbers to calculate voltage, current, and power ratios. Following these simple rules allows you to come up with electronic engineering math solutions quickly and accurately as well as demonstrate professional competencies.

Quiz

1) What is the submultiple name of 1×10^{12}?

2) A DMM can display digits down to one thousandth of a volt. What is the smallest change in decimal value this DMM can display?

3) Convert 2.5 uA to nA.

4) Convert 120 ns to frequency.

5) Using mental math, calculate $XL = 2\pi \times 2$ MHz $\times 2$ uH.

6) A common emitter amplifier delivers to a resistive load draws 2 mA at 12 V to ground rails. The power measured at the load (Pout) is 10 mW. What is the power efficiency in dB?

7) Convert 0.707 to a percentage.

8) An emitter follower's voltage changes by 0.5 V while input changes by 1 V. What is the voltage loss in dB?

9) 1 angstrom is equal to 10^{-10} meters, which is often used to describe the thickness of CMOS transistor gate oxide. If the FET's gate oxide is 50 angstrom, what is the value in nanometers (nm)?

10) If an amplifier's open-loop gain is 80 dB, what is the gain ratio of V / mV?

Abbreviations and Acronyms

< (less than)	Br (boron)
% (percentage)	C (capacitance or coulomb)
(a)(b) (multiply a and b)	C_eq (equivalent capacitance)
/ (divide)	CA (common anode)
|| (parallel)	CAD (computer-aided design)
> (greater than)	CAN (Control Area Network)
Δ (delta)	CC (common cathode)
≈ (approximately equal to)	CDMA (Code Division Multiple Access)
≤ (less than or equal to)	CDS (drain-to-source capacitance)
≥ (greater than or equal)	CDSub (drain-to-substrate capacitance)
°C (degrees Celsius)	CGD (gate-to-drain capacitance)
∞ (infinity)	CGS (gate-to-source capacitance)
A or Amp (ampere)	CLoad, Cload (capacitive load)
A/C (air-conditioning)	CMOS (complementary metal oxide semiconductor)
AC (alternating current)	CMRR (common mode rejection ratio)
Acm (common mode gain)	COM (common potential)
ADC (analog-to-digital converter)	Cox (gate-oxide capacitance per unit area)
ADD (add instruction)	CPU (Central Processing Unit)
Adm (differential gain)	CSSub (source-to-substrate capacitance)
AM (amplitude modulation)	Cu (Copper)
AMD (Advanced Micro Device)	D (digital input code)
ARM (Advanced RISC Machines)	DAC (digital-to-analog converter)
ASIC (Application Specific Integrated Circuit)	dB (decibel)
BiCMOS (Bipolar and CMOS)	DC (direct current)
BNC (Bayonet Neill-Concelman)	Diff amp (differential amplifier)
BOR (Brown-Out Reset)	DIV (divide instruction)
Bps (bit per second)	DMM (digital multi-meter)

DN (done bit)	**GPR** (general purpose register)
DRC (design rule check)	**GRT** (greater than)
DSP (digital signal processing)	**GSM** (Global System for Mobile)
e (exponential)	**H** (Henry)
E (voltage potential)	**hfe** (voltage gain)
e-, E- (electron)	**HMI** (human machine interface)
EC (quartz crystal resonators)	**Hz** (hertz)
ECL (emitter-coupled logic)	**I/O** (input output)
ELI (voltage-inductor-current)	**I_A** (current A)
Emax (maximum peak-to-peak level)	**I_B** (current B)
Emin (minimum peak-to-peak level)	**I_C** (current C)
EN (enable bit)	**I_total** (total current)
EQ (equal to)	**I2C** (Inter-Integrated Circuit)
EQU (equal to)	**Ib, IB** (base current)
ESL (equivalent series inductance)	**IBM** (International Business Machine)
ESR (equivalent series resistance)	**IC** (collector current)
F, f (frequency or farad)	**ICD3** (In-Circuit Debugger 3)
Fab (fabrication)	**ICE** (current-capacitor-voltage)
FCC (Federal Communications Commission)	**ICs** (integrated circuits)
FCY (instruction frequency)	**ID** (drain current)
FM (frequency modulation)	**ID** (identification)
FOSC (oscillator frequency)	**IDE** (integrated development environment)
FPGA (Field Programmable Gate Array)	**IE** (emitter current)
fresonant (resonant frequency)	**IEEE** (Institute of Electrical and Electronics Engineers)
Gbps (gigabit per second)	**Iin** (input current)
GEQ (greater than or equal to)	**Iload** (load current)
GHz (gigahertz)	**Iout** (output current)
gm, GM (transconductance)	**IR drop** (voltage drop across resistor)
GPIO (general purpose Input Output)	**IS** (saturation current)

IS (source current)	**MIPS** (million instructions per second)
Isense (sense current)	**mm** (millimeter)
ISR (interrupt service routine)	**MOSFET** (metal oxide semiconductor field effect transistor)
I-V curve (current vs. voltage curve)	**MOV** (move instruction)
JMP (jump instruction)	**MSB** (most significant bit)
JSR (jump to subroutine instruction)	**MSPS** (mega-sample per second)
K (degree Kelvin)	**MUL** (multiply instruction)
KB (kilobyte)	**MUX** (multiplexer)
KCL (Kirchhoff's current law)	**MUL** (multiply instruction)
kg (kilogram)	**n** (bit number)
KVL (Kirchhoff's voltage law)	**N** (negative type)
L (inductor or transistor length)	**NC** (normally-closed)
L_eq (equivalent inductance)	**NEQ** (not equal to)
LBL (label instruction)	**NFET** (N-typed field effect transistor)
lbs (pounds)	**NMOS** (N-typed metal oxide semiconductor)
LCD (liquid crystal display)	**NO** (normally-open)
LDO (low drop-out regulator)	**Op-amp** (operational amplifier)
LED (light emitting diode)	**Op-code** (operation code)
LEQ (less than or equal to)	**OST** (Oscillator Start-up Timer)
LES (less than)	**P** (power or Phosphorus or positive type)
LTE (Long Term Evolution)	**PAC** (Programmable Automation Controller)
ln (natural logarithm)	**Parasitic cap** (parasitic capacitance)
LO (local oscillators)	**PB** (push button)
LP, XT, HS (lower speed, external, high speed)	**PCB** (printed circuit board)
LSB (least significant bit)	**PFD** (phase frequency detector)
mA (milliampere)	**PFET** (P-type field effect transistor)
mAh (milliampere-hour)	**PIC** (peripheral interface controller)
MCU (microcontroller unit)	**PLC** (programmable logic controller)

PLL (phase lock loop)	**Rms** (root mean square)
PMOS (P-type metal oxide semiconductor)	**TT** (timing bit)
POR (Power-On-Reset)	**ROM** (read only memory)
POT (potentiometer)	**Rout** (output impedance)
ppm (part-per-million)	**Rs** (source resistor)
PSRR (power supply rejection ratio)	**RS-232** (recommended standard 232)
PWM (pulse width modulation)	**Rvin** (input impedance)
Q factor (quality factor)	**Rz** (zener impedance)
Q# (transistor number)	**SAR** (successive approximation)
q, Q (electron charge)	**SBR** (subroutine)
Q_bar (Q bar)	**SCADA** (Supervisory Control And Data Acquisition)
R (resistance)	**SFR** (special-function registers)
R leakage (leakage resistance)	**Si** (silicon)
r π (intrinsic base resistance)	**SiGe** (silicon germanium)
R_eq, R_equivalent (equivalent resistance)	**Sine** (sinusoidal)
R_total (total resistance)	**SOC** (system-on-chip)
RAM (read access memory)	**Spec** (specification)
RC mode (Resistor-Capacitor mode)	**SPI** (synchronous peripheral interface)
RD (drain resistor)	**SQO** (sequencer output instruction)
Rdson (drain-to-source on-resistance)	**S-R** (set, reset)
RET (return)	**SUB** (subtract instruction)
Rf (feedback resistor)	**SW** (switch)
RF (radio frequency)	**T0CK1** (external clock source)
RFID (radio frequency ID)	**T0XCS** (clock select bit)
Rgate (gate resistance)	**TC** (temperature coefficient)
Ri (input terminal resistor)	**TCY** (instruction period)
RISC (Reduced Instruction Set Computing)	**TND** (temporary end)
RJ-45 (registered jack 45)	**Toff, toff** (off time)

RLoad or RL (resistive load)	**Ton, ton** (on time)
TTL (transistor-transistor logic)	**TSMC** (Taiwan Semiconductor Mftg. Corp.)
U (effective mobility)	**VHDL** (very high level descriptive language)
um (micrometer)	**Vin, VIN** (input voltage)
UMC (United Microelectronics Corporations)	**Vin_diff** (input voltage difference)
USART (Universal Synchronous Asynchronous Receive Transceiver)	**Vout, VOUT** (output voltage)
USB (universal serial bus)	**Vout_diff** (output voltage difference)
V− (negative terminal)	**Vpeak** (peak voltage)
V (voltage)	**Vpeak-to-peak** (peak-to-peak voltage)
V_cap (capacitor voltage)	**Vref** (reference voltage)
V+ (positive terminal or positive voltage supply)	**Vrms** (root mean square voltage)
V++ (positive voltage supply)	**VS** (source voltage)
VB (base voltage)	**Vsense** (sense voltage)
VBE (base to emitter voltage)	**VT** (threshold voltage or thermal voltage)
VC (collector voltage)	**W** (watt or transistor width)
VCC (positive power supply)	**WDT** (watchdog timer)
VCE (collector to emitter voltage)	**WiFi** (wireless fidelity)
VCEsat (collector to emitter saturation voltage)	**X** (multiply)
VCO (voltage controlled oscillator)	**Xc** (capacitive reactance)
VD (drain voltage)	**XL** (inductive reactance)
VDD (positive voltage supply)	**XLP** (extra Low Power)
Vdiff (voltage difference)	**α** (alpha)
Vdiode (diode voltage)	**β** (beta)
VDS (drain-to-source voltage)	**λ** (wavelength)
VE (emitter voltage)	**π** (pi or 3.14)
VFB (feedback voltage)	**Σ-Δ** (sigma-delta)
VG (gate voltage)	**Ω** (ohm, unit of resistance)
VGS (gate-to-source voltage)	**ω** (omega)

Index

− 20 dB per Decade, 65
− 3 dB, 68, 83

Δ

ΔQ = C (ΔV), 71

1

1 / SC, 57
1 / ωC, 57
120 V, 94
1N4001, 48

2

2 π, 70

3

3-dimentional cross section model, 131

5

555-timer, 226
    Hans Camenzind, 226
    one-shot timer, 228
    precision timing, 226
    Pulse Width Modulation (PWM), 226

6

60 Hz, 53

7

7-segment display, 256

A

AA, 7
AAA, 7
AC analysis, 63
AC choke, 74
AC parameters, 49
AC short, 56
Acm (common mode gain), 150
active loads, 151
active low-pass filter, 174
ADC
    gain error, 222
    offset error, 222
Adm (differential mode gain), 150
aerospace, 106
Agilent, 27
alkaline household battery, 7
Alpha, 112
alternating current, 49
aluminum, 1
AM demodulation circuit, 240
AM detector, 240
AM transmitter, 240
amplifier, 106, 140
Amplitude modulation (AM), 238
Analog Devices, 3, 24, 219
analog electronics, 105
analog IC vendors
    Analog Devices, 106
    Infineon Technologies, 106
    Qualcomm, 106
    STMicroelectronics, 106
    Texas Instruments, 106
analog market, 106
analog signals, 105
Analog-to-Digital Converter, 25, 106
    ADC, 219
AND gate, 205
angstrom, 131
angular velocity, 70
anode, 41
arc, 70
assembler, 262
Asynchronous Receiver Transceiver (USART), 247
atoms, 37

atoms
    electrons, 37
    neutrons, 37
audio amplifier, 157

B

band-pass filter, 233
band-stop filter, 233
bandwidth, 103
base, 108
Baud, 236
Bayonet-NeillConcelman (BNC), 178
Bessel chart, 239
Beta, 112
bill-of-materials, 68
Bipolar versus CMOS, 147
bit rate, 236
bit-oriented operations, 255
bode plot,, 63
body diode, 133
Boltzmann's constant, 113
BOMs, 68
Boolean algebra, 207
boron, 37
bounce, 86
break-before-make, 199
breakpoint, 264
Brown-Out-Reset (BOR), 261
buck regulator, 97, 156
buffer, 119, 171
built-in diode voltage, 40
byte-oriented instructions, 255

C

C = F λ, 237
Capacitive load (CLoad), 182
capacitor, 136
capacitor
    capacitance, 55
    capacitive reactance, 55
    dielectric, 55
    electric field, 55
    electrolytic, 55
    passive electronic device, 55
    polyester, 55
    tantalum, 55
    Xc, 55
carbon resistors, 2
carrier concentration, 38
cascode, 167
cathode, 41, 108
cell phone battery chargers, 93
cellular bands
    Code Division Multiple Access (CDMA), 241
    Global System for Mobile (GSM), 241
    Long Term Evolution (LTE), 241
ceramic resonators, 257
charge pump, 103
charging, 60
circuit simulation software
    Multisim, 176
class A amplifier, 129
class AB, 129
class B amplifier, 129
closed-loop, 153
closed-loop voltage gain, 161
CMOS, 130
CMOS cacosde, 170
collector, 108
collector current equation, 152
color bands, 3
combinational logic, 206
common base amplifier, 120, 128
common collector amplifier, 118
common emitter amplifier, 115, 123
common gate amplifier, 139, 145
common mode rejection ratio (CMRR), 150
common mode voltage, 149
common source amplifier, 139
communications, 231
    full duplex, 231
    half duplex, 231
    Radio Frequency (RF), 231
    simplex, 231
commutating diode, 88

comparators, 106
compiler, 262
computer-aided design (CAD), 135
copper, 1
core, 93
coulomb, 8
CPU, 110
Cu, 1
current, 1, 8, 329
current divider rule, 15
current lead, 89
current mirror, 152, 165
current source, 8

D

Darlington pair, 168
data compare instructions, 308
data manipulation instructions, 304
data memory, 251
dB, 64
DC, 5
DC block, 56
dead zone, 199
debugger, 263
   ICD3, 264
   PICKIT3, 263
decade, 65
decibel, 63
decimal numbers, 320
design example
   comparator, 265
   timer, 269
Design Rule Checking (DRC), 135
design systems, 135
desktop computer, 110
D-flip-flop, 211
die, 130
diff amp, 148
differential amplifiers, 153
diffusion, 110
digital circuits
   flip-flop, 208
   latch, 208

digital electronics, 195
digital voltage levels
   CMOS, 219
   Emitter-Coupled-Logic (ECL), 219
   Transistor-Transistor-Logic (TTL), 219
Digital-to-Analog Converter
   DAC, 224
   System-On-Chip (SOC), 224
digital-to-analog converters, 106
diode circuits, 43
diode clamp, 88
diode parameters
   diode output current, 42
   maximum forward voltage, 42
   maximum power dissipation, 42
   maximum reverse current, 42
   reverse voltage, 42
diodes, 37, 108
direct current, 1, 5
discharging, 60
distributors
   Arrow electronics, 24
   Digikey, 24
   Future electronics, 24
   Mouser electronics, 24
DMM, 27, 28, 33, 48, 320, 328, 329
dominant pole, 183
doping levels, 38
drain current, 132
drain resistor (RD), 140
duty cycle, 52, 99
dynamic gate current, 138

E

Early effect, 115
electric power generation, 93
electrical
   current, 51
electrical engineering, 3, 4
electrical isolation, 93
electrical outlet, 51
electromagnetic theory, 93
electron charge, 42

electronic load, 7
electrons, 8, 9
ELI, 77, 90
emitter, 108
emitter follower, 118, 127
Energizer, 7
engineering, 3, 323, 325, 328
equivalent series inductance, 83
ESL, 83
ESR, 83
Ethernet, 106

F

f −3dB., 68
fab, 130
fabrication, 130
FCY, 257
Federal Communications Commission
    Frequency spectrum, 231
feedback, 99
file registers, 251
finFET, 131
flip-flop, 210
    edge-triggered flipflop, 210
Fluke, 27
FM, 85
FM noise clipper, 85
Forward-biased, 40
FOSC, 257
Fraction-to-nonfraction conversions, 324
Freescale, 24
frequency, 50
frequency divider, 211
frequency domain, 63, 64
frequency domains, 232
Frequency Modulation, 85
Frequency Modulation (FM), 241
Fritzing, 55
full-wave rectifier, 102
function generator, 27, 179

G

General Motor, 273
General Purpose Registers (GPRs), 253
germanium, 8, 37, 110
Gm, 121
ground, 8
Gummel-Poon model, 125

H

half-wave rectifier, 95
Harvard architecture, 251
height, 105
high- and low-level languages, 263
high-pass filter, 80, 101
hole, 37
humidity, 105
hybrid π model, 122
hysteresis, 179
    hysteresis zone, 179

I

I (Δt) = C (ΔV), 57
IBM, 3, 110
IC design and simulation software
    Cadence Design Systems, 178
    Mentor Graphics, 178
    Synopsis, 178
IC foundries
    Global Foundries, 130
    Samsung Semiconductor, 130
    Taiwan Semiconductor Manufacturing Company (TSMC), 130
    United Microelectronics Corporations (UMC), 130
IC layout, 134
IC package manufacturers
    Advanced Semiconductor Engineering, 25
    Amkor, 25
    Siliconware Precision Industries, 25
IC package types
    dual inline package, 25
    flip-chip, 25
    wire-bond, 25

IC Package types
    ball-grid-array, 25
IC Packages, 24
IC versus VCE Curve, 114
ICE, 71, 90
ideal diode, 42
Ideal voltage source, 22
IE = IC + IB, 113
impedance, 54
induce, 93
inductive load, 87
inductor
    copper, 73
    ferrite, 73
    Henry, 73
    inductance, 73
    inductive reactance (XL), 73
    iron, 73
    magnetic field, 73
    pass device, 73
    XL = 2 π f L, 73
inductor schematic symbol, 73
Institute of Electrical and Electronics Engineers
    (IEEE), 231
instruction clock, 257
instrumentation amplifier (INA), 184
Integrated Development Environment (IDE), 262
Intel, 134
intensity, 105
Inter-Integrated Circuit (I2C), 247
internal oscillator, 258
internal regulators, 261
International Rectifier, 24
interrupt, 260
Interrupt Service Routine (ISR), 260
Intersil, 24
intrinsic resistance, 125
inverter, 196
    NOT gate, 196
inverting amplifier, 158
ion implantation, 38, 110
iPod, 93

J

James Early, 115
JK flip-flop, 211
Jump to Label instructions, 300
Jump to Subroutine Instructions, 301

K

KCL, 11
Kelvin (K), 113
Kirchhoff's Current Law, 11
Kirchhoff's Voltage Law, 9
KVL, 9

L

ladder logic, 273
laptop, 93
latch, 208
    memory, 208
    memory seats, 208
    sequential logic, 208
Layout versus Schematic (LVS), 135
LDO, 97
leakage current, 42
leakage currents, 138
leakage resistor, 83
Least Significant Bit
    LSB, 204
LED, 44
level shifter, 217
light, 105
light emitting diode, 44
linear regulator
    low dropout regulator, 186
    zener regulator, 97, 185
Linear Technology, 24
literal control, 257
load line, 114
log to number, 326
logarithm, 63
lots, 135
low drop-out regulator
    error voltage, 186

negative feedback, 186
low-pass filter, 68

M

mAh, 7
mathematics, 3, 4, 49, 61
Maxim Integrated Circuits, 24
MCU instructions, 255
MCU parameters, 248
MCU peripherals
    ADCs, 247
    comparators, 247
    DACs, 247
    timers, 247
MCU vendors
    Atmel, 247
    Freescale Semiconductor, 247
    Fujitsu, 247
    Infineon Technologies, 247
    Microchip Technology, 247
    NXP, 247
    Renesas Electronics, 247
    Samsung, 247
    STMicroelectronics, 247
    Texas Instruments, 247
medical equipment, 106
mental math, 319
Mentor Graphics, 135
Metal Oxide Semiconductor Field Effect Transistor, 107
Microchip Technology, 3, 24, 25, 217, 219, 248, 249, 250
microcontroller Units
    embedded systems, 247
Microcontroller Units (MCUs), 247
microelectronics, 3
microscope, 134
milliamphour, 7
Millions of instructions per second (MIPS), 258
mixed-signal
    ADCs, 216
    DACs, 216
modulation, 236

modulation index, 239
monostable, 228
MOSFET cross section, 136
MOSFET parasitic
    drain-to-source (CDS), 142
    drain-to-substrate (CDSub), 142
    gate-to-drain capacitor (CGD), 142
    gate-to-source capacitor (CGS), 142
    source-to-substrate (CSSub), 142
MOSFETs, 130
Most Significant Bit (MSB), 204
motor control applications, 106
MOV instruction, 306
MOV instruction application, 307
MPLABX, 262
multimeter, 27
multiples, 319
multiples and submultiples conversions summary, 322
multiples number conversion, 320
multiplexer, 215
multiplication and division with multiples and submultiples, 325
multistage amplifiers, 155

N

NAND gate, 205
National Semiconductor, 3, 176
National Instruments, 176
natural log, 113
negative temperature coefficient, 46
nested subroutines, 303
neutrons, 8
NFET, 130
NFET and PFET Inverter, 197
N-junctions, 131
NMOS, 130
NMOS Inverter, 197
non-ideal capacitor, 83
non-ideal diode, 42
non-ideal voltage source, 22
non-inverting amplifier, 160

NOR gate, 204
normally-closed (NC), 278
normally-open (NO), 278
NPN, 108
NPN schematic symbol, 109
Nyquist frequency, 221

O

off-time, 52
Ohm's Law, 6
omega, 57
On Semiconductor, 24
One-Over Reciprocal, 323
on-time, 52
op-amp, 99
op-amp
   LM741, 164
op-amp parameters
   supply and input voltage, 162
   supply current, 162
   Common Mode Rejection Ratio (CMRR), 162
   input impedance, 162
   input offset current, 162
   input offset voltage, 162
   open-loop gain, bandwidth, 162
   output source and sink current, 162
   output voltage swing, 162
   power consumption, 162
   Power Supply Rejection Ratio (PSRR), 162
op-amp rules
   input impedance, 155
   input offset voltage, 155
   output impedance, 155
open-collector, 193
open-drain, 193
operand, 255
operation code (opcode), 255
OR logic gate, 202
Oscillator Startup Timer (OST), 261
oscilloscope, 27, 177
oscilloscopes
   attenuation ratio, 178
   gigabits per second, 178
   gigahertz, 178
output code, 230
output impedance, 124
output symbol, 280
oxide, 131

P

parallel capacitor rule, 63
parallel circuit, 11
parallel data transmission, 214
parallel inductor rule, 78
parallel LC, 92
parallel resistor rule, 12
parasitic, 83, 84
passive electronic device, 1
passive load resistors, 151
peak voltage, 52
Peak-to-Peak Voltage, 52
percentage to real number, 326
percentage-decimal conversion, 319
period, 50
periodic waveform, 49
PFET, 130
Phase Lock Loop (PLL), 242
   feedback loop, 242
   low-pass filter, 242
   phase detector, 242
   Voltage Controlled Oscillator (VCO), 242
phase shift, 69, 71, 119, 120, 129, 145
phosphorus, 38
PLC benefits, 275
PLC components, 276
   CPU, 276
   I/O modules, 276
   input modules, 276
   memory (program, data), 276
   power supply, 276
   programming device, 276
PLC conveyor system, 290
PLC counter, 297
PLC counter application, 298
PLC data structure, 305
PLC math instructions, 311

PLC off-timer, 295
PLC off-timer application, 296
PLC on-timer, 293
PLC on-timer application, 294
PLC periodic clock signal generator, 318
PLC program control instructions, 300
PLC programming
    ladder logic, 278
PLC programming example, 283
PLC programming syntax, 286
PLC sequencer instructions, 315
PLC suppliers
    AllenBradley, 273
    Bosch, 273
    General Electric, 273
    Mitsubishi Electric, 273
    Panasonic, 273
    Siemens, 273
PLC timer, 292
PLC trends, 317
PLL frequency multiplier, 244
    Analog Devices, 244
    Low noise digital Phase frequency Detector (PFD), 244
PLL implementation
    crystal oscillator, 243
PMOS, 130
P-N junction
    carrier concentration, 39
    concentration imbalance, 39
    depletion, 39
    diffusion, 39
    equilibrium, 39
P-N junctions, 37
PNP, 108
PNP schematic symbols, 109
polysilicon, 131
positive feedback
    bode plot, 182
positive feedback
    oscillations, 182
power, 7
power efficiency, 129
power management, 76, 106
power ratio, 64
Power-On-Reset (POR), 261
pressure, 105
printed circuit board, 24
program memory (flash), 251
Programmable Logic Controllers (PLCs), 273
programming languages
    assembly, 247
    C, 247
    C++, 247
protons, 8
Psubstrate, 131
pull-up resistor, 193
Pulse Width Modulation (PWM) channel, 247
push button, 279
Pythagorean Theorem, 104

Q

Q factor, 77

R

R C circuit, 104
r π, 123
radian, 70
radio, 85
radio frequency, 106
Random Access Memory (RAM), 247
RC low-pass filter, 69
RC time constant, 62, 96
reactance, 54
Read Only Memory (ROM), 247
reference current, 152
relay, 274
renewable energy, 106
resistance, 1
resistivity, 1
resistor, 1
resonant frequency, 90, 91
reverse-biased, 40
RFID (radio frequency ID), 261
ring oscillator, 200

ringing, 86
ripple voltage, 95
RLoad, 129
room temperature, 123
rotation degree, 70

S

saturation, 115
saturation current, 113
sawtooth wave, 49
schematics, 8
Schottky diode, 192
scope probe, 178
seal-in circuit, 288
semiconductor, 4, 5, 38, 106, 218, 317, 318
semiconductor fab conveyor system, 318
semiconductor package, 24
Serial Peripheral Interface (SPI), 247
series capacitor rule, 63
series circuit, 9
series inductor rule, 79
series resistor rule, 13
settling time, 157
sheet rho, 4
shift register, 213
shoot-through current, 199
SiGe, 110
silicon, 37
silicon dioxide (SiO2), 131
silicon germanium, 110
sine wave, 49
single-ended amplifier, 115, 129
sinusoidal, 49
sleep mode, 261
slew rate, 157
small-signal analysis, 124
small-signal model, 121, 128, 140
smartphone car chargers, 93
SOC, 106
solid-state, 38
sound, 105
source follower, 139, 143
Special-Function Registers (SFRs), 253

spectrum analyzer, 234
speed, 105
square wave, 49
step response, 86
submultiples, 319
submultiples number conversion, 321
summing amplifier, 172
superposition, 35
Superposition theorems, 19
surface-mount
    resistors, 4, 46
switching regulators, 97
Synopsis, 135
system-on-a-chip, 106

T

tank circuit, 91, 103
tapeout, 135
TCY, 257
Tektronix, 27
telecommunication applications, 106
temperature, 112
temperature coefficient, 4, 23
temporary end, 304
Texas Instruments, 24
common collector amplifier, 127
thermal voltage, 113
thermocouple, 191
threshold voltage, 132
time, 50
timer, 259
touch screen, 106
transconductance (Gm), 121
transfer function, 64
transformer, 93
transistor Beta, 116
transistor types, 107
    BiCMOS, 107
    CMOS, 107
transistors, 107
    MOSFET, 107
    NFET, 107
    NPN, 107

PFET, 107
PNP, 107
trigonometry, 49
truth table, 196

U

unity gain amplifier, 171
Universal Serial Bus (USB), 247
USB, 236

V

V (Δt) = L (ΔI), 75
variable-gain op-amp, 214
VBE, 113
VBE equation, 113
vector diagram, 80, 104
Verilog, 135
Very High Level Descriptive Language (VHDL), 135
VFB, 99
virtual ground, 158
voltage, 5
voltage divider, 16, 64, 99
voltage follower
    buffer, 171
voltage gain, 126
voltage gain (hfe), 117
voltage leads, 77
voltage source, 7
voltage-doubler circuit, 103
Vpeak, 52

Vpeak-to-peak, 52
Vrms, 54
VT, 113

W

wafer, 130
watchdog timer (WDT), 261
waveform, 5
weight, 105
whole numbers, 320
Wilson current mirror, 166
wireless network, 106
wireless smoke detector, 248

X

XOR gate, 206

Z

zener diode, 47
zener regulator, 96
ZigBee®, 261

Ω

Ω per square, 4

ω

ω, 57

www.ingramcontent.com/pod-product-compliance
Lightning Source LLC
Chambersburg PA
CBHW081233180526
45171CB00005B/417